THE PHYSICS OF HIGH PRESSURE

THE PHYSICS
OF HIGH PRESSURE

BY

P. W. BRIDGMAN, Ph.D.

PROFESSOR OF PHYSICS, HARVARD UNIVERSITY

DOVER PUBLICATIONS, INC.

NEW YORK

Published in Canada by General Publishing Company, Ltd., 30 Lesmill Road, Don Mills, Toronto, Ontario.

Published in the United Kingdom by Constable and Company, Ltd., 10 Orange Street, London WC 2.

This Dover edition, first published in 1970, is an unabridged and unaltered republication of the work first published by G. Bell and Sons, Ltd. in 1931. This edition is reprinted by special arrangement with G. Bell and Sons, Ltd.

International Standard Book Number: 0-486-62712-8
Library of Congress Catalog Card Number: 76-139974

Manufactured in the United States of America
Dover Publications, Inc.
180 Varick Street
New York, N. Y. 10014

PREFACE

This book is intended primarily as a summary of work in which I have been personally engaged in the last twenty-five years in the field of high pressure. It is also intended, however, that the book shall give a fairly complete survey of all the important work in this field; to this end the first chapter is devoted exclusively to historical matters, a historical introduction giving the previous state of the art is appended to many of the later chapters, and in the final chapter on miscellanies the work of others on a number of topics which I have not yet had a chance to touch is discussed in some detail. The references and the index will also facilitate a comprehensive grasp of the whole field.

There are comparatively few previous books in this field; of these the book of Cohen and Schut, *Piezochemie Kondensierter Systeme* (Akademische Verlagsgesellschaft, Leipzig, 1919), is especially to be mentioned. This was written in 1914 and published in 1919. It aims to give a complete summary of work in the field of high pressure physics and chemistry up to 1914, with some of the work of the period 1914–1919 treated in appendices. So much has been done since that time that no apology is necessary for a new book. In addition to this principal source of information in book form, the second part of Cohen's Cornell lectures, *Physico-Chemical Metamorphosis and Problems in Piezo-Chemistry* (M'Graw-Hill Book Co., New York, 1926), deals with certain aspects of his pressure work at Utrecht, and is a valuable source of information. Practically the only other work is that of Tammann, *Kristallisieren und Schmelzen* (Barth, Leipzig, 1903), republished with some changes under the title *Aggregatszustände* (Voss, Leipzig, 1922). This deals with the special topics indicated by the titles, and is concerned mostly with Tammann's own work.

The reader who is interested in the details of much of the early work should consult these books. Finally, there is my own recent article in the *Handbuch der Experimental Physik*, vol. 8^2, dealing with a limited aspect of the subject—namely, the effect of high pressure on various thermal properties of matter. The information contained in that article is, of necessity, similar to that in the corresponding sections of this book; the arrangement and general method of treatment are, however, different.

The titles of my own papers are collected in an appendix; in the body of the book reference is made to these by number, prefixed by a B. References to other work are numbered consecutively through each chapter, and are collected at the end of the chapter. At the end of the first chapter all these references to other work are given by title, but this was less necessary in the later chapters where the subject is more definitely indicated, and there the reference is to the periodical only. Many of these later references are duplicates of those given in the first chapter, which may therefore often be consulted for the titles.

It has not been possible in a book of this size to reproduce all the numerical data, but the attempt has been made in the various tables scattered through the text to give enough data to fix the broad features of the effects of pressure and temperature on the more important substances, particularly the elements. I believe that the data given here will be found sufficient for most theoretical use for which we are at present prepared in this field. For finer details and for many of the more complex substances the original papers must be consulted. Many of my own detailed results have been published in the *Proceedings of the American Academy of Arts and Sciences*, a somewhat inaccessible publication. It may be that some libraries which do not have the *Proceedings* may have the *Contributions* from the Jefferson Physical Laboratory, in which all my papers are collected.

CONTENTS

THE
PHYSICS OF HIGH PRESSURE

CHAPTER I

HISTORICAL INTRODUCTION

IN this historical introduction only those experiments are to be especially considered which are the natural forerunners of experiments to high pressures. Many experiments dealing with pressure effects do not need the emphasis given by high pressures, as, for example, those dealing with the critical point liquid-vapour; such experiments will be considered only briefly if at all.

Perhaps the earliest experiment on our theme was the celebrated attempt of the Florentine Academy to find whether water is compressible. Water was sealed into a sphere of lead, which was then flattened between the jaws of a press until the water exuded through the lead walls. A measurement of the distortion of the lead showed no measurable loss of volume of the cavity containing the water. Apparently the tinsmith who soldered together the lead hemispheres was as confident of the integrity of his work as are present-day smiths, for the experiment was interpreted as meaning that the water had been forced through the pores of the homogeneous lead, and the conclusion was drawn that water must be incompressible, because it suffered no measurable loss of volume under a pressure so tremendous as to force it through the pores of solid lead. In view, however, of the very low strength of lead, it is improbable that the pressure in this experiment was more than a small fraction of 100 kg./cm.², so that the resulting change of volume would have been much less than 1 per cent., evidently too small to detect with such crude methods of measurement.

The next experiments take us to the years 1762 and 1764,

when Canton [1] published experiments in the *Transactions of the Royal Society* to prove that water is compressible, often paradoxically misquoted because of a misprint in the index as experiments to prove that water is incompressible.. Canton placed his liquids in a thermometer-like arrangement with a large bulb and fine capillary, anticipating the later piezometers, and with the great magnification obtained in this way was able. to produce measurable changes of volume in water and other liquids with the small changes of pressure realisable under the receiver of an air pump.   The phenomena of volume changes in liquids remained the principal object of high-pressure investigations for a long time, the era which was principally concerned with this theme culminating perhaps with the classical paper of Amagat in 1893.

The next name is that of Perkins,[2] an ingenious Yankee, who made his first experiments in America, and then later, after the example of his illustrious predecessor Rumford, emigrated to England, where he very much extended his original experiments.   Perkins must have been a mechanical genius, for his experiments were cast in a mould heroic for those days.   His first experiments were made to a pressure of 100 kg./cm.², using a cannon as the containing vessel.   His pressure measurements were checked by sinking the apparatus in the sea to known depths, this being the first use of this method of producing high pressures.   Later he reached pressures of 2000 kg./cm.², an enormous extension of range, and a pressure that was not again reached in accurate experiments until Amagat.   His apparatus was so accurately made that no soft packing was necessary on the plunger with which pressure was produced.   His pressures were measured with a sort of safety-valve arrangement, the precursor of the free piston gauge; a number of his other devices were later reinvented.   He observed the decrease of compressibility of water with increasing pressure, but his absolute values of compressibility were four times too small.   He also observed the raising of the melting-point of acetic acid with pressure, a phenomenon forgotten until the thermodynamic discussions of the Thomsons, and not even then connected with the name of Perkins

Following Perkins, one branch of pressure investigation was a long series of experiments concerned chiefly with the attempt to get consistent values for the compressibilities of liquids. There was nothing noteworthy about these experiments from the high-pressure point of view, the pressures often being of the low range obtainable in the receivers of air pumps, and the joints often being made with sealing-wax. The crucial point in all these experiments was to properly correct for the distortion of the containing vessel, which was a very appreciable fraction of the whole effect. All the early attempts to correct for the distortion of the containing vessel were indirect, based on calculations from elastic theory, using data obtained from deformations under stresses other than hydrostatic pressure. In those days when elastic theory was not well understood, there was much disagreement and serious errors were made even by men as skilful as Jamin. The principal names associated with this series of experiments are: Oersted,[3] 1823 and 1828; Colladon and Sturm,[4] 1838; Aimé,[5] 1843, who followed Perkins's scheme of using the ocean as a source of pressure, but who got very bad numerical results; Grassi,[6] 1851; Jamin, Amaury, and Descamps,[7] 1868–69; and Dupré and Page,[8] 1868.

The period of fifty years from 1819, the year of the first paper of Perkins, to 1869, the year of the first paper of Amagat, constitutes a self-contained period in high-pressure history, in which a good deal of the ground was mapped out, but few final or even good numerical results were obtained. Besides the papers mentioned as being concerned chiefly with the compressibility of liquids, the following also belong in this period. In 1833 Parrot and Lenz [9] in St. Petersburg observed several miscellaneous pressure effects to 100 kg./cm.^{2}; they showed that glass is compressible, checked Boyle's law for air, and measured pressure for the first time with a free piston gauge. Natterer [10] published three papers in 1850, 1851, and 1854, dealing with the attempted liquefaction of gases by high pressures. Being unable to obtain any apparatus capable of standing the desired pressures without leak, he was forced to become his own mechanic, and made apparatus capable of withstanding 3600 kg./cm.^{2}. He introduced several tricks of technique, among others that of making a tight joint with a hardened cone forced against a right-angled edge, which still is frequently used. With this apparatus he was unable to liquefy N_2, CO, H_2, air, and illuminating gas at ordinary temperatures by pressures up to 3600 kg./cm.^{2}, and he drew the conclusion that it was impossible to liquefy these gases by any pressure at this temperature. He attempted the liquefaction at $-80°$ C., but here his apparatus leaked because of the freezing of his oil. He made approximate measurements of the volume to these pressures, and observed the very great departures from Boyle's law. In 1857 and 1858 Jamin [11] observed with his interferometer the change of index of refraction of water at a few atmospheres, probably the first experiment on optical properties under pressure. Wartmann in 1859 [12] observed the change of electrical resistance of metals under pressure. These experiments were very crude, the metal was imbedded in gutta-percha, which was squeezed between the jaws of a press. The pressure, which could have been only very roughly hydrostatic, was estimated to be 400 kg./cm.^{2}. Joule in 1859 [12a] measured quantitatively the adiabatic temperature changes accompanying the sudden application of about 25 kg./cm.^{2} to water and sperm-oil, and obtained agreement with the theoretical values within about 5 per cent. The effect had been previously sought unsuccessfully by Regnault.

Andrews [13] in 1861 discovered the critical phenomena in gases, and immediately wide attention was attracted to this field. Although we have perhaps arbitrarily refused to consider these phenomena as part of our subject, nevertheless the work of Andrews had an important effect on high-pressure work proper. The pressures which he had to reach were several hundred kg., and were therefore much higher than the ordinary run of pressures at which the compressibilities of liquids had been measured, although low compared with the pressures reached by Perkins and Natterer, so that the technique which Andrews developed for reaching his pressures reacted on the whole high-pressure technique. Furthermore, the perspective opened by the discovery of critical phenomena embraced subjects proper to our range, such as the behaviour of gases at pressures much higher than the critical pressures, and the impetus given by the discoveries of Andrews lasted for a number of years.

The next important period is of nearly twenty-five years, terminated by the classical paper of Amagat in 1893 on the compressibility of gases and liquids over a pressure range of 3000 kg./cm.^{2} and a temperature range of 200°. This period is dominated by two Frenchmen; in the early part of the period Cailletet,[14] and, later, pre-eminently Amagat.[15] The direct inspiration of this work was the discovery of Andrews. Cailletet was mostly interested in the liquefaction of gases, although his measurements included also the compressibility of liquids. He considerably extended the pressure range of Andrews, up to something of the order of 1000 kg./cm.^{2}, and devised a convenient pump for reaching these pressures, which is still manufactured in nearly its original form and which still goes by his name. Cailletet also played a large part in devising accurate methods for measuring these pressures, including the use of compressed air manometers and an anticipatory form of the free piston gauge of Amagat, but his chief reliance was on long mercury columns in open steel tubes, with which he reached 300 or 400 kg./cm.^{2}.

The first paper of Amagat, in 1869, was on the departure of gases from Boyle's law ; his papers continued with ever-increasing scope to the number of about thirty, until they reached their climax in the 1893 paper, the subject of which has been indicated above, and which is an epitome of nearly all his work. Amagat developed a special packing technique by which he was able consistently to reach pressures of 3000 kg./cm.^{2} or more. He devised special methods of measuring compressibility, including an arrangement utilising glass windows up to 1000 kg., and an arrangement with a series of electric contacts, adapted from Tait, to 3000 kg./cm.^{2}. He devised methods for measuring high pressures, improving the open air manometer, and for the maximum pressures devised his celebrated free piston gauge with large and small pistons. The difficult question of the proper method of correcting for the distortion of the containing vessel became finally understood from the theoretical side in this period, and Amagat took an experimental step which might have led to the practical resolution of this question as well, although he never carried the work through.

Amagat's experimental work came to a nearly abrupt stop in 1893, there being only one later paper dealing with the melting of ice when the pressure to which it has been exposed is released, and from this date on he contented himself with theoretical calculations and speculations on the large amount of experimental material that he had collected.

The period of Cailletet and Amagat was one of intense activity by other experimenters also, and there are a great many titles in this period. Regnault [16] published one of his latest papers in 1871 on a form of gas manometer to measure high pressures, seeing that the subject of high pressures was one of growing importance. Mascart [17] in 1874 had two papers on the index of refraction of compressed water, and in 1877 one on the index of compressed gases. Buchanan [18] in 1880 made the first experiments capable of giving the linear compressibility of a solid, without using any of the equations of elasticity, by observing with microscopes through a heavy glass tube the change of length of a rod of metal within the tube exposed to hydrostatic pressure. His numerical values were not good, probably due to irregular refraction effects in the glass. In 1880 Spring [19] published the first of a series of experiments dealing in general with cohesion effects and chemical reactions when dry powders of various solids were subjected to high pressure. Spring did not have the problem of leak of the transmitting liquid to trouble him, and so was able to reach pressures materially higher than previous observers, up to 6000 or 7000 kg. There has been considerable controversy about the results of Spring, many of which were highly spectacular, and it now seems certain that many of his results were not due to pressure alone, but involve in addition the rubbing motion of one particle on another with intense shearing stress. In 1880 Dewar [20] observed the lowering of the freezing point of ice by pressure up to 700 kg./cm.^{2}. In 1881 Tait [21] published a paper on the pressure errors of the thermometers of the *Challenger*. The temperature of the ocean at various depths was measured on this famous expedition with mercury-in-glass thermometers exposed directly to the action of the sea water. It was obvious enough that under the great pressure in the ocean errors were to be expected in such thermometer readings, and for a number of years after the return of the expedition Tait was occupied with reproducing in the laboratory the conditions met in the depths of the sea. His pressure range was about 600 kg./cm.². He measured the compressibility of pure water, sea water, and a number of solutions. He proposed that the pressure effect on the exterior of a thermometer should be used as a method of measuring pressure, and in a number of subsequent investigations the so-called Tait gauge was used, in which pressure was indicated by the change of internal volume of a cylinder exposed to external hydrostatic pressure.

In 1881 Roentgen and Schneider [22] determined the compressibility of dilute solutions and of solid NaCl, and followed this in 1886 with a paper in which a connection was sought between the compressibility and the surface tension of a number of liquids. Chwolson [23] in 1881 observed for apparently the first time the decrease of resistance of a metal when exposed to a truly hydrostatic pressure, the magnitude of which was only 60 kg./cm.². He attempted to avoid the errors arising from the temperature effects of compression by operating at the maximum density point of water. He also compared the change of resistance produced by hydrostatic pressure with the change to be expected from the change of dimensions, and showed that there is a specific pressure effect. Marshall and others [24] in 1882 observed the depression of the maximum density point of water with pressure. In 1883 Pagliani and Vicentini,[25] and Pagliani and Palazzo [26] determined the compressibility of a number of new liquids in a small range of pressure. Quincke [27] studied the compressibility and the change of index of refraction of a number of liquids, Chappuis and Rivière [28] measured the index of refraction of gases under pressure, and Tomlinson [29] published a most elaborate account of the effects of all kinds of mechanical stress on a wide variety of physical properties, including a few measurements of liquid compressibility and an unsuccessful attempt to detect an effect of hydrostatic pressure on magnetic permeability.   In 1885 Creelman and Crockett [30] extended Joule's quantitative study of the changes of temperature accompanying an adiabatic application or release of pressure to several hundred kg./cm.[2] and a miscellaneous collection of solids and liquids.   There have been comparatively few repetitions of the experiments of Creelman and Crockett; it would appear that there may be useful possibilities here, particularly as a more direct method than any yet used for the determination of specific heats at high pressures.   In 1888 Parsons [31] published the first accounts of investigations, which were later very much extended, of the behaviour of carbon at high pressures and temperatures, the ultimate interest being the artificial formation of diamond, a problem not yet solved.   Hallock [32] in 1888, in the attempt to confirm or interpret the experiments of Spring, found that solids like wax or lead do not actually become fluid under pressure, as had been many times erroneously stated, but the appearance of flow arises from the resistance to yield being overcome by the enormous stresses.

Barus [33] in 1889 published the first account of work covered in great detail later in two long papers from the Geological Survey in Washington in 1892.   He was interested in studying the question of rock formation and similar geological problems by reproducing in the laboratory, as far as possible, actual terrestrial conditions of pressure and temperature. He attained pressures of 2000 kg. and temperatures of 400° C., and drew a number of interesting conclusions.   There is an individual quality about his work that makes interesting reading.   Among other things he observed the enormous solvent action of water on glass at high pressures and temperatures, measured the electric conductivity of several solutions of electrolytes, studied in detail the hysteresis of several forms of pressure gauge, made several interesting additions to technique, observed the enormous increase of viscosity of a substance like marine glue under pressure, and made a number of observations of melting phenomena under pressure, the interpretation of which was unfortunately obscured by the effect of impurities.

Damien [34] in 1889 published his first paper on the melting of solids up to 2000 kg., and followed it with a second paper in 1891. De Metz [35] in 1890 investigated the compressibility of oils and colloids, and in 1892 the much more difficult question of the compressibility of mercury. The greater difficulty arises from the fact that the compressibility of mercury is so low compared with the compressibility of the container that an accurate evaluation of the correction for the distortion of the container becomes much more necessary. De Metz obtained values not far from the present accepted values. Roentgen and Zehnder [36] in 1891 measured the effect of pressure on the index of refraction of water, CS_2, and other liquids. In 1892 Roentgen [37] followed this with a paper dealing with a variety of new phenomena to a pressure of several hundred kg./cm.[2], including the conductivity of solutions of electrolytes, the velocity of inversion of sugar in an acid solution, the velocity of diffusion in a liquid, and a confirmation of the observation of Barus of the increased viscosity of glue. In 1891 Des Coudres [38] measured the effect of pressure up to two atmospheres on the thermo-elective quality of mercury and several dilute amalgams, the first observations of their kind. Galopin [39] in 1892 made one of the few repetitions of the experiments of Creelman and Crockett, observing the adiabatic changes of temperature in water up to 500 kg. In 1893 Voigt [40] attempted to find the effect of a hydrostatic pressure of 60 kg./cm.[2] on the breaking strength of NaCl crystals. No effect was found on the differential stress required to produce rupture, which means that if a rod is stretched by hanging weights on it, it will rupture at the same weight, whether the rod together with the weights is immersed in an atmosphere under pressure or not. This is an important question, and the experiments should be repeated at a higher pressure, where probably some effect will be found.

Since the termination of Amagat's work in 1893 there have been a very large number of high-pressure investigations, mostly groups of a few papers by various investigators, who evidently have not made this work their chief activity, and which will be referred to in more detail later, but there have in addition been a few foci in which several individuals or institutions have devoted an important part of their effort to high-pressure work.

Without doubt the most important work done in the period immediately following that of Amagat was by Tammann and his pupils [41]; Tammann's first paper in this field was in 1893, dealing with internal pressures in solutions.   His main thesis was that dilute solutions, under ordinary conditions, behave approximately like pure water under an external pressure equal to that part of the internal pressure in the solution due to the action of the dissolved substance.   In support of this thesis Tammann published, in the few years following 1893, a number of papers dealing with such topics as the compressibility and expansion of solutions, specific heats, heats of neutralisation, and pressure effects on electrical conductivity of electrolytes.   A number of interesting correlations were brought to light, and the thesis gives a qualitative account of a considerable range of phenomena.   The idea must be applied with caution, however, in any new field; there are a number of phenomena known in dilute alcohol solutions which are the exact opposite of what would be expected.   Tammann later became interested in phenomena of solidification, such as the velocity of solidification, and from this it was a natural step to determine the effect of pressure on melting temperature and on the transition temperature from one solid phase to another.   In this field his work was by far the most extensive and systematic of anything that had yet been done.   His pressure range was 3000 kg./cm.^{2}, the same as that of Amagat.   Tammann did not make as many improvements in technique as might be expected of one occupied so extensively in this field; this is doubtless explained by the fact that his mechanical facilities were always limited, so that he had to content himself with more or less stock apparatus that could be purchased from instrument dealers.   Tammann first extensively used the method of discontinuity of volume in locating a melting or transition point.   He observed the universal direction of curvature of the melting curves, which is the same as that of the liquid-vapour curves.   He followed the change of volume to high pressures along the melting curve, from this calculated the latent heat, and from this in turn drew certain conclusions about the general character of the melting curve, which will be the subject of much more detailed discussion later. He observed a number of new transitions between solids possible only under pressure, including the very interesting two varieties of ice, Ice II and III. Tammann's chief activity in this field terminated with the publication of his book *Kristallisieren und Schmelzen* in 1903, of which the second edition in 1922 is called *Aggregatszustände*. Here will be found a more complete reference list to his papers than that given here. He has, however, published sporadic papers dealing with this general topic down to the present day.

Next after Tammann in point of time is Lussana,[42] whose first paper was published in 1895, and who has published some twenty papers at irregular intervals ever since. This work has apparently not had the influence which would be the natural reward of the industry and ingenuity displayed. It is evident that much of the work has been done under heavy material handicaps, which have sometimes led to the adoption of methods not capable of the highest accuracy, and in fact sometimes effects have been found not verified even qualitatively by other observers. The pressure range of much of Lussana's later work is 3000 kg./cm.², the range of Amagat. Lussana began in 1895 with three papers: one on the resistance of solutions of electrolytes at various pressures and temperatures, one on the effect of pressure on the maximum density point of water, and one on the effect of pressure on the transformations of NH_4NO_3 and HgI_2. In 1897 there was a very extended paper on the resistance of electrolytes. In 1899 and 1903 there were papers on the resistance of metals under pressure, in which, in addition to the permanent effects, temporary effects were found immediately after the application of pressure which other observers have not been able to confirm. In 1903, 1904, and 1910 there were several papers entitled, "Thermal Properties of Solids and Liquids," dealing with the compressibility, thermal expansion, and melting of pure metals, alloys, and a number of ordinary liquids.   In 1908, 1912, and 1914 there are papers dealing with the effect of pressure on the specific heat of liquids.   The attempt was made to measure this effect directly, a most difficult thing to do; this is one of the few, if not the only, attempts at a direct attack on this problem. In 1918 and 1923 there are papers on the effect of pressure on thermal conductivity of metals and the law of Wiedemann-Franz.   This again is a difficult subject, and Lusanna was apparently the first to attempt measurements.

Next should be mentioned the work of T. W. Richards.[43] His first work on compressibility was published in 1903, and continued, usually in collaboration with some pupil, up to his death in 1928.   Richards's work is important, not for the range of pressure and temperature covered (nearly all his work was confined to a pressure range of 500 kg./cm.[2] and to room temperature), but for the great range of materials investigated, and for the theoretical significance which he saw in his results.   Richards was the first to make a systematic investigation of the compressibility of a large number of the elements and to notice their periodic relations.   His measurements also embraced a large number of solutions and organic liquids.   He introduced a much improved method of measuring compressibility, later used by a number of other investigators, by which accurate and reproducible values for the difference of compressibility between any substance and a reference substance, such as mercury, could be determined. His efforts to convert differential compressibility into absolute compressibility were not so successful.   When Richards first began his work the common view of the atom was of a rigid billiard ball-like object.   Richards had the insight to see that much physical and chemical evidence was opposed to this view, and his chief concern in making his measurements and discussing their significance was to emphasise the necessity for a "compressible" atom.   Unfortunately this was a conception on which physics was already converging from many other points of view, so that historically the work of Richards did not have as much effect in moulding the conception of the atom as it might under more favourable conditions.

At Utrecht Cohen [44] and his collaborators have been

engaged in high-pressure work since 1894. The pressure range of this work is 1500 or 2000 kg./cm.². Many sorts of question have received attention, mostly of a chemical character, with comparatively little work on such physical questions as compressibility. The results have been mostly published in the *Zeitschrift für Physikalische Chemie*, in a series of papers entitled *Piezo-Chemische Studien*. Investigations have been made of the effect of pressure on the E.M.F. of a number of reversible cells, and the first satisfactory proof has been given that the effect is actually that deduced theoretically by thermodynamic reasoning. An exhaustive investigation was made of the validity of Faraday's law under pressure. Reaction velocities have been measured, and the effect of pressure on chemical equilibrium studied. An extensive study has been made of the effect of pressure on solubility, and an experimental demonstration given of the law of Braun, which applies to these phenomena. Finally, the diffusion velocity of cadmium into mercury has been measured under pressure, and also the effect of pressure on the viscosity of mercury. Cohen has published an important book entitled *Piezochemie* dealing with high-pressure effects mostly up to 1914, and has also discussed some of his own more recent work in his Cornell lectures on the Baker Foundation. Extended references to Cohen's papers will be found in the two books.

In the United States, the Geophysical Laboratory in Washington [45] has been systematically engaged in high-pressure work ever since the foundation of the Laboratory in the early years of this century. The investigations have been mostly concerned with problems of geophysical interest, such, for example, as a determination of the compressibility of a large number of minerals and rocks. In actual geophysical applications high pressures are encountered simultaneously with high temperatures, and a great deal of the effort at the Geophysical Laboratory has been devoted to devising methods of extending experiments at pressures of several thousand kg./cm.² to temperatures of the order of 1000° C., a problem of great difficulty. Bombs have been developed by which a great many of the reactions involved in rock formation have been followed under terrestrial conditions of pressure and temperature.

In addition to these well-established centres of high-pressure investigation, A. Michels [46] in Amsterdam is engaged in building up an elaborate high-pressure plant, and has already published several important papers, in which special emphasis is laid on the precision of the work. Much more work may be expected from this laboratory in the next few years. At Iowa Wesleyan College, Poulter [47] is now engaged in constructing apparatus with which pressures are reached limited only by the strength of the steel. He is beginning to publish results, and already has several papers describing special parts of the technique.

Finally, there is my own work done at Harvard University since 1906.[48] A good part of the remainder of this book is to be devoted to it, so that no more need be said here except to give the one characterisation that all this work grew out of the development of a packing technique which makes it possible to reach without leak any pressure allowed by the mechanical strength of the walls of the containing vessels.

Returning now to the more sporadic high-pressure investigations since 1893, in 1894 there were papers by Fanjung [49] and Piesch [50] on the electric conductivity and polarisation effect in solutions of electrolytes under pressure. Roentgen [51] in this year apparently made the first measurement of dielectric constant under pressure. Up to 500 kg. Roentgen could find no change in the dielectric constant of water of as much as 1 per cent., although a much greater increase would be indicated by the Lorenz-Lorentz formula. In the following year Ratz [52] was able to measure the effect of pressure on the dielectric constant of a number of organic liquids up to 250 kg., and found that it was about ten times smaller than would be indicated by the formula.

In 1896 and 1897 Demerliac [53] published two papers on the effect of pressure up to 300 kg./cm.^{2} on the melting points of several organic substances. At low pressures he found agreement with the Clapeyron formula, but at higher pressures very great disagreement. This disagreement played a considerable part in the development of theories of the melting curve; it was later shown by Tammann to be an effect of the rapidly increasing solubility of the air with which the pressure was transmitted. In 1896 Von Stackelberg [54] published on the effect of pressure on solubility, Rothmund [55] on the effect of pressure on reaction velocity, and Gilbaut [56] on the compressibility of salt solutions. In 1898 Mack [57] followed the melting curve of naphthalin to 2100 kg./cm.^{2}, measuring pressure with a free piston gauge, and determining the melting-point by the change of volume indicated by the motion of the piston of the gauge, thus anticipating the method used so extensively by Tammann, and later by me. Mack had previously observed melting in heavy glass capillaries to 900 kg./cm.^{2}; all his experiments gave to the melting curve curvature in the same direction as the liquid-vapour curves, but no trace of the effects claimed by Damien and Demerliac could be found. Heydweiller [58] also in 1898 observed melting phenomena in heavy glass tubes, reaching pressures in the neighbourhood of 3000 kg./cm.^{2}, probably the highest pressures that have ever been reached in glass capillaries. He could find no evidence of the phenomena of Damien and Demerliac, nor did he find evidence of a critical point between liquid and solid. Neither was there any evident connection between melting phenomena and the phenomena of continuous passage to vapour. Hulett [59] in 1899 measured the effect of pressure up to 300 kg./cm.^{2} on the melting of a number of liquid crystals, hoping that if there were critical phenomena between liquid and solid this would be the class of substance that would show them most easily, but no difference in behaviour could be found between liquid crystals and ordinary substances.

Majorana [60] in 1900 attacked the diamond problem by the application of high pressure and temperature with the usual result, and in 1907 he made a second attempt. In 1901 Hauser [61] measured the effect of pressure up to 500 kg./cm.^{2} on the viscosity of water over a temperature range of 100°; here he found a reversal of sign of the pressure coefficient, which below 32° is negative and above 32° positive. In 1902 Lampa [62] published work on two component systems, determining the effect of pressures up to 200 kg./cm.^{2} in depressing the freezing-point of various salt solutions. In 1902 Agricola [63] extended the work of Des Coudres on thermo-electric phenomena in mercury and amalgams from 2 to 100 kg./cm.^{2} The effect of pressure on thermo-electric properties was again attacked by Wagner [64] in 1908, who measured the effect of pressure up to 300 kg./cm.^{2} on a number of solid metals, by Siegel [65] in 1912, who worked up to 400° C. with liquid metals, and by Hörig [66] in 1909, who applied 1400 kg./cm.^{2} to mercury and the Na—K eutectic.

Carnazzi [67] in 1903 made an extended study of the volume of mercury between 0° and 200° up to 3000 kg./cm.^{2}. He states that it was possible to reach 5000 kg./cm.^{2} with his apparatus, but since his pressure measurements were based on the results of Amagat, quantitative measurements could not be made beyond Amagat's limit. In 1905 he followed this with a study of the compressibility up to 1000 kg./cm.^{2} of a number of mixtures of organic liquids.

In 1907 Eve and Adams [68] made an attempt to find an effect of pressure on radioactive disintegration. The radioactive material was imbedded in lead, which was exposed directly, without the intervention of a liquid, to a pressure of 30,000 kg./cm.^{2}, the yield point of the steel vessel. No effect was found. Since theory does not suggest any large effect, and since it would obviously not be possible to greatly exceed this pressure, the inducement to repeat the experiment has been small.

Also in 1907 Ladenberg [69] measured the effect of pressure on the viscosity of very viscous liquids, such as Venice turpentine. He found increases of as much as 100 per cent. for 100 kg./cm.^{2}, confirming the earlier results of Barus and Roentgen. Williams [70] observed for the first time the anomalous increase of resistance of bismuth under pressure.

Montén [71] in 1909 observed very large and irregular effects of pressure on the electrical resistance of Se and AgS. Lafay [72] measured the effect of pressure on the resistance of several metals, including liquid mercury, to a pressure of 4500 kg./cm.^{2}, and discussed the results particularly with a view to using the decrease of resistance with pressure as a method of measuring pressure.

In 1911 Parsons and Cook [73] measured the compressibility of a number of liquids up to 4500 kg./cm. by measuring the motion of the plunger with which pressure was produced. Beckman and Onnes [74] in 1913 observed the change of resistance of lead under pressure at the temperature of liquid hydrogen, and found a pressure coefficient much higher than at room temperature. This increase of coefficient has been confirmed in 1929 by measurements of Seemann [75] at liquid-air temperatures.

The outstanding recent development in the combination of high pressure with low temperature is the solidification of helium by Keesom [76] at several hundred kg./cm.^{2}, and the very important work of Simon and collaborators,[77] who have investigated the freezing of a number of the "permanent" gases under pressure, and in particular has followed the melting curve of helium to nearly 6000 kg./cm.^{2}.

This historical survey has not aimed to be in any sense complete, but has endeavoured to mention only the more important investigations. Other references will be given as occasion arises in the body of the text.

[1] JOHN CANTON, "Experiments to prove that Water is not Incompressible," *Trans. Roy. Soc.*, pp. 640–643 (1762).
"Experiments and Observations on the Compressibility of Water and some other Fluids," *ibid.*, pp. 261–262 (1764).

[2] JACOB PERKINS, "On the Compressibility of Water," *ibid.*, pp. 324–329 (1819–20).
"On the Progressive Compression of Water by High Degrees of Force, with some Trials of its Effect on other Fluids," *ibid.*, pp. 541–547 (1826).

[3] OERSTED, "Sur la compressibilité de l'eau," *Ann. Chim. Phys.*, **22**, 192–198 (1823).
"Sur la compression de l'eau dans des vases de matières différentes," *ibid.*, **38**, 326–330 (1828).

[4] D. COLLADON et C. STURM, "Mémoire sur la compression des liquides," *Mem. Sav. Etr. Inst. France*, **5**, 267–347 (1838).

[5] G. AIMÉ, "Mémoire sur la compression des liquides," *Ann. Chim. Phys.*, **8**, 257–280 (1843).

[6] GRASSI, "Recherches sur la compressibilité des liquides," *ibid.*, **31**, 437–478 (1851).

[7] Jamin, Amaury, et Descamps, " Sur la compressibilité des liquides," *C.R.*, **66**, 1104–1106 (1868).

Amaury et Descamps, " Sur la compressibilité des liquides," *ibid.*, **68**, 1564–1565 (1869).

[8] A. Dupré et F. J. M. Page, " On the Specific Heat and other Physical Properties of Mixtures of Ethylic Alcohol and Water," *Trans. Roy. Soc.*, **159**, 591–614 (1869).

Section VI, p. 610, deals with compressibility.

[9] Parrot et Lenz, "Experiences de fortes compressions sur divers corps," *Mem. St. Pet. Acad.*, **2**, 595–630 (1833).

[10] Natterer, " Gas-verdichtungs-Versuche," *Wien. Ber.*, **5**, 351–358 (1850) ; **6**, 557–570 (1851) ; **12**, 199–208 (1854).

[11] J. Jamin, " Recherches sur les indices de réfraction," *C.R.*, **45**, 892–894 (1857).

" Mémoire sur la mesure des indices de réfraction des gaz," *Ann. Chim. Phys.*, **49**, 282–303 (1857).

" Mémoire sur les variations de l'indice de réfraction de l'eau à diverses pressions," *ibid.*, **52**, 163–171 (1858).

[12] E. Wartmann, "On the Effect of Pressure on Electric Conductibility in Metallic Wires," *Phil. Mag.*, **17**, 441–442 (1859).

[12a] J. R. Joule, " On the Thermal Effects of Compressing Fluids," *Trans. Roy. Soc.*, **149**, 133 (1859).

[13] Andrews, " On the Effect of Great Pressure combined with Cold on the Six Non-Condensable Gases," *B.A. Rep.*, p. 76 (1861).

[14] L. Cailletet, " Compressibilité des gaz à hautes pressions," *C.R.*, **70**, 1131–1134 (1870).

" Compressibilité des liquides sous des hautes pressions," *ibid.*, **75**, 77–78 (1872).

" Sur la résistance des tubes de verre à la rupture," *ibid.*, **78**, 411–413 (1874).

" Manomètre destiné à mesurer les hautes pressions," *ibid.*, **83**, 1211–1213 (1876).

" Nouveau manomètre destiné à mesurer les hautes pressions," *Jour. de Phys.*, **5**, 179–181 (1876).

" Sur la construction des manomètres à air libre, destinés à mesurer les hautes pressions," *C.R.*, **84**, 82–83 (1877).

" Recherches sur la liquéfaction des gaz," *Ann. Chim. Phys.* (5), **15**, 132–144 (1878).

" Recherches sur la compressibilité des gaz," *C.R.*, **88**, 61–65 (1879).

" Recherches sur la compressibilité des gaz," *Jour. Phys.*, **8**, 267–274 (1879).

" Sur la mesure des hautes pressions," *Ann. Chim. Phys.* (5), **19**, 386 (1880).

[15] E. H. Amagat, " De l'influence de la temperature sur les écarts de la loi de Mariotte," *C.R.*, **68**, 1170–1173 (1869).

" Sur la compressibilité et la dilatation des gaz," *ibid.*, **73**, 183–186 (1871).

" Mémoire sur la compressibilité des liquides," *Ann. Chim. Phys.*
   (5), **11**, 520–549 (1877).
" Recherches sur la compressibilité des liquides," *C.R.*, **85**, 27–29,
   139–142 (1877).
" Sur la compressibilité de gaz à des pressions élevées," *ibid.*, **87**,
   432–434 (1878).
" Recherches sur la compressibilité des gaz à des pressions élevées,"
   ibid., **88**, 336–338 (1879).
" Recherches sur la compressibilité des gaz à des pressions élevées,"
   ibid., **89**, 437–439 (1879).
" Mémoire sur la compressibilité des gaz à des pressions élevées,"
   Ann. Chim. Phys., **19**, 345–385 (1880).
" Sur la compressibilité de l'oxygène, et l'action de ce gaz sur la
   mercure dans les experiences où ces corps sont mis en contact,"
   C.R., **91**, 812–815 (1880).
" Sur la deformation des tubes de verre sous de fortes pressions,"
   ibid., **90**, 863–864 (1880).
" Influence de la temperature sur la compressibilité des gaz sous
   de fortes pressions," *ibid.*, **90**, 995–997 (1880).
" Sur la dilatation et la compressibilité des gaz sous de fortes
   pressions," *ibid.*, **91**, 428–431 (1880).
" Sur la compressibilité des gaz de fortes pressions," *Ann. Chim.
   Phys.*, **22**, 353–398 (1881).
" Sur les experiences faites pour déterminer la compressibilité du
   gaz azote," *C.R.*, **95**, 638–641 (1882).
" Recherches sur la compressibilité des gaz," *Ann. Chim. Phys.*,
   28, 456–464 (1883).
" Résultats pour servir aux calculs des manometers à gaz com-
   primés," *C.R.*, **99**, 1017–1019, 1153–1154 (1884).
" Sur la densité et le volume atomique des gaz, et en particulier de
   l'oxygène et de l'hydrogène," *ibid.*, **100**, 633–635 (1885).
" Sur la mesure des très fortes pressions et la compressibilité des
   liquides," *ibid.*, **103**, 429–432 (1886).
" Sur la dilatation des liquides comprimés, et en particulier sur la
   dilatation de l'eau," *ibid.*, **105**, 1120–1122 (1887).
" Dilatation et compressibilité de l'eau et déplacement du maxi-
   mum de densité par la pression," *ibid.*, **104**, 1159–1161 (1887).
" Compressibilité des gaz : oxygène, hydrogène, azote et air
   jusqu'a 3000 atmos.," *ibid.*, **107**, 522–524 (1888).
" Recherches sur l'elasticité du cristal," *ibid.*, **107**, 618–620 (1888).
" Sur la verification expérimentale des formules de Lamé, et la
   valeur du coefficient de Poisson," *ibid.*, **106**, 479–482 (1888).
" Compressibilité du mercure et elasticité du verre," *ibid.*, **108**,
   228–231 (1889).
" Détermination directe (c'est-à-dire sans faire usage d'aucune
   formule) de la compressibilité du verre, du cristal, et des métaux,
   jusqu'à 2000 atmos.," *ibid.*, **108**, 727–730 (1889).

" Recherches sur l'élasticité des solides," *ibid.*, **108**, 1199–1202 (1889).

" Variation de l'élasticité du verre et du cristal avec la température," *ibid.*, **110**, 1246–1249 (1890).

" Nouvelle méthode pour l'étude de la compressibilité et de la dilatation des liquides et des gaz. Resultats pour les gaz : oxygène, hydrogène, azote, et air," *ibid.*, **111**, 871–875 (1890).

" Mémoire sur l'élasticité et la dilatabilité des fluides jusqu'aux très hautes pressions," *Ann. Chim. Phys.*, **29**, 68–136 (1893).

" Dilatation et compressibilité de l'eau," *C.R.*, **116**, 41–44 (1893).

" Sur la cristallisation de l'eau par décompression au-dessous de zéro," *ibid.*, **117**, 507–509 (1894).

[16] V. REGNAULT, " Sur un nouveau monomètre pour mesurer les hautes pressions des gaz," *Ann. Chim. Phys.*, **24**, 258 (1871).

[17] M. E. MASCART, " Über die Brechung des zusammen gedruckten Wassers," *Pogg. Ann.*, **153**, 154–158 (1874).

" Sur la réfraction de l'eau comprimée," *C.R.*, **78**, 801–805 (1874).

" Sur la réfraction des gaz," *Ann. de l'École Nor. Sup.*, **6**, 9–78 (1877).

[18] J. Y. BUCHANAN, " Preliminary Note on the Compressibility of Glass," *Trans. Roy. Soc. Edin.*, **29**, 589–598 (1880).

[19] W. SPRING, " Recherches sur la propriété que possèdent corps solides de se souder par l'action de la pression," *Bull. Roy. Acad. Belg.*, **49**, 323–379 (1880).

" Sur les quantités de chaleur dégagées pendant la compression des corps solides," *Bull. Soc. Chem. Paris*, **41**, 488–492 (1884).

" Sur la diminution de densité qu'éprouvent certain corps à la suite d'une forte compression et sur la raison probable de ce phénomene," *Jour. Chim. Phys.*, **1**, 593–606 (1903).

" Ueber einen Fall durch Druck bewirkter chemischer Zersetzung," *Zeit. Phys. Chem.*, **1**, 227–230 (1887).

" Ueber die Kompression von feuchtem Pulver fester Körper, und die Formbildung des Gesteine," *ibid.*, **2**, 532–535 (1888).

" Ueber die chemische Einwirkung der Körper im festen Zustande," *ibid.*, **2**, 536–538 (1888).

" The Compression of Powdered Solids : A Note," *Amer. Jour.*, **36**, 286–289 (1888).

[20] J. DEWAR, " On the Lowering of the Freezing Point of Water by Pressure," *Proc. Roy. Soc.*, **30**, 533–538 (1880).

[21] P. G. TAIT, " The Pressure Errors of the *Challenger* Thermometers : Report of the Voyage of H.M.S. *Challenger*," Narrative **2**, Appendix A, 42 (1881).

" Report on some of the Physical Properties of Fresh Water and Sea Water : Report of the Voyage of H.M.S. *Challenger*," *Phys. and Chem.*, **2**, 1–76 (1889).

[22] W. C. ROENTGEN und J. SCHNEIDER, " Über die Compressibilität von verdünnten Salzlösungen und die des festen Chlornatriums," *Wied. Ann.*, **31**, 1000–1005 (1881).
" Über Compressibilität und Oberflächenspannung von Flüssig-keiten," *ibid.*, **29**, 165–213 (1886).

[23] O. CHWOLSON, " Über den Einfluss des Druckes auf den electrischen Leitungswiderstand von Metall-Drähten," *Bull. St. Pet. Acad.*, **27-28**, 187–212 (1881–1882).

[24] D. H. MARSHALL, C. M. SMITH, and R. T. ORMUND, " Experiments to determine the Lowering of the Maximum Density Point of Water by Pressure," *Proc. Roy. Soc. Edin.*, **11**, 626–629, 809–815, (1882).

[25] S. PAGLIANI et VICENTINI, " Sur la compressibilité des liquides," *Jour. Phys.*, **2**, 461–462 (1883).

[26] S. PAGLIANI e L. PALAZZO, " Sulla compressibilita dei liquidi," *Mem. R. Acc. Lin.*, **19**, 273–300 (1883–1884).

[27] G. QUINCKE, " Über die Änderung des Volumens und des Brechungs-exponenten von Flüssigkeiten durch hydrostatischen Druck," *Wied. Ann.*, **19**, 401–438 (1883).

[28] CHAPPUIS et RIVIÈRE, " Sur les indices de réfraction des gaz à des pressions élevées," *C.R.*, **96**, 699–701 (1883).

[29] H. TOMLINSON, " The Influence of Stress and Strain on the Action of Physical Forces," *Trans. Roy. Soc. Lond.*, **174**, 1–172 (1883).

[30] II. G. CREELMAN and J. CROCKETT, " On the Thermal Effects pro-duced in Solids and in Liquids by Sudden Large Changes of Pressure," *Proc. Roy. Soc. Edin.*, **13**, 311–319 (1885).

[31] C. A. PARSONS, " Experiments on Carbon at High Temperatures and under Great Pressures, and in Contact with other Sub-stances," *Proc. Roy. Soc.*, **44**, 320–323 (1888).
" Some Notes on Carbon at High Temperatures and Pressures," *Proc. Roy. Soc.*, **79**, A, 532–535 (1907).

[32] W. HALLOCK, " The Flow of Solids : A Note," *Amer. Jour.*, **36**, 59–60 (1888).

[33] C. BARUS, " A Method of Obtaining and of Measuring very High Pressures," *Proc. Amer. Acad.*, **25**, 93–109 (1889–1890).
" Note on the Relation of Volume, Pressure, and Temperature in the Case of Liquids," *Amer. Jour.*, **38**, 407–408 (1889).
" The Effect of Pressure on the Electrical Conductivity of Liquids," *Amer. Jour. Sci.*, **140**, 219–222 (1890).
" The Compressibility of Liquids," *Bull. U.S. Geol. Sur.*, No. 92 (1892).
" The Volume Thermodynamics of Liquids," *ibid.*, No. 96 (1892).

[34] B.-C. DAMIEN, " Sur la variation du point de fusion avec la pression," *C.R.*, **112**, 785–788 (1891).
" Appareil pour la determination du point de fusion dans les conditions ordinaires et sous des pressions variables," *ibid.*, **108**, 1159–1161 (1889).

[35] G. DE METZ, " Über die Compressibilität der Öle und Colloide," *Wied. Ann.*, **41**, 663–674 (1890).
" Über die absolute Compressibilität des Quicksilbers," *ibid.*, **47**, 706–742 (1892).

[36] ROENTGEN und S. ZEHNDER, " Über den Einfluss des Druckes auf die Brechungsexponenten von Wasser, Schwefelkohlenstoff, Benzol, Aethyläther, und einigen Alkoholen," *ibid.*, **44**, 24–51 (1891).

[37] ROENTGEN, " Kurze Mittheilung von Versuchen über den Einfluss des Druckes auf einige physikalische Erscheinungen," *ibid.*, **45**, 98 (1892).

[38] T. DES COUDRES, " Über thermoelectrische Eigenschaften des Quecksilber und der sehr verdünnten Amalgame," *ibid.*, **43**, 673–699 (1891).

[39] P. GALOPIN, " Sur les variations de température de l'eau comprimée subitement à 500 atm. entre 0° et 10°," *C.R.*, **114**, 1525–1528 (1892).

[40] W. VOIGT, " Beobachtungen über die Festigkeit bei homogener Deformation," *Gött. Nach.*, pp. 521–533 (1893).

[41] G. TAMMANN, " Über die Binnendrucke in Lösungen," *ZS. Phys. Chem.*, **11**, 676–692 (1893).
" Über die Wärmeausdehnung und Kompressibilität von Lösungen," *ibid.*, **13**, 174–187 (1894).
" Korrespondierende Lösungen," *ibid.*, **14**, 163–173 (1894).
" Über den Einfluss des Druckes auf die Eigenschaften von Lösungen," *ibid.*, **14**, 286–300 (1894).
" Über die Abhängigkeit der Volumina von Lösungen von Drück," *ibid.*, **17**, 620–637 (1894–1895).
Über die spezifischen Wärmen der Lösungen," *ibid.*, **18**, 625–644 (1895).
" Über den Einfluss des Druckes auf das elektrische Leitvermögen von Lösungen," *ibid.*, **17**, 725–736 (1895).
" Über die Volumenänderungen bei der Neutralisation," *ibid.*, **16**, 139–146 (1895).
" Über die Volumenänderungen bei der Neutralisation verdünnter Lösungen," *ibid.*, **16**, 91–96 (1895).
" Über die Lage der thermodynamischen Flächen eines Stoffes im festen und flüssigen Zustande," *ibid.*, **21**, 17–34 (1896).
With K. ROGOYSKI, " Über adiabatische Volumenänderungen an Lösungen," *ibid.*, **20**, 1–18 (1896).
With A. BOGOJAWLENSKY, " Über den Einfluss des Druckes auf die Reaktiongeschwindigkeit in homogenen flüssigen Systemen," *ibid.*, **23**, 13–23 (1897).
With J. FRIEDLÄNDER, " Ueber die Krystallisationsgeschwindigkeit," *ibid.*, **24**, 152–159 (1897).
" Ueber die Erstarrungsgeschwindigkeit," *ibid.*, **23**, 326–328 (1897).
With A. BOGOJAWLENSKY, " Ueber den Einfluss des Druckes auf

das elektrische Leitvermögen von Lösungen," *ibid.*, **27**, 457–474 (1898).

" Ueber die Abhängigkeit des elektrischen Leitvermögens vom Druck," *Wied. Ann.*, **69**, 767–780 (1899).

" Ueber den Einfluss des Druckes auf den Schmelzpunkt des Zinns und des Wismuts," *ZS. Anorg. Chem.*, **40**, 54–60 (1904).

With A. D. COWPER, " Ueber die Änderung der Kompressibilität bei der Erweichung eines amorphen Stoffes," *ZS. Phys. Chem.*, **68**, 281–288 (1909).

" Das Eis III," *ZS. Anorg. Chem.*, **63**, 285–305 (1909).

" Ueber das Verhalten des Wassers bei hohen Drucken und tiefen Temperaturen," *ZS. Phys. Chem.*, **72**, 609–631 (1910).

" Die Stabilitätsbedingungen der beiden Kristallarten des Phenols," *ibid.*, **75**, 75–80 (1910–11).

" Das Zustandsdiagramm des Iodsilbers," *ibid.*, **75**, 733–762 (1910–1911).

" Über Zustandsgleichungen im Gebiete kleiner Volumen," *Ann. Phys.*, **37**, 975–1013 (1912).

" Das Zustandsdiagramm der Kohlensäure," *ZS. Phys. Chem.*, **80**, 737–742 (1912).

" Die Bestimmung der Schmelzkurven einiger bei tiefen Temperaturen schmelzender Stoffes," *ibid.*, **81**, 187–203 (1912).

" Die Methode der Bestimmung von p-T-Linien zur Feststellung von Zustandsdiagrammen," *ibid.*, **80**, 743–754 (1912).

" Über das Zustandsdiagramm des Wassers," *Gött. Nachr. Math.-phys. Klasse*, 38 pp. (1913).

" Über das Zustandsdiagramm des Wassers," *ZS. Phys. Chem.*, **84**, 257–292 (1913).

" Die Beziehungen der Volumfläche zum Polymorphismus des Wassers," *ibid.*, **84**, 293–312 (1913).

" Die Änderung der Temperatur des Volumenminimums von Wasser in Abhängigkeit vom Druck," *ZS. Anorg. Chem.*, **174**, 216–224 (1928).

" Die Volumenisobaren des Wassers bis zur Schmelzkurve," *ibid.*, **174**, 225–230 (1928).

" Über die Abhängigkeit des inneren Druckes in Lösungen von der Natur des gelösten Stoffes," *ibid.*, **174**, 231–243 (1928).

[42] S. LUSSANA, " Contributo allo studio della resistenza elettrica delle soluzioni considerata come funzione della pressione e della temperatura," *Nuov. Cim.*, **2**, 263–271 (1895).

" Influenza della pressione sulla temperatura del massimo di densità dell'acqua e delle soluzioni acquose," *ibid.*, **2**, 233–252 (1895).

" Influenza della pressione sulla temperatura di trasformazione," *ibid.*, **1**, 97–108 (1895).

" Contributo allo studio della resistenza ellettrica delle soluzioni considerata come funzione della pressione e della temperatura (Memoria I)," *ibid.*, **5** (1897), 49 pp.

" Influenza della pressione sulla resistenza elettrica dei metalli,"
ibid., **10** (1899), 14 pp.

" Influenza della pressione sulla resistenza elettrica dei metalli,"
ibid., **5** (1903), 12 pp.

" Proprietà termiche dei solidi e dei liquidi," *ibid.*, **5** (1903),
29 pp.

" Proprietà termiche dei solidi e dei liquidi," Memoria II, Parte I,
ibid., **7** (1904), 22 pp.

" Sul calore specifico liquidi a pressione costante sotto varie
pressioni," *ibid.*, **16** (1908), 4 pp.

" Sull'influenza della pressione e della temperatura sulla resistenza
elettrolittica," *ibid.*, **18**, 170–172 (1909).

" Proprietà termiche dei solidi e dei liquidi," Memoria II, Parte II.
" Compressibilità, coefficiente de dilatazione e calori specifici dei
metalli e delle leghe," *ibid.*, **19**, 182 (1910).

" Sul calore specifico dei liquidi a pressione costante per pressioni
e temperature diverse," *ibid.*, **4** (1912), 28 pp.

" La termodinamica dei gas e dei liquidi in rapporto alle appli-
cazioni pratiche," *Atti Soc. Ital. Progresso Sci.*, VII Riunione
(1913), 25 pp.   (Printed in 1914.)

" Sul calore specifico dei liquidi a pressione costante per pressioni
e temperature diverse," *Nuov. Cim.*, **7** (1914), 11 pp.

" Variation de la chaleur spécifique des gaz avec la pression,"
Annales de Physique, **6**, 344–356 (1916).

" Influenza della pressione sulla conducibilità calorifica ed
elettrica dei metalli e la legge de Wiedmann-Franz," *Nuov.
Cim.*, **15**, 130–170 (1918).

" Influenza della pressione sulla conducibilità calorifica ed
elettrica dei metalli e la legge Wiedemann-Franz," *ibid.*, **25**,
115–130 (1923).

[43] T. W. RICHARDS, with W. N. STULL, " New Method for Determining Compressibility," *Carnegie Inst.*, *Washington* (1903),
No. 7, 45 pp.

With W. N. STULL, F. N. BRINK, and F. BONNET, jun., " The Compressibilities of the Elements and their Periodic Relations,"
ibid., No. 76 (1907), 67 pp.

With G. JONES, " The Compressibilities of the Chlorides, Bromides,
and Iodides of Sodium, Potassium, Silver, and Thallium," *Jour.
Amer. Chem. Soc.*, **31**, 158–191 (1909).

" The Fundamental Properties of the Elements," Faraday Lecture
(1911), 18 pp.

With W. N. STULL, J. H. MATTHEWS, and C. L. SPEYERS, " Compressibilities of certain Hydrocarbons, Alcohols, Esters, Amines,
and Organic Halides," *Jour. Amer. Chem. Soc.*, **34**, 971–993
(1912).

" The Critical Point, and the Significance of the Quantity b in the
Equation of Van der Waals," *ibid.*, **36**, 617–634 (1914).

With C. L. SPEYERS, " The Compressibility of Ice," *ibid.*, **36**, 491–494 (1914).

" The Present Aspect of the Hypothesis of Compressible Atoms," *ibid.*, **36**, 2417–2439 (1914).

With E. P. BARTLETT, " Compressibilities of Mercury, Copper, Lead, Molybdenum, Tantalum, Tungsten, and Silver Bromide," *ibid.*, **37**, 470–481 (1915).

" Concerning the Compressibilities of the Elements, and their Relations to Other Properties," *ibid.*, **37**, 1643–1656 (1915).

With J. W. SHIPLEY, " The Compressibility of certain Typical Hydrocarbons, Alcohols, and Ketones," *ibid.*, **38**, 989–999 (1916).

With SVEN PALITZSCH, " Compressibility of Aqueous Solutions, especially of Urethane, and the Polymerisation of Water," *ibid.*, **41**, 59–69 (1919).

With J. SAMESHIMA, " The Compressibility of Indium," *ibid.*, **42**, 49–54 (1920).

" The Magnitudes of Atoms," *ibid.*, **43**, 1584–1591 (1921).

With E. P. BARTLETT and J. H. HODGES, " The Compressibility of Benzene, Liquid and Solid," *ibid.*, **43**, 1538–1544 (1921).

" Compressibility, Internal Pressure, and Atomic Magnitudes," *ibid.*, **45**, 422–437 (1923).

With E. P. R. SAERENS, " The Compressibilities of the Chlorides, Bromides, and Iodides of Lithium, Rubidium, and Cæsium," *ibid.*, **46**, 934–952 (1924).

" Compressibility, Internal Pressure, and Change of Atomic Volume," *Jour. of The Franklin Institute*, pp. 1–27 (July 1924).

" The Internal Pressures of Solids," *Jour. Amer. Chem. Soc.*, **46**, 1419 1436 (1924).

" Internal Pressures produced by Chemical Affinity," *ibid.*, **47**, 731–742 (1925).

With H. M. CHADWELL, " The Densities and Compressibilities of several Organic Liquids and Solutions, and the Polymerization of Water," *ibid.*, **47**, 2283–2302 (1925).

" A Brief History of the Investigation of Internal Pressures," *Chem. Reviews*, **2**, 315–348 (1925).

" Further Evidence concerning the Magnitude of Internal Pressures, especially that of Mercury," *Jour. Amer. Chem. Soc.*, **48**, 3063–3080 (1926).

With L. P. HALL and B. J. MAIR, " The Compressibility of Sodium, Barium, and Berylium," *ibid.*, **50**, 3304–3310 (1928).

" A Brief Review of a Study of Cohesion and Chemical Attraction," *Trans. Faraday Soc.*, No. 81 ; **24**, Pt. 2, 111–120 (1928).

With J. D. WHITE, " The Compressibility of Thallium, Indium, and Lead," *Jour. Amer. Chem. Soc.*, **50**, 3290–3303 (1928).

" Affinité chimique, cohésion, compressibilité et volume atomique. Étude des effets des pressions internes," *Jour. Chim. Phys.*, **25**, 83–119 (1928).

[44] E. COHEN, *Piezochemie Kondensierter Systeme*, Akademische Verlagsgesellschaft, Leipzig (1919).

Physico - Chemical Metamorphosis and Problems in Piezo-Chemistry, M'Graw Hill Book Co., New York (1926).

[45] J. JOHNSTON and L. H. ADAMS, " The Influence of Pressure on the Melting Points of certain Metals," *Amer. Jour. Sci.*, **31**, 501–517 (1911).

J. JOHNSTON, " A Correlation of the Elastic Behaviour of Metals with certain of their Physical Constants," *Jour. Amer. Chem. Soc.*, **34**, 788–802 (1912).

J. JOHNSTON and L. H. ADAMS, " On the Density of Solid Substances with especial Reference to Permanent Changes produced by High Pressures," *ibid.*, **34**, 565–584 (1912).

J. JOHNSTON and L. H. ADAMS, " On the Effect of High Pressures on the Physical and Chemical Behaviour of Solids," *Amer. Jour. Sci.*, **35**, 205–253 (1913).

J. JOHNSTON, " Some Aspects of Recent High Pressure Investigation," *Jour. Frank. Inst.* (January 1917), 32 pp.

G. W. MOREY, " Solubility and Fusion Relations at High Temperatures and Pressures," *Jour. Engineers' Club of Phila.* (Nov. 1918), 11 pp.

L. H. ADAMS and E. D. WILLIAMSON, " Some Physical Constants of Mustard ' Gas,' " *Jour. Wash. Acad. Sci.*, **9**, (19th Jan. 1919).

L. H. ADAMS, E. D. WILLIAMSON, and J. JOHNSTON, " The Determination of the Compressibility of Solids at High Pressures," *Jour. Amer. Chem. Soc.*, **41**, 12–42 (1919).

L. H. ADAMS, " The Compressibility of Diamond," *Jour. Wash. Acad. Sci.*, **11**, 45–50 (1921).

E. D. WILLIAMSON, " Change of the Physical Properties of Materials with Pressure," *Jour. Frank. Inst.*, **193**, 491–513 (1922).

L. H. ADAMS and E. D. WILLIAMSON, " The Compressibility of Minerals and Rocks at High Pressures," *ibid.*, **195**, 475–529 (1923).

G. W. MOREY, " The Development of Pressure in Magmas as a Result of crystallization," *Jour. Wash. Acad. Sci.*, **12**, 219–230 (1922).

L. H. ADAMS and R. E. GIBSON, " The Compressibilities of Dunite and of Basalt Glass, and their bearing on the Composition of the Earth," *Proc. Nat. Acad. Sci.*, **12**, 275–283 (1926).

L. H. ADAMS, " A Note on the Change of Compressibility with Pressure," *Jour. Wash. Acad. Sci.*, **17**, 529–533 (1927).

G. W. MOREY and N. L. BOWEN, " The Decomposition of Glass by Water at High Temperatures and Pressures," *Trans. Soc. of Glass Technology*, **11**, 97–106 (1927).

R. E. GIBSON, " The Influence of Pressure on the High-Low Inversion of Quartz," *Jour. Phys. Chem.*, **32**, 1197–1205, 1206–1210 (1928).

L. H. ADAMS and R. E. GIBSON, " The Elastic Properties of certain Basic Rocks and of their Constitutent Minerals," *Proc. Nat. Acad. Sci.*, **15**, 713–724 (1929).

46 A. MICHELS, " Einfluss der Rotation auf die Empfindlichkeit einer absoluten Druckwage," *Ann. Phys.*, **72**, 285–320 (1923).

" Genauigkeit und Empfindlichkeit einer Druckwage mit einem sogenannten Amagatzylinder," *ibid.*, **73**, 577–623 (1924).

" Mechanische Einflüsse auf die elektrische Leitfähigkeit von Metallen," *ibid.*, **85**, 770–780 (1928).

" The Behaviour of Thick Walled Cylinders under High Pressures," *Proc. Roy. Acad. Amst.*, **31**, 552 (1928), 8 pp.

" Isothermenmessungen bei höheren Drucken," *Ann. Phys.*, **87**, 850–876 (1928).

" The Use of the Effect of Pressure on the Electrical Resistance of Manganin as a Method of Measuring Pressure," *Proc. Roy. Acad. Amst.*, **32**, 1379 (1929), 7 pp.

47 T. C. POULTER, " A Glass Window Mounting for withstanding Pressures of 30,000 Atmospheres," *Phys. Rev.*, **35**, 297 (1930).

48 Complete List of References in the Appendix.

49 I. FANJUNG, " Über den Einfluss des Druckes auf die Leitfähigkeit von Elektrolyten," *ZS. Phys. Chem.*, **14**, 673–701 (1894).

50 B. PIESCH, " Änderungen des elektrischen Widerstandes wässeriger Lösungen und der galvanischen Polarisation mit dem Drucke," *Sitzber. Wien. Akad.*, **103**, 2a, 784–808 (1894).

51 W. C. ROENTGEN, " Über den Einfluss des Druckes auf die Dielectricitäts-constante des Wassers und des Aethylalkohols," *Wied. Ann.*, **52**, 593 (1894).

52 F. RATZ, " Über die Dielektrizitätskonstante von Flüssigkeiten in ihrer Abhängigkeit von Temperatur und Drück," *ZS. Phys. Chem.*, **19**, 94–113 (1895).

53 R. DEMERLIAC, " Sur l'application de la formule de Clapeyron à la température de fusion de la benzine," *C.R.*, **122**, 1117–1118 (1896).

" Sur la variation de la température de fusion avec la pression," *ibid.*, **124**, 75–76 (1897).

54 E. F. v. STACKELBERG, " Ueber die Abhängigkeit der Löslichkeit von Druck," *ZS. Phys. Chem.*, **20**, 337–358 (1896).

55 V. ROTHMUND, " Über den Einfluss des Druckes auf die Reactionsgeschwindigkeit," *Öfver. af Kong. Vet. Akad. Förhand. Stock.*, **53**, 25–40 (1896).

56 H. GILBAUT, " Untersuchungen über die Kompressibilität der Salzlösungen," *ZS. Phys. Chem.*, **24**, 385–441 (1897).

57 E. MACK, " Températures de fusion de quelques corps à des pressions élevées," *C.R.*, **127**, 361–364 (1898).

58 A. HEYDWEILLER, " Über Schmelzpunkterhöhung durch Druck und den continuirlichen Übergang vom festen zum flüssigen Aggregatzustand," *Wied. Ann.*, **64**, 725–734 (1898).

28 THE PHYSICS OF HIGH PRESSURE

" Erwiderung auf Herrn. G. Tammann's Bemerkung," *ibid.*, **66**, 1194–1195 (1898).

[59] G. A. HULETT, " Der stetige Übergang fest-flussig," *ZS. Phys. Chem.*, **28**, 629–672 (1899).

[60] Q. MAJORANA, " Comportamento del carbone sotto alte pressioni e temperature," *Rend. R. Acc. Lin.*, **9**, 2 Sem., 224–232 (1900).

[61] L. HAUSER, " Über den Einfluss des Druckes auf die Viscosität des Wassers," *Ann. d. Phys.*, **5**, 597–633 (1901).

[62] A. LAMPA, " Der Gefrierpunkt von Wasser und einigen wasserigen Lösungen unter Druck," *Wien. Ber.*, **111**, 2 A., 316–332 (1902).

[63] H. AGRICOLA, " Die thermoelektromotorische Kraft des Quecksilbers und einiger sehr verdünnter Amalgame in ihrer Abhängigkeit von Druck und Temperatur," *Diss. Erlangen.* (1902), 27 pp. ; *Beibl.*, **27**, 277 (1903).

[64] E. WAGNER, " Über den Einfluss des hydrostatischen Druckes auf die Stellung der Metalle in der thermoelektrischen Spannungsreihe," *Ann. d. Phys.*, **27**, 955–1001 (1908).

[65] E. SIEGEL, " Über den Einfluss des Druckes auf die Stellung flüssiger Metalle in der thermoelektrischen Spannungsreihe," *Ann. Phys.*, **38**, 588–636 (1912).

[66] H. HÖRIG, " Über den Einfluss des Druckes auf die thermoelektrische Stellung des Quecksilbers und der eutektischen Kalium-Natrium Legierung," *ibid.*, **28**, 371–412 (1909).

[67] P. CARNAZZI, " Influenza della pressione e della temperatura sul coefficiente di compressibilita del mercurio," *Nuov. Cim.*, **5**, 180–189 (1903).
" Dilatazione e Compressibilità delle Miscele," *ibid.*, **9**, 161–174 (1905).

[68] A. S. EVE and F. D. ADAMS, " Effect of Pressure on the Radiation from Radium," *Nature*, **76**, 269 (1907).

[69] R. LADENBURG, " Über die inere Reibung zäher Flussigkeiten und ihre Anhängigkeit vom Druck," *Ann. d. Phys.*, **27**, 287–309 (1907).

[70] F. MONTÉN, *Om Tryckets Inflytande på det Elektriska Ledningsmotståndet hos Selen ock Svavelsilver*, Upsala (1909), Almquist & Wiksells Boktryckeri.

[71] W. E. WILLIAMS, " On the Influence of Stress on the Electrical Conductivity of Metals," *Phil. Mag.*, **13**, 635–643 (1907).

[72] A. LAFAY, " Sur la mesure des pressions élevées déduite des variations de resistivité des conducteurs soumis à leur action," *C.R.*, **149**, 566–569 (1909).

[73] C. A. PARSONS and S. S. COOK, " Experiments on the Compression of Liquids at High Pressures," *Proc. Roy. Soc.*, **85**, 332–349 (1911).

[74] H. K. ONNES and B. BECKMAN, " On the Change induced by Pressure in Electrical Resistance at Low Temperatures : I. Lead," *Proc. Amst. Acad.*, **15**^{2}, 947–952 (1913).

[75] H. J. SEEMANN, " Der Einfluss allseitigen Druckes auf die metallische Leitfähigkeit in tiefer Temperatur (Kupfer)," *Phys. ZS.*, **30**, 256–258 (1929).

[76] W. H. KEESOM, " Solid Helium," *Comm. Phys. Lab. Leiden*, **184**, 9–20 (1926).

[77] F. SIMON, M. RUHEMANN, and W. A. M. EDWARDS, " Untersuchungen über die Schmelzkurve des Heliums," *ZS. Phys. Chem.*, **2**, 340–344 (1929) ; **6**, 62–77 (1930).

" Die Schmelzkurven von Wasserstoff, Neon, Stickstoff und Argon," *ibid.*, **6**, 331–342 (1930).

CHAPTER II

GENERAL TECHNIQUE

THE problems of pressure technique may be divided into two groups: general problems which must be solved in any piece of high-pressure work, such as problems of packing or of measurement of pressure, and special problems peculiar to different specific investigations. The special problems will be discussed later in the chapters dealing with special topics; only the general topics will be discussed here. The object of this discussion of technique is primarily to give what, in my opinion, is most useful for present-day laboratory practice, with special emphasis on methods of reaching the highest pressures, and only secondarily to present the historical aspects of the subject.

Packing. Evidently the most vital problem which must be faced in one form or another in practically every high-pressure investigation is the prevention of leak, and this again means in nearly every case some form of packing. The packing problem may take a great many forms; let us discuss by way of introduction the problem of packing a pipe connection. Three distinct stages in the evolution of such a packing may be distinguished. Fig. 1 sufficiently characterises the most primitive method, in which a heavy flange is attached to the end of the pipe, and a pliable packing is compressed by heavy bolts between the flange and the vessel to which connection is made. Packing of this sort is often used in steam connections.

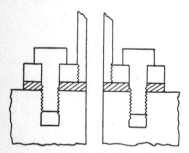

FIG. 1.—Most primitive form of packing, liable to be blown out either by the pressure or by tightening the bolts.

30

An obvious limitation of this method is that if the packing is too tightly compressed in the effort to make it proof against high pressure it will flow sidewise, either into the interior of the pipe or to the outside. Its range of use may be increased by incorporating into the flexible packing some more resistant material, such as metal gauze or cotton fabric. The design may further be improved by prolonging the end of the pipe beyond the flange sufficiently to enable it to enter the hole in the block. In this way the squeezing of the packing into the interior hole is avoided, but it is obvious that at sufficiently high pressure the packing will be blown out by the pressure itself, even if it is not squeezed out by the initial pressure of the bolts.

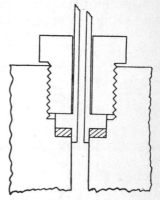

FIG. 2.—Amagat's type of fully enclosed packing. This leaks when the pressure gets as high as the initial pressure exerted by the screw.

The second stage in evolution is shown by fig. 2, which carries to its logical conclusion the idea suggested by prolonging the end of the pipe into the hole in fig. 1. In fig. 2 we have essentially a flange on the end of the pipe, but the flange enters a recess in such a way that the soft packing is entirely surrounded by rigid metal walls, and cannot flow out under the initial pressure of the screw or be blown out by the pressure. This is the essential motif of most of Amagat's packing; it enabled him to reach consistently very much higher pressures than had been commonly used before, up to a maximum of 3000 or even 4000 kg./cm.² An upper limit to which this packing may be used is evidently set by the compressing action of the pressure on the packing. When the hydrostatic pressure has reached an intensity equal to that initially exerted on the packing by the screw, there is balance, so that the packing exactly fills the surrounding space, but above this the packing shrinks away from the retaining walls and leak occurs. Leak will occur below the theoretical upper limit if there are geometrical irregularities, and the limit reached by Amagat is about the maximum that has been found practical.

The range of the packing of fig. 1 may often be increased by a crude application of the principle of fig. 2, by slipping a metal ring over the outside of the flange, thus retaining the packing on the outside.

The third stage in evolution (B. 12) is shown in fig. 3. At first sight this appears to be merely a modification of fig. 2, in that the location of the packing has been changed from the front to the back side of the flange, but this apparently trivial change has introduced a most important change of principle. It will be seen that the tendency of the pressure to blow the pipe out of the hole is transmitted to the screw through the packing. The force tending to blow the pipe out is the force exerted by the pressure on the annular packing area plus that exerted on the cross-section of the pipe itself. The total force exerted by the packing must be equal and opposite to this, but the force exerted by the packing is applied only to the annular area of the flange, which is less than the area on which the hydrostatic pressure acts by the cross-section of the pipe. It follows that the intensity of pressure in the packing in pounds per square inch must be greater than the intensity of the fluid pressure in the ratio of the area of the flange plus pipe to the area of the flange alone. That is, the geometrical design is such that the hydrostatic pressure in the packing is automatically maintained at a fixed percentage higher than the pressure in the liquid, and leak of the liquid cannot occur as long as the packing remains soft or as long as the retaining walls hold.

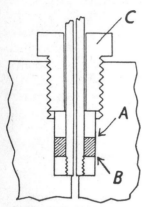

FIG. 3.—Packing utilising the 'unsupported area' principle, in which the pressure is automatically maintained a fixed percentage above the pressure in the liquid.

There is still the problem of preventing the packing from leaking out past the outer washer at the point indicated by the arrow A, or inward against the fluid pressure at the point marked B. This latter is an amusingly paradoxical possibility, but it may, nevertheless, be actually realised in practice, as one can see on considering that under actual conditions the pressure in the packing may be 5000 kg./cm.^{2} higher than the pressure in the liquid, an excess pressure very much above the flow pressure of the packing. However, the escape of the packing at these two places is easily prevented, as will be explained below, and the only real limit set to the pressures attainable with this type of packing at ordinary temperatures is set by the strength of the metal parts. There is a very definite limit to the strength attainable by the metal parts, which depends on an effect which was not generally known when I began my experiments, which I have called the "pinching-off" effect; this is described in greater detail in the next chapter. That there is such an effect can be shown by simple considerations. It is obvious that if the retaining screw C of fig. 3 is screwed in more and more tightly, the flexible packing is forced against the walls of the pipe, tending to collapse them, and that if sufficient force is exerted, collapse will take place, and the pipe will be "pinched off," the protruding part being expelled by the pressure of the packing. Very much the same sort of thing takes place when the packing is strongly compressed by the action of a high hydrostatic pressure; the pipe is eventually pinched off and expelled. It does no good to increase indefinitely the thickness of the walls of the pipe; a solid rod replacing the pipe of fig. 3 is pinched off in the same way. The maximum pressure that a solid rod can withstand without pinching off is numerically somewhere in the neighbourhood of the maximum tensile strength of the steel as measured by ordinary tensile tests. This is also approximately the maximum pressure that a pipe in a state of ease can stand without bursting, but in practice the pinching-off limit is always somewhat lower than the bursting limit. In designing apparatus it is wise to stay well below the expected upper limit, for there is considerable danger when pinching-off occurs, the velocities of the expelled parts sometimes reaching those of rifle bullets. In that part of my work which deals with pressures approaching the tensile strength of the steel, an entirely different sort of connection is used, which will be described later.

In practice I have used continuously the type of packing of fig. 3 with heavy copper tubing (6 mm. outside diameter and 1·5 mm. inside diameter) up to pressures of 1000 kg./cm.², and with hard-drawn carbon steel tubing of about 0·50 per cent. carbon content and with an outside diameter about five times the inside diameter up to 3000 or 4000 kg./cm.². It is now possible to obtain tubing of alloy steel of similar dimensions of higher tensile properties, but I do not believe that there is any steel that could be safely used in this way up to 12,000 kg./cm.².

For ordinary service conditions, the most convenient packing material is soft rubber. Lead is considerably more permanent, but must be compressed more tightly initially to prevent leak. Rubber is possible only in a moderate temperature range; at low temperatures it gets hard and cracks, and at high temperatures it decomposes.

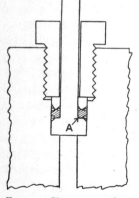

FIG. 4.—Shows a method of retaining the packing with metal rings.

Returning now to the matter of preventing the packing itself from leaking, various devices are possible. It is paradoxical that rubber packing can be prevented from leaking more easily than lead packing. If the pressure is not too great, it is sufficient merely to make the metal parts a good fit; this is adequate up to 2000 kg./cm.². For higher pressures, one method is to taper the metal washers to knife-edges; this method is illustrated in my paper on technique (B. 12). Later, however, I have found more convenient the use of rings of soft steel in a way suggested clearly enough in fig. 4. This method is particularly useful in closing holes; in this case the pipe with flange screwed on the end is replaced by a single piece of alloy steel, as shown in fig. 4, heat treated to give a high tensile strength combined with considerable elongation. The corner at A must be rounded, otherwise pinching-off is likely to start at this place.

The fundamental principle illustrated by fig. 3 I have called the principle of the unsupported area. The pipe projecting through the screw is the unsupported area in this case. Wherever there is such an unsupported area, the pressure in the packing is greater than the pressure in the fluid, and leak cannot occur. The principle is capable of a great many adaptations, as will appear in the following. I believe that one form or another of this principle will be essential in reaching the absolutely highest pressures; it is, at any rate, the only thing that has made possible my own high-pressure work.

If one is not trying for the maximum pressure, there are a good many possibilities in the way of connections that it is convenient to have in one's repertory. The ordinary plumber's tapered thread can, with care, be made to hold up to several thousand kg./cm.^{2}. Such connections demand considerable care, however, and are likely to go wrong in use. I personally have seldom been in a situation where I have felt that the initial saving of effort has justified their greater liability of failure. Barus[1] invented an interesting modification of this idea, his so-called "tinned screw." A carefully cut steel screw is coated with soft solder. If the screw is turned home with great force, the solder is sufficiently forced into the cracks between the threads to withstand pressures up to 2000 kg./cm.^{2} for a long time.

A few general comments on the use of solder in making connections may be interjected here. Brazed or silver soldered connections have a limited usefulness, because of the necessary softening of the steel incident to the red heat necessary in such soldering. This restricts the use of such connections to 3000 or perhaps 4000 kg./cm.^{2}. Soft soldered connections have a somewhat wider application in the low-pressure parts of the apparatus. The strength of the soft solder is itself very low, so that the parts must be so designed that the solder is contained in narrow cracks, supported as far as possible by the surrounding strong metals. In making such a connection, the pipe and the part into which it screws are first both carefully tinned, and then screwed together hot, so that the solder fills all the empty spaces. Such a joint will eventually fail, perhaps after continued use for many years, by the solder pushing out of the crack like a viscous liquid. The length of time that such joints will last at 1000 kg./cm.² is so great, however, and the incon-

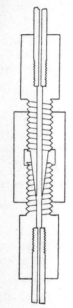

venience of renewing the connection so inconsiderable, that I use this type of connection regularly with brass and copper parts for this low-pressure range.   If one goes to 2000 kg./cm.², however, the chance of failure is so much greater that some more permanent type of connection is usually justified.   For example, there is on the market in America a Bourdon pressure gauge made by a well-known manufacturer, calibrated and designed for 2000 kg./cm.².   The connections to the spring of the gauge are made with soft solder, and although the gauge always passes the test of the manufacturer, it invariably fails after a few weeks of use, so that I have always had to replace the solder with a packing, using the principle of the unsupported area.

FIG. 5.—Pipe connection for quick assembly at low pressures.

Returning now to the question of convenient forms of pipe connection, the internal and external cone arrangement, pulled together with a right- and left-handed screw, shown in fig. 5, is useful for connections up to 1000 kg./cm.² that have to be taken apart rather often.

The double-cone arrangement of fig. 6 is one of the most convenient methods of making connections to steel pipe. The double cone of hardened steel A is pressed firmly into the ends of the pipe by a suitable screw arrangement. The pipe need receive no treatment except to be dressed flat with a file and threaded to receive a flange. The hole in the double cone must be considerably smaller than the hole in the pipe, in order that the cone may actually enter the pipe. The pressure applied with the screw drawing the pipes together is great enough to cause considerable flow around the tip of the cone, so that the local stresses are high. This practically limits the application of the method to steel tubing. The small areas of contact involve the advantage that the screw parts need not be as heavy as in many other schemes. Theoretically this connection has the disadvantage that the higher the pressure the greater the tendency to leak; but, nevertheless, because of the unusual smallness of the bearing surfaces, it is capable of use over an extended pressure range. Personally I have not used it perhaps as much as I might, because I have not been invariably successful with it. I have lost track of the origin of this device; it probably originated in the Fixed Nitrogen Laboratory after the war, at any rate it has

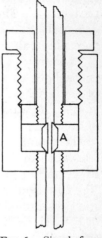

Fig. 6.—Simple form of pipe connection for steel tubing.

been extensively used there for gas pressures up to 2000 kg./cm.[2].

Another connecting device, with which I have had no direct experience but which should be convenient under many conditions, is a sort of cross between the arrangements of figs. 5 and 6, and is described in a paper from the Geophysical Laboratory by Johnston and Adams.[2]

The method of pipe connection which I believe is capable of the greatest range, and which I have consistently used above 6000 kg., applies the principle of the unsupported area in a somewhat modified

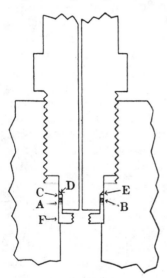

Fig. 7.—Pipe connection for the highest pressures.

form. This is shown in fig. 7. It does not apply to drawn tubing, but the tube must be drilled from the solid rod and machined to the required dimensions. Only in this way is it possible to make tubing of the highest grades of alloy steel and give them the most suitable heat treatment. The packing consists of three rings, two of steel, A and C, and one of lead, B. The ring C pushes against a conical shoulder D. The annular space of triangular section E is empty. The 'unsupported area' in this case is the entire back surface of the ring C. The ring is crowded by the pressure into contact with the shoulder and out against the containing walls, in such a way that the intensity of pressure in the regions of contact is greater than the intensity of pressure in the liquid. This evidently demands some flow in the steel ring, and in fact this part of the packing does not begin to function until a pressure of 5000 or 6000 kg./cm.^{2} is reached. The function of the lead ring B is to provide tightness until the pressure of flow of the steel is reached. With continued use the ring C is gradually forced into the space E, until it is squeezed out to a knife edge. When flow to this extreme has taken place there is no unsupported area left; the packing may now leak and the rings should be renewed. It is obvious that the rings must be of softer steel than the pipe, that the conical surface must be smooth, and that the rings must be a good enough fit so that the lead washer may receive a sufficient amount of initial compression. These conditions are easy to attain, and I have found the method very convenient in practice. To facilitate the removal of the rings, they are compressed against the edge of the steel cup F, which is threaded to allow its removal, when it draws the rings out with it.

The discussion thus far has been concerned only with the packing of stationary parts, but somewhere in a high-pressure apparatus there must always be movable parts to be packed. If the pressure is produced by a hydraulic intensifier, there is in the first place the plunger which operates the low-pressure end of the intensifier. This plunger may perhaps best be packed with a rather conventional stuffing-box arrangement, in which the stuffing material is leather washers made a close fit for the hole and the plunger, and restrained on all sides by metal parts after the manner of Amagat.

These washers must be impregnated initially with a heavy grease, such as mutton tallow, until they are soft and flexible. In use the grease gets squeezed out gradually, and the washers eventually get very rigid and must be replaced by fresh ones. The life of a packing like this depends very much on the conditions of use. If the liquid to be pumped is, for example, a glycerine and water mixture, which does not attack the grease in the leather, and if the pressure range is below 1000 kg./cm.^{2}, such packing may be used for years. But if the liquid (such as kerosene) attacks the grease, and if the pressure gets up to 2000 kg./cm.^{2}, renewal must be made very much oftener. I have under extreme conditions obtained 4000 kg./cm.^{2} with a plunger packed in this way. Of course such a packing always leaks to a certain extent, and parts of the apparatus containing such plungers must be cut off from the rest of the apparatus by suitable valves, so that they are exposed to pressure only during the actual operation of changing pressure.

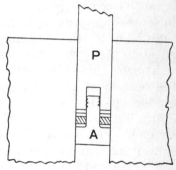

FIG. 8.—Method of packing the moving piston.

To reach the highest pressures the principle of the unsupported area must be used. Fig. 8 illustrates an arrangement that I have found very convenient. The essence of it is the mushroom-shaped plug A, which is pushed into the high-pressure vessel by the plunger of hardened steel P. The stem of the mushroom projects freely into a hollow in the end of the piston, and it is this stem that constitutes the unsupported area. It is at once evident that the pressure in the soft packing, indicated by the shading, is greater than the pressure in the liquid in the ratio of the area of the head of the mushroom to the area of the annular space on which the packing acts. The soft packing, for which soft rubber is convenient, is prevented from leaking by discs of soft steel and copper, as indicated. The friction in the packing is considerable, so that its thickness should be kept low. On a diameter of 0·5 in. I usually use a total thickness of rubber of $\frac{1}{8}$ inch. If this is made in two thicknesses of $\frac{1}{16}$ in., a lubricant of a mixture of graphite and vaseline may be smeared between the two rubber washers, considerably reducing friction. This packing is so very easy to renew that it does not pay to try to use it for more than two or three strokes of the piston. It is absolutely without leak, and has been used in measuring compressibilities, when readings consistent to 0·0001 in. have been obtained. In case pressure is lost and the plug sticks, as is often the case, removal of the plug is facilitated by a thread on the end of the stem of the mushroom. This method of extracting the plug when stuck is much superior to another method which I have described in several previous accounts of high-pressure technique. The limit to which the plug can be used is set by the pinching-off effect on the stem of the mushroom. For this reason the heat treatment has to be carefully adjusted, and the shoulder at the bottom of the stem must not be sharp. With suitable grades of alloy steel the pinching-off effect is no serious limitation up to pressures of 20,000 kg., and perhaps materially beyond. For the very highest pressures I have used on the end of the piston a modification of the ring packing of fig. 7. This is still in the experimental stage, however, and at present suffers from the disadvantage that only the first forward stroke can be made without leak.

Tubing. Most of the essential discussion of tubing has already been covered incidentally in connection with the packing, but it will be convenient to collect it into one place.

For pressures up to 1000 kg./cm.², such as are employed in the low-pressure end of the intensifiers with which the higher pressures are attained, copper tubing of 6 mm. outside diameter and 1·5 mm. inside diameter, or with a similar ratio of inside to outside diameter, is most convenient, because of its flexibility. For pressures from 1000 up to perhaps 4000 kg./cm.² a hard-drawn carbon steel tubing of approximately 0·50 per cent. carbon content, and a ratio of outside to inside diameter of about 5 to 1 may be used with the type of connection of fig. 3, and with the connection of fig. 6 to perhaps 6000 or possibly 7000 kg./cm.². Higher pressures may doubtless be attained with some of the drawn tubing of alloy steel now available, but my direct experience with such tubing is limited. For considerably higher pressures the tube must be of alloy steel, drilled from the solid rod, and heat treated. For security the ratio of outside to inside diameter should be as large as possible; in my most usual tubing the ratio is ten to one, the outside diameter being 1·25 inches, and the inside 0·125. It is not nearly as difficult as might be supposed to drill such long holes as are required for this tubing. The rod to be drilled is rotated in the lathe against the stationary drill. The speed of rotation must be high. Very accurate starting of the drill is, of course, essential. After the drill has penetrated a sufficient distance to guide itself, the back stop of the lathe should be entirely removed, and the drill fed in by hand with a suitable handle. Only in this way is it possible to remove the drill often enough to keep the hole free from chips. If the drill is kept well lubricated and the hole free from chips, a $\frac{1}{8}$-in. hole may be easily drilled at the rate of 6 or 7 in. per hour, and a hole 0·040 in. in diameter at the rate of 3 in. per hour.

Glass tubing has often been used in pressure measurements at comparatively low pressures. Several hundred kg./cm.^{2} may be attained in heavy glass capillaries without difficulty, and there is the extreme record by Heydweiller of a pressure above 3000 kg./cm.^{2}. Glass suffers, however, from the disadvantage of great unreliability; it will often break for no apparent reason at a low pressure after its successful use at a much higher pressure. Its use should be avoided wherever possible. When necessary, connections to glass tubing may be made with some cement, such as sealing-wax or marine glue, a bulge of some sort being preferably blown on the tube to withstand part of the tendency of the pressure to expel the tube from the high-pressure vessel to which it is attached.

Miscellaneous Problems of Technique. A convenient way of systematising the remaining problems of technique will be to describe a typical high-pressure apparatus, referring in detail to the various parts. A section of the principal parts of the high-pressure part of such an apparatus is shown in

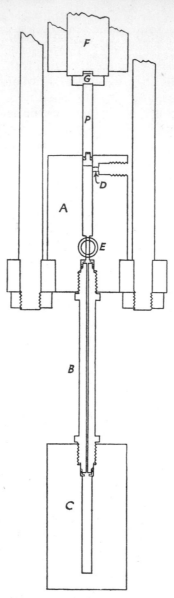

FIG. 9.—General assembly, showing the cylinder, above, in which pressure is produced by the advance of a piston driven by a hydraulic press, the connecting pipe, and the lower cylinder adapted to the particular experiment.

PLATE I

FIG. 9A.—Photograph of a typical high-pressure installation. The tank, motor and stirrer, bunsen burners, etc., are appurtenances for maintaining the lower cylinder at constant temperature. In use, the tank surrounds the lower cylinder.

fig. 9, and a photograph of a typical high-pressure installation, with various appurtenances connected with the thermostated bath, in fig. 9A (Plate I). The part of the apparatus subject to the high pressure consists of three principal parts— an upper cylinder A, in which pressure is produced by the descent of a plunger P, a connecting pipe B, and a lower cylinder C, in which the experiment of the day is set up. The lower cylinder changes from experiment to experiment, the rest remains fixed. The upper cylinder contains two important attachments. Near the upper end there is a side connection to an auxiliary pressure generator by which an initial pressure of 2000 or 3000 kg./cm.^{2} may be produced in the apparatus before the high-pressure piston starts to descend. The function of the initial pressure is to permit the attainment of the maximum pressure in a single stroke, with a larger capacity in the lower cylinder than would otherwise be possible. The apparatus is limited to a single stroke, and the compressibility of the transmitting liquid sets a very definite limit to the capacities that can be employed, the volume loss of the transmitting liquid under 12,000 kg./cm.^{2} varying from 25 to 30 per cent. The initial pressure at the upper end is led in through a very narrow by-pass, made either by drilling a very small hole in the screw plug D, or else by making a fine longitudinal scratch across the threads of this plug. After the initial pressure is produced the main plunger is actuated, moving over the by-pass and cutting it off.

The second attachment in the upper cylinder is the pressure-measuring gauge, which is attached through E, and which will be described in greater detail later.

The upper cylinder has to be subjected to a special treatment before it is suited to continued use at high pressure. The scale is sufficiently shown in the figure. The outside diameter is of the order of nine times the inside diameter. The dimensions which I usually use are about 0·5 in. for the inner diameter and 4·5 in. for the outer. The cylinder is subjected to a preliminary application of pressure higher than that at which it is intended to use it in order to season it. This involves making the inner hole at first somewhat smaller than the final size. The procedure is as follows: The cylinder is machined from an annealed piece of a suitable alloy steel. I have used a chromium-vanadium steel of the type known in America as type D. A typical analysis is: Cr 0·93, Mn 0·68, C 0·51, Si 0·23, Va 0·18, S 0·010, P 0·008. The steel which I have found most successful is a product of the electric furnace. This makes it somewhat more expensive, but the chances against internal flaws are so much greater that the difference in cost is worth while. There is nothing more exasperating than to spend a great deal of time and labour in machining and treating a piece of high-pressure apparatus, and then to have it fail because of some minute flaw deep in the body of the ingot which cannot be detected by any ordinary means. The first machining of the cylinder should leave the inside diameter somewhat under size, perhaps $\frac{15}{32}$ in. if the final size is to be $\frac{1}{2}$ in. The steel is then hardened and drawn by a suitable amount, so as to leave the steel with high tensile strength, but still capable of a fair degree of elongation before rupture. The steel that I use is drawn back to a Brinell hardness of 400. After drawing, the upper hole in the cylinder is subjected to a seasoning application of pressure by filling the hole with lead, and exerting pressure on it with a plunger until the interior takes a perceptible set. The purpose of the preliminary stretching of the cylinder is to increase the elastic limit by producing in it a distribution of internal stress, which will be explained in greater detail in the next chapter. The amount of stretch must be judged somewhat by experience, and the pressure for which the cylinder is ultimately destined. If the pressure range contemplated is 12,000 kg./cm.², it is not necessary to go beyond 20,000 kg./cm.²; under these conditions the stretch of the interior is such as to leave the inside diameter still well below the desired $\frac{1}{2}$ in. If the cylinder is to be used to a higher pressure, greater initial pressures must be applied. The cylinders which I have intended for 20,000 kg./cm.² have been stretched to 30,000, but there is considerable uncertainty of getting the desired results at these very high pressures, and one must be prepared for disappointment. Such high preliminary pressures are likely to start obscure flaws in the interior, and even if no such flaws are encountered, the raising of the elastic limit by such a pressure is only more or less temporary. The final experiments must be made as soon as possible after the treatment, and one must be prepared to find that the cylinder has a short life. After stretching, the hole is enlarged to the final size, $\frac{1}{2}$ in. for the comparatively low pressure ranges, or the minimum necessary to leave the hole completely cylindrical if the cylinder has been considerably stretched by a materially higher pressure. Even for use at 12,000 kg./cm.^{2}, one cannot expect the stretching process to be absolutely permanent, but the cylinder very slowly becomes larger in continued use, and at long intervals the hole must be further enlarged. I have for this purpose a set of reamers in steps of 0·010 in.

Sometimes, after stretching the upper part of the cylinder, which during the stretching is shut off from the other parts by a plug in the bottom of the hole, it may be desirable to stretch the two lower holes for the pressure gauge and the bottom pipe connection. Lead cannot be used for the second stretching process, but a liquid, such as a mixture of water and glycerine; furthermore, the regular packing must be used on the piston, whereas for the original stretching with lead a simple cupped packing of steel is sufficient to prevent the lead from leaking. This stretching of the two lower holes is not usually necessary, however, since obviously the demands of accuracy in the hole which contains the moving plunger are much more exacting than in those holes which carry stationary parts. If the lower holes should stretch in use, it is easy to enlarge them by a small amount if necessary. The only real difficulty that arises from the stretching of the holes for stationary attachments is that if the stretch goes too far the connecting pipe is not a close enough fit, and the steel packing rings are in danger of blowing out through the annular space. However, there is considerable latitude here. I have one cylinder in which the stuffing gland for the pipe has gradually stretched to 0·800 in., whereas the end of the pipe which should fit it is 0·750 in., making a crack 0·025 in. wide. A packing ring of ordinary mild steel would be blown out through a crack of this size, but by making the rings of alloy steel, and drawing back to a temperature considerably in excess of that to which the cylinder or pipe is drawn, I have been able to reach 12,000 kg./cm.^{2} without difficulty. There is, however, a limit to the use which can be made of hardened packing rings, in that the rings must obviously be softer by a very material degree than the pipe against which they are forced.

The connecting pipe and the lower cylinder are not usually subjected to a preliminary stretching, although occasionally such treatment may be desirable for the lower cylinder. The connecting pipe has already been sufficiently discussed, and will not be described further here.   The lower cylinder varies somewhat with the particular experiment, and will be discussed later under the special technique required for the different sorts of measurement.

The high-pressure plunger has to be made of an entirely different grade of steel from the rest of the apparatus, since it is subject to a different sort of stress.   This plunger is best made of some steel capable of extreme hardness, such as the steel from which ball-bearings are made, and should be left glass hard.   It is surprising how much compressive stress such a steel in the glass-hard condition can stand if it is so supported that the compression is applied absolutely uniformly, with no tendency to buckle or side motion.   I have with one steel reached compressive stresses of 750,000 lb./in.^{2}.   In order that the force on the plunger should be without a sidewise component, it is necessary that the press by which the plunger is actuated should be made with as great accuracy as possible, taking great care that the centre lines of all parts are true.   If by any chance the plunger should break, the rupture will occur with great suddenness, and parts are likely to be projected with considerable violence. Once a small piece flew out of such a plunger with sufficient velocity to drill a clean hole through an adjacent window, with no cracking of the surrounding glass, exactly as a rifle bullet might have done.   For this reason it is always advisable to surround the hardened piston with a sheet-metal shield of sufficient thickness to guard against such flying fragments.

The appliance by which the plunger is pushed into the

high-pressure chamber may be made to suit the personal taste of the user. In the early days of high-pressure work, the plunger was almost always driven by some sort of a screw, and pictures may be found in the literature of enormous capstan-like arrangements which required the whole force of one or two men to operate. These arrangements are very cumbersome, however, and, furthermore, if the screw is used to a pressure approaching the capacity of the steel, are very inefficient. My early plunger was driven by a screw, and when near the limit of the apparatus, which was 6000 kg./cm.^{2}, the screw was delivering to the plunger only 5 per cent. of the energy applied. There can be no question that some sort of a hydraulic press is to be preferred from every point of view for driving the plunger. The figure illustrates the arrangement which I have myself found convenient. The main piston of the press F is 2·5 in. in diameter, which, operating on a high-pressure plunger P of 0·5 in. diameter, gives a multiplication of twenty-five fold. This 2·5-in. piston is driven by a hand-pump of the Cailletet type capable of 1000 kg./cm.^{2}, so that with no allowance for friction the pressure attainable in the high-pressure part is 25,000 kg./cm.^{2}. As a matter of fact, the friction with the design of packing shown later in fig. 14 cuts this maximum down by about 2000 kg./cm.^{2}. The 2·5-in. piston is packed with the same type of mushroom packing as the high-pressure piston. The only particularly novel feature about the large piston is that it is threaded for its entire length, and provided with a nut heavy enough to stand the maximum pressure. For simplicity in drawing, the nut is not shown in fig. 9. This nut can be set when the piston has reached any desired point, and in this way the high pressure is retained even if the low pressure should be lost. It is somewhat paradoxical that the low pressure is always more difficult to retain without leak than the high pressure. The reason for this is that the low-pressure part of the apparatus must always have certain valves for convenience of operation, to which the principle of the unsupported area cannot be easily applied, so that there is almost always some slight leak in the low-pressure end. Because of the thread, it is necessary to guide the end of the low-pressure piston to prevent side motion, with consequent buckling of the glass-hard, high-pressure piston. This guiding is conveniently done with a plate of brass, also not shown in fig. 9, attached to the lower end of the piston, and sliding on the tie rods of the press, which must evidently also be accurately aligned. All the low-pressure parts of the press may be made of ordinary mild steel. To prevent the glass-hard piston P being forced into the end of the soft steel piston, an interposed block of hard steel G is necessary, sufficiently indicated in the figure. A further attachment to the low-pressure part of the press, which is so convenient as to be almost indispensable, is some method of withdrawing the piston after the end of the stroke. This is particularly necessary if by some accident the pressure should be lost from the high-pressure part and the piston left in the hole with no pressure to drive it out. The withdrawal of the piston is conveniently accomplished with another smaller cylinder, not shown, in my case 0·875 in. inside diameter, mounted coaxially with and back to back to the 2·5-in. cylinder. The piston of the smaller cylinder is connected with tie rods and yokes to the large piston. After the termination of the forward stroke, the pressure from the hand-pump may be transferred by a suitable arrangement of valves to the smaller cylinder, while the larger cylinder is open to the atmosphere, and the large piston withdrawn by the advance of the small piston.

Low-pressure Pumps. If the high-pressure piston is advanced by a hydraulic press instead of a screw, some sort of pump for its operation will be necessary. There is not space in this book to go into the design of low-pressure pumps in any detail. There are a number on the market sufficiently good, such, for example, as the Cailletet pump of the Société Genevoise, capable of 1000 kg./cm.². There are, however, a few general remarks worth making on the subject. The connections which are usually found with commercial high-pressure outfits are some sort of soft-soldered joint, or else some variety of the brute-force connection. These will last for a time, but it has been my experience that they eventually give out, and that I have had to replace them with "unsupported area" packings. In the pumps of my own design, made in the shops of the laboratory, this design of packing and connection is used wherever possible. There are, however, a couple of places where this is not possible. The moving plunger must be used so often that some other packing of the stuffing-box variety must be used. There is also the question of the valves, an inlet valve to the chamber in which the plunger operates, and the outlet valve from this same chamber. I have never had a valve that is entirely satisfactory. The valve supplied by the pump with the Société Genevoise is a conical valve of hard rubber. This works well for a time, but after much use gradually gets squeezed down into the lower end, where it sticks, and eventually breaks. I have used a conical valve of copper in place of this with rather good results; or a ball valve works fairly well. Probably everyone who has much experience with a pump of this kind will have his own try at designing a valve, and he will doubtless succeed in getting many forms that appear for a long time to have solved the problem, and then suddenly something inexplicable will go wrong, and the pump will be out of commission for days. Probably the greatest single enemy of an efficient valve is dirt in the pump liquid. The most extravagant precautions in straining the pump liquid will be well rewarded. But even perfect straining will not completely solve the problem, because it is not possible to entirely prevent fragments detaching themselves from the leather packing of the plunger and sticking sometimes to the valve seats. The inlet valve is the one which I have found to make almost all the trouble.

For a pump liquid I have found a mixture of two parts glycerine and one part water satisfactory. It has the advantage over a lubricating oil, which is also used by many experimenters, that it does not attack any soft rubber packings, which are rather convenient to use in applying the principle of unsupported area.

Valves. Besides the automatically operating valves necessary in the pump discussed in the last paragraph, other valves in the low-pressure part of the apparatus are very desirable for deflecting the pressure first to one part of the

apparatus and then to another, as from the lower cylinder
to the upper cylinder of the hydraulic press, in this way
avoiding the necessity for a plurality of pumps. A very
convenient principle to use in such valves is that of a hardened
steel point pressed into a small hole in softer steel. An
example of such is shown in fig. 10. Such a valve is obviously
bilateral. The hardened point with which the hole is cut off
has to be packed with some packing that will permit its
motion. The simple arrangement of leather washers indicated in the figure is one that I have found convenient. Such

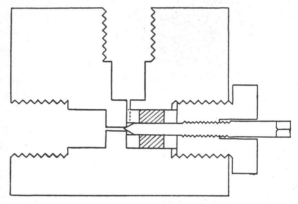

Fig. 10.—Needle valve.

a packing can be made to stand pressures of several thousand
kg. The point may be forced into the hole with a force
up to the elastic limit of the steel, so that on the other side
of the point, when the valve is shut, it is possible to confine
considerably higher pressures than can be used on the low
side. I have used such valves up to 7000 kg./cm.^{2} on the
high-pressure side, and have no reason to think that they
may not be carried considerably farther, since similar devices
have been carried higher. For instance, a hardened point
pressed against the open end of a pipe is a simple and rapid
method of plugging off part of the apparatus which might
happen temporarily not to be wanted. Combinations of
valves of this type may be mounted together in the same
block, connected in various ways according to the job in

hand. I have a combination of three in a single block, which is very convenient for operating with a single pump the two cylinders of the hydraulic press and also the intensifier, by which pressure is fed into the by-pass at the upper end of the high-pressure cylinder, and which will be described in a later paragraph.

Technique of Insulation. A great many times it is desirable to get electrically insulated leads into the pressure chamber. It is obvious that in order to do this the insulating substance must somewhere be exposed to a stress gradient, determined by the maximum pressure, and that therefore rather severe demands are put on the strength of the insulating material. Insulating materials in general do not possess a high degree of mechanical strength, having either a low flow point, if they are like rubber, or else being brittle, if like glass. It has proved to be a matter of considerable difficulty to make an insulating connection that will satisfactorily withstand the highest pressures; everything depends on the design, which must be such that the steel parts afford the maximum assistance to the insulating material. On the other hand, insulating connections for low pressures can be made with comparative ease.

For pressures up to as much as 1000 kg./cm.^{2} or so, a perfectly satisfactory insulating electrode can be made with any of the ordinary cements, such as sealing-wax or de Khotinski cement. The insulated rod may most conveniently be surrounded with a bushing of glass tubing, which is thrust over the rod when filled with melted cement. The rod must also be prevented from pushing out of the pressure chamber by some sort of a head. Since resistance to leak is here afforded primarily by the resistance of the cement to shearing stresses, the thickness of the layer of cement should be as little as possible, and its length as great as possible. There are doubtless many other arrangements convenient for a few hundred kg./cm.^{2}. Professor Keyes of the Massachusetts Institute of Technology has a simply made device, using a low melting glass instead of the softer cement.

A connection capable of reaching considerably higher pressures is the inverted cone connection of Amagat, shown in fig. 11.   The insulation is here a thin conical shell of ivory, shown shaded in the drawing, between the electrode and the body of the pressure vessel.   The principle of the unsupported area is effective here to a certain extent, because the protruding electrode is unsupported, and the pressure in the

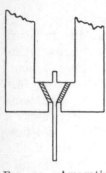

ivory is greater than the pressure in the liquid.   The parts of the apparatus must evidently be machined with considerable exactness to prevent initial leak, and for this reason it is somewhat difficult to make. Amagat used it successfully to 3000 kg./cm.².   This is about the limit of its use, however, because at materially higher pressure the ivory is extruded between the electrode and the containing vessel into a tube.   Considerable search on my part has failed to

FIG. 11.—Amagat's method of insulating electrodes with an ivory cone.

disclose any substance easy to machine and capable of reaching materially higher pressures than ivory.   I have used most of the things that readily suggest themselves, such as various bakelite and celluloid materials, and also a great variety of bones, including the anvil bone of the ear of the manatee or sea-cow, which a physiologist assured me is the hardest substance in the animal kingdom.   Bone, by the way, is not suitable for use as an insulator at the highest pressures, even in parts of the apparatus where it is exposed only to hydrostatic pressure all over, and therefore is not called on to stand any mechanical strain, because the small amounts of water which it contains, and which cannot be removed by any process of heating in vacuum, are expelled by the pressure and introduce a small but appreciable amount of conductivity.

After much experimenting I have finally adopted the form shown in fig. 12, or some modification of it, for the highest pressures.   The essential parts of the packing are the washers of mica A, which are the nearest to the outside, and are called on to stand the brunt of the mechanical stress.   Tightness against leak is provided by the packing of soft rubber B above the mica.   This rubber is made somewhat too large for the hole initially, and for this reason is tight, in addition

to being compressed by the unsupported area of the stem
of the electrode. The area of the stem is not a very large
fraction of the area of the mica washers, however, so that
the advantage gained from the principle of the unsupported
area is not as great here as in most other places. In fact, the
advantage at the highest pressures might not be great enough
to overcome the greatly increased rigidity of the rubber at
high pressure, so that if the rubber
were not made initially enough larger
than the hole to allow for the volume
compression, it would shrink away
from the walls of the hole and leak
would ensue. Above the soft rubber
there is another thickness of mica C,
or any other conveniently worked insulating material, by which the force
tending to expel the electrode is transmitted to the rubber. A very important feature is the thin washer of steel
S at the bottom of the rubber packing; the function of this is to prevent
the rubber from extruding the mica as
a cylindrical tube. A convenient ring
of steel R, fitting into a conical seat
on the washer S, prevents the rubber
from squeezing between the stem and
S. It is evident that the greatest
stresses in the mica are at the unsupported annular ring D; this ring should

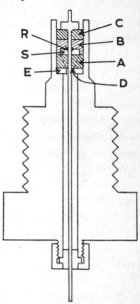

Fig. 12.—Insulated elect-
rode for the highest pressures.

be made as narrow as possible consistent with safe insulation.
This may involve drilling the steel plug with an inconveniently
small hole; to avoid this difficulty in construction the plug
may be drilled with a hole of convenient size, and the annular space made smaller with a disc of steel E placed in the
bottom of the hole, as indicated.

Such a plug, carefully constructed, is capable of a great
many applications of 12,000 kg./cm.^{2}, and of a fair number
of applications of 20,000 or even more, but failure eventually
takes place along the cone of shear indicated by the dotted

lines, the mica is reduced to a powder, which is expelled through the annular space, and insulation fails, usually by bending of the electrode, so as to come into contact with the steel walls. This insulating plug is the least satisfactory part of the high-pressure apparatus. It may be possible to improve it by the use of some insulating substance materially stronger than mica. In small sizes it may be possible to use such things as sapphires, and I am at present working on such a design, but it is too early to know the results.

In fig. 12 the electrode shown is turned in a single piece from alloy steel and heat treated. This is the most convenient method if the electrode is of some size, such, for example, as 0·06 in. in diameter at the stem. But if the electrodes must be made smaller, as is necessary when mounting three or more electrodes in a single plug, machining the slender stem of the electrode is not convenient, and I have used instead the proper size of piano wire, which is suitable because of its great mechanical strength. To prevent the wire from being expelled it must be attached to a head of some kind, analogous to the head turned on the solid electrode of fig. 12. If the wire is too large to bend readily to a small radius, the head may be a split cone affair. I have used this arrangement with wire of 0·035 in. in diameter. If the wire is small enough to be more readily bent, 0·015 in. in diameter, for example, it may be more easily attached to a head by threading through holes or notches in a steel piece in a way that will readily suggest itself. Before use in the high-pressure apparatus, the piano-wire stems should be tested in tension; the head should be able to stand up to 90 or 95 per cent. of the tensile strength of the steel without pulling off. If the head is not firmly attached the wire will be pushed through it by the pressure, and expelled with considerable violence. There is always danger of the stem of the insulating plugs being expelled, even if the original assembly is without fault, because as the mica washers gradually shear through, the stem becomes exposed to bending and shearing strains, under which it sometimes breaks loose from the head and is then expelled. It is well to protect all such stems with a heavy plate of steel, so placed as to catch them in case of expulsion.

There is another source of failure in the insulating plugs—namely, failure of the rubber packing. The highest pressure must be transmitted by a liquid which does not freeze or become too viscous under pressure, and, in addition, wherever electrical measurements are involved, must be an electrical insulator. There seem to be no such liquids which at the same time do not exert some solvent action on the rubber. The transmitting liquids which I usually use are kerosene, if a moderate amount of viscosity is no disadvantage, or petroleum ether, if freedom from viscosity is more essential. Either of these attack rubber; the rubber packing in fig. 12 softens and swells under the action of the liquid, so that sometimes the electrode rises above the edge of the plug, or else the deteriorated rubber fails mechanically. It is more or less of a gamble as to whether the rubber or the mica will fail first. Attack of the liquid on the rubber may be minimised by melting around the upper part of the stem some thick grease, such as a mixture of vaseline and beeswax, by which access of the transmitting liquid to the rubber is hindered. But any such protecting coating goes into solution in the transmitting liquid, so that if the pressure run is at all long the protection disappears. This is particularly the case at temperatures materially higher than room temperature. A great deal can be done by selecting the best grades of rubber. Recently there has been developed in the automobile industry a rubber for the tread of tires which has considerable resistance to the action of oils, and at the same time has sufficient softness; this is perhaps the best rubber that I have found. There are other grades made especially to resist the action of oil by impregnating the rubber with glue, which are also good, but these are likely not to be as flexible as desirable.

At high temperatures I have not succeeded in devising any direct method of getting electrically insulated connections into the pressure chamber; the device above must evidently fail at temperatures above 150° C., where the rubber chars, and in fact the design of fig. 12, cannot be carried much above 100°. For use at higher temperatures, some indirect device is necessary in which the insulating plug is in a separate chamber, kept cool, and the leads pass from this cool chamber through the connecting pipe to the hot chamber. At high temperature the lead rings of the pipe packing may be successfully replaced by copper.

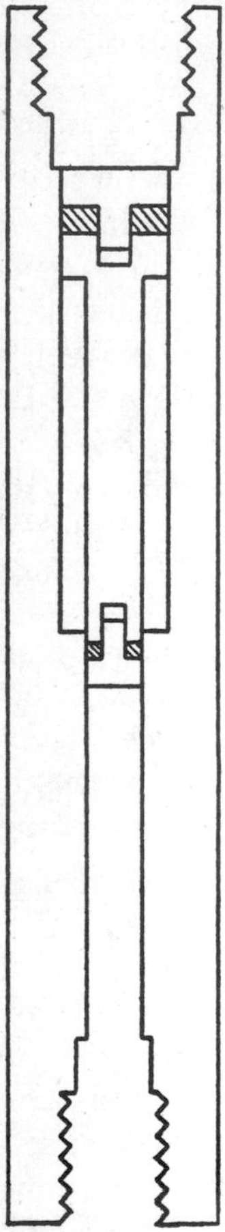

Intensifier. A very convenient appliance, useful in many manipulations, is some compact and light device by which pressures can be attained above the range of the hand-pump, and below the pressure of the high-pressure part of the apparatus. For example, such an appliance can be used to prime the high-pressure cylinder through the by-pass at the top to an initial pressure of 2000 kg. or 3000 kg./cm.² For this I have used a simple arrangement which I have called an intensifier, which gives pressures up to 4000 kg./cm.² It is nothing but a hydraulic press, with two pistons of four to one ratio, mounted together in a single cylinder. The design is sufficiently indicated in fig. 13; no extended description will be given. The packing of both pistons is the conventional unsupported area piston packing. An inconvenience of the intensifier is that no specific provision is made for the return stroke. This may be done in several ways; a simple method is to unscrew the connection at the small end and force the plunger back in an arbor press. Some such arrangement as this is almost necessary in those cases in which the high-pressure apparatus is to be filled with some liquid different from the liquid of the hand-pump. It is, for example, impractical to use such a liquid as petroleum ether in the hand-pump, although such a liquid is necessary in the high-pressure part of the apparatus if pressures as high

Fig. 13.—The intensifier, a compact arrangement for stepping up the pressure fourfold.

as 12,000 kg./cm.^{2} are to be reached at a temperature of 0° C.

Glass Windows. For many purposes it is desirable to be able to examine optically the inside of the pressure chamber. The problem, however, is a difficult one, because of the very low strength of glass, or any other available transparent material, compared with steel, and the fact that glass is likely to receive internal strains, so that after several applications of pressure the breaking stress may be a small fraction of the initial value.

If it is possible to work on a small scale, the whole apparatus may sometimes be mounted in heavy glass capillaries, but the pressures so attainable are limited to a few hundred kg./cm.^{2}. Madelung and Fuchs [3] used an ingenious modification of the capillary by so mounting the apparatus that the capillary was exposed to pressure on the outside rather than the inside, and the interior of the pressure apparatus was viewed through a system of mirrors in the interior of the capillary. The crushing strength of glass tubing to external pressure is much higher than the bursting strength to internal pressure. However, the difficulties of making a tight connection between the glass and the steel are serious, so that in actual use Madelung and Fuchs did not go higher than a few hundred kg./cm.^{2}.

For pressures of a 1000 kg./cm.^{2} or more some form of window, supported as well as possible by the steel parts of the apparatus, appears to be necessary. The first arrangement was due to Amagat, who used a cone, packed with ivory, in very much the same way as his insulated electrodes. The pressure range of this was not much over 1500 kg./cm.^{2}; at this pressure the windows fail in a curious way by separating into thin plates parallel to the plane faces. Wahl [4] improved the design by inverting the cone, so that the large end is outside, and using a double cone packing of red fibre, forced by the pressure into the space between the window and the steel walls. With this arrangement he reached pressures as high as 4000 kg./cm.^{2}; the manner of failure was the same as in Amagat's arrangement.

Higher pressures may be reached, and the laminar method of failure avoided, if the window is given a straight cylindrical form rather than the form of a cone. The problem of packing the glass cylinder then must be reopened. In the apparatus used by Miss Wick in my laboratory for measuring optical absorption, the packing was done by an arrangement of soft rubber tubing. With this, pressures as high as 12,000 kg./cm.² were reached, although the actual experiments were to much lower pressures. After this, Poulter [5] made the interesting discovery that no packing at all is necessary, provided the plane surface of the glass window is made sufficiently flat. Fig. 14 illustrates an arrangement used by Collins [6] in my laboratory employing the packing-less window of Poulter. The abutting surfaces of steel and glass are made optically flat; evidently under these conditions there can be no leak, because the intensity of stress on the bearing surface is greater than the hydrostatic pressure, by the principle of the unsupported area. In practice it is convenient in handling to stick the window to the steel initially with a very thin layer of Canada balsam. With this arrangement Poulter states he has reached 30,000 kg./cm.², but the most that I have been able to reach is 12,000, the same as with the rubber-packed arrangement, and more usually something in the neighbourhood of 6000. Rupture occurs in a curious way on hemispherical surfaces indicated by the dotted line in the figure. Reflection shows that the distribution of stresses in the rubber-packed and the packing-less windows is the same, so that the packing-less arrangement should have only the advantage of greater simplicity. There is a source of uncertainty not yet cleared up in the fact that all of Poulter's high pressures were reached in an oil as transmitting medium, which may freeze under pressure. Poulter reached no higher pressures than I when the transmitting medium is water.

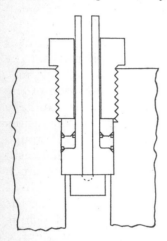

FIG. 14.—Poulter's packing-less glass window as used by Collins.

The strongest material yet found for the windows is an ordinary greenish plate-glass containing considerable iron. Collins, however, found this opaque in the infra-red, and for his purpose was forced to use pyrex, which, unfortunately, has much less mechanical strength. Quartz, either crystalline or amorphous, is distinctly inferior to plate-glass.

[1] C. BARUS, *Proc. Amer. Acad.*, **25**, 93–109 (1889–1890).
[2] JOHN JOHNSTON and L. H. ADAMS, *Amer. Jour. Sci.*, **31**, 505 (1911).
[3] E. MADELUNG und R. FUCHS, *Ann. Phys.*, **65**, 289 (1921).
[4] W. WAHL, *Trans. Roy. Soc. Lond.*, A, **212**, 117–148 (1912).
[5] T. C. POULTER, *Phys. Rev.*, **35**, 297 (1930).
[6] J. R. COLLINS, *ibid.*, **36**, 305 (1930).

CHAPTER III

THE MEASUREMENT OF HIGH PRESSURE

METHODS of measuring pressure might have been described in the chapter on technique, but so much effort has been given to this particular subject, and it is so extensive, that it seems better to treat it separately.

Pressure gauges may be conveniently classified into primary gauges—that is, gauges so constructed that the absolute pressure can be at once approximately found from the construction of the instrument itself; and secondary gauges, the readings of which can be interpreted into absolute pressure only after a proper calibration.

Primary Gauges.—By far the simplest and the first to be used is the open mercury column, in which the pressure at any point in a liquid is given directly by the height above that point of an open mercury column. The pressures measurable with such means are limited by the height of the column, and have been restricted in practice to a few hundred kg. Perhaps the highest pressures reached in practice with gauges of this type have been attained by Cailletet [1] and Amagat,[2] who reached 300 or 400 kg. by utilising a deep mine or the Eiffel Tower. There are obviously serious inconveniences in such a gauge if it is to be used over a range of pressure great enough to necessitate mounting the tube that carries the mercury in several sections or the equivalent. Cailletet, who used in his early work a high hill, had an ingenious scheme by which the mercury was carried in a flexible steel tube wound on a reel, which could be unrolled by different amounts for different heights. There are various corrections to be applied to the direct reading of such a gauge—for temperature differences at different parts of the column, and also for the compressibility of the mercury. This latter correction must be determined by independent experiment, so that in a strict sense this is not a primary gauge; indeed, there is no primary gauge in the strict sense, for corrections, even though slight, which demand an approximate knowledge of the pressure, must be applied to the readings of all such gauges. The corrections to an open mercury column gauge are not difficult to apply, however, and this sort of gauge is probably capable of greater percentage accuracy than any of the others.

The open mercury column evidently becomes impractical when extended to pressures of several thousand kg., and some radically different method must be designed. The device of a number of mercury columns in series, which readily suggests itself, and which has been used at Leyden and other places in work on critical phenomena at lower pressures, is evidently too clumsy for such high pressures. There is practically only one type of primary gauge that has been successfully applied to pressures above 1000 kg., namely, some form of the so-called free piston gauge. This consists merely of a piston directly exposed on one end to the pressure to be measured, and so accurately fitted to its cylinder that leak is unimportant. The force required to keep the piston in place against the expelling effect of the pressure is measured in any convenient way, and so the pressure is obtained directly in terms of the force and the measured cross-section of the piston. The first suggestions of this form of gauge go back a hundred years or more to Perkins [3] and Parrot and Lenz. [4]. Natterer, [5] in the 1850's, used essentially this. The first extensive use of the gauge was by Amagat, [6] who gave it a very convenient form. A novelty of Amagat's gauge consisted in the use of a second free piston to measure the force expelling the high-pressure piston. The instrument is shown in fig. 15. A large and small piston are mounted coaxially; the small piston is exposed to the high pressure, and the total thrust on it is transmitted directly to a much larger piston, which is in connection with a free mercury column. It is evident that when the system is in equilibrium the pressure under the large piston is less than the pressure acting on the smaller piston in the ratio of the areas, and this ratio may be made so large that a pressure of several thousand kg. may be measured with a mercury column a few meters high. One great convenience of the apparatus is obvious in that the mercury column comes at once automatically to the height

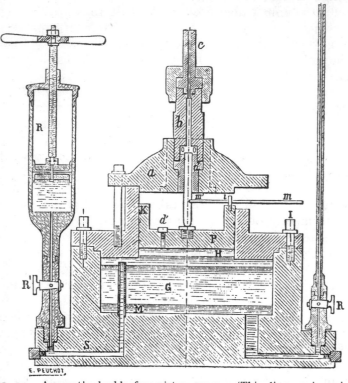

FIG. 15.—Amagat's double free piston gauge. (This diagram is copied from Amagat's 1893 paper.)

required to balance the high pressure. In use, the friction of the pistons is eliminated as far as possible by giving them a rotary motion immediately before each reading. There are evidently corrections to be applied to the cross-sections of the pistons for the distortion produced by the pressure, but these corrections are small. Amagat used this gauge to 3000 kg./cm.^{2}; this was about the maximum possible, as at higher pressures leak past the high-pressure piston became prohibitively great.   Leak was cut down as far as possible by using a very viscous liquid—molasses.   The gauge never came into extensive use; one reason, doubtless, is that it demanded unusually accurate construction.   I know of one attempt to reproduce the gauge in this country, which was not successful above 2000 kg./cm.[2] because of leak.

There have been a large number of different designs for the free piston gauge, and several very careful examinations of the theory of their performance,[7] nearly all, however, to pressures of only a few hundred kg.   Free piston gauges are a commercial article, and are most commonly used in the calibration of the more convenient forms of secondary gauge, such as the Bourdon spring.   Practically all these free piston gauges exert the force on the piston by hanging weights on it, thus accounting for the commercial name of such gauges, "dead-weight testers."   For the most part the differences of design are concerned with the method of suspending the weights from the piston, and of giving it a rotating motion.   This latter feature, following the lead of Amagat, has now become universal in all free piston gauges making any pretence to accuracy or sensitiveness.

There are two main variations of the free piston gauge as ordinarily made; the first kind is the simplest, in which a single piston is used and weights are directly applied to it. In the second type, two pistons are used, of slightly different diameters, and so mounted as to oppose each other, so that the total force measured is a differential effect of pressure due to the difference of the two areas.   It is immaterial whether the two pistons are joined in a single piece, as in fig. 16, the appearance being then that of a single piston with a shoulder, or whether the pistons are separate pieces of steel, joined by an external framework as in fig. 17.   The manifest advantage of such an arrangement is that the equilibrating force for a given pressure is very much reduced compared with what would be demanded by a single piston. If one tries to achieve the advantage of small weight with a small piston by reducing its diameter, two difficulties are encountered; the mechanical stiffness of a small piston becomes smaller, so that it is difficult to hang weights on it without flexure, and the difficulty of measuring the effective area of the piston with a given percentage accuracy increases as the diameter of the piston becomes smaller.

The most recent work on the free piston gauge is by A. Michels.[8] A certain gauge investigated by Michels up to 200 kg. has a sensitiveness of 1/120,000; in order to attain this sensitiveness the piston must be rotated with a speed

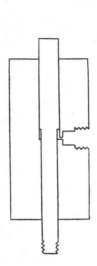

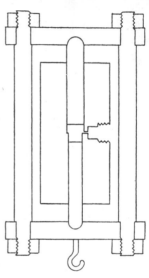

FIG. 16.—Differential free piston gauge, in which the two diameters are on the same piston. The weights are placed in a pan hung from the piston by means of the screw thread at its lower end.

FIG. 17.—Differential free piston gauge, the two diameters being on two separate pistons, united by an external frame. The pan for the weights is hung from the hook on the lower part of the frame.

above a certain critical speed, depending on the viscosity of the oil and the dimensions, among other things. Within the accuracy of the measurements, which was about 1/10,000, the effective diameter was the mean of the diameter of the piston and the cylinder. Within this limit there was no appreciable change of the effective area over the pressure range, which a theoretical calculation indicated should not be more than 1/23,000.

The free piston gauge, as conventionally constructed, can

probably be used to several thousand kg., the limit being imposed by leak past the piston.   To reach higher pressures, the leak must be avoided by special design.   The free piston gauge that I have used to pressures of 13,000 kg./cm.², shown in fig. 18, utilises two methods of reducing leak (B. 4).   In the first place, the piston P is made small, only $\frac{1}{16}$ in. in diameter, which has the incidental advantage of cutting down the

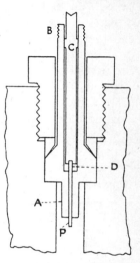

necessary equilibrating force, and in the second place the cylinder in which the piston plays is so mounted that the pressure acts on its external surface A as well as in the annular space between piston and cylinder.   The external pressure contracts the cylinder around the piston, and so prevents as great an enlargement of the annular space as would otherwise take place.

For convenience in operation, the weights for measuring the force on the piston are replaced by stiff springs, not shown in the diagram, attached by the screw thread B.   The force tending to expel the piston is transmitted to the springs by the plunger C.   The springs consist of two discs of sheet steel about 6 cm. in diameter and 2 mm. thick, dished to the form of saucers, and mounted face to face, in contact around

Fig. 18.—Form of free piston gauge for high pressures. The action of the pressure on the outside of the cylinder carrying the piston diminishes leak.

the edges.   By this method of mounting, relative motion of the springs when deflected is avoided, and so frictional effects also.   The stiffness is such that the deflection under a force of 600 pounds, corresponding to the maximum pressure, is only about 1 mm.   This small deflection may be read to an accuracy considerably better than 0·1 per cent. by a simple mechanical and optical magnifying device.   In addition to the advantage of freedom from clumsiness, the use of stiff springs very much lessens the necessary stroke of the piston, thereby permitting the unsupported part D of the piston to be made so short that there is no danger of buckling under the end thrust, a danger which would be very great because of the unusually small diameter of the piston if weights had to be employed.   But there is a disadvantage in the use of springs instead of weights, in that any force determined from the elastic deformation of a spring is subject to error from hysteresis and other effects.   These errors may be cut down by various devices.   In the present case the design of the springs themselves was arrived at only after considerable experiment.   Furthermore, the best procedure in using the springs was carefully worked out by studying the deflection of the springs under dead weights in a specially constructed apparatus.   It was found that if the weights are applied or removed monotonically, a hysteresis loop of considerable width is described, but that if before any reading in a series with increasing weights the load is run somewhat beyond the intended value and then released, and if the inverse procedure is adopted for a series with decreasing weights, hysteresis effects may be eliminated, the relation between force and deflection becoming linear within the error of the readings, which was somewhat less than 0·1 per cent.

· The piston is so small that the effective area could not be determined with sufficient accuracy by direct measurement of the geometrical dimensions.   The area was determined indirectly by the use of a second free piston gauge with piston large enough so that its area could be determined directly by geometrical measurement.   The two pistons were exposed to the same low pressure, and the effective area of the small piston determined from the ratio of the forces on the two pistons and the known area of the larger piston.

At high pressures the deformation of the piston produced by pressure becomes appreciable, and some correction must be applied.   There seems no direct way of determining this correction, which I found by calculation from elastic theory. In view of the smallness of this correction, which at the maximum pressure of 13,000 kg./cm.² was only 0·23 per cent., the correction may probably be accepted with considerable confidence, although if it had been higher there might have been more uncertainty because of the increasingly important part played by hysteresis and elastic after-effects at high stresses.

The liquid by which pressure is transmitted to the high pressure piston must be carefully selected. The ordinary oil suitable for free piston gauges designed for low pressures freezes at high pressures. If a hydrocarbon, such as kerosene or petroleum ether is used, which does not freeze at the highest pressure, the leak in the first few thousand kg. will be prohibitive. I found suitable a mixture of equal parts water and glycerine, to which was added enough glucose to make a rather thick syrup. Most liquids increase enormously in viscosity at high pressures, so that even if they do not freeze they become so stiff that it is impossible to rotate the piston to secure freedom from leak. The effect of pressure on the viscosity of water is comparatively small, however, so that with the liquid mentioned the piston could still be rotated at 13,000 kg.

Employing all precautions, readings good to 0·1 per cent. were obtained over a pressure range of 13,000 kg. It would doubtless be possible to materially increase this accuracy. However, accuracy of this amount seemed to me all that was required in the virgin field of pressures above 3000 kg., and I have since then devoted all my efforts to the measurement of various effects referred to pressure readings on this gauge for their ultimate absolute measurement. Recently, however, Michels has taken up the problem of increasing the accuracy of high-pressure measurement by a method involving the step-wise extension by means of a free mercury column of the range of the free piston gauge, and doubtless results will soon be published by him, if indeed they do not appear before this book.

Secondary Pressure Gauges. In almost all high-pressure investigations the direct measurement of pressure with a free piston gauge would be inconveniently clumsy, and there would be the further disadvantage that such a gauge always has a slight amount of leak, so that some form of secondary gauge is nearly always desirable. Any reproducible and conveniently measured high-pressure effect may be made the basis of a secondary gauge. Secondary gauges may be classified into those which depend on some specific property of matter, so that after the property has once been measured no calibration of the secondary gauge is necessary, and those which depend on some more complicated property of the special gauge, which must therefore be calibrated individually. The dividing line between the two classes is not sharp but depends on the desired accuracy.

The most common and one of the most convenient of secondary gauges is the Bourdon spring, which is a tube flattened and bent to an arc of a circle. When exposed to internal pressure the tube uncoils by an amount measured with a suitable multiplying mechanism, so arranged that the pressure is shown by the position of a pointer on a dial. Such a gauge obviously belongs to the second group, and each gauge must be individually calibrated, usually against a dead-weight tester. The pressure range of such a gauge is limited by the fact that the walls of the tube must be thin enough to allow the requisite flexibility. The maximum range at present claimed for any such gauge is 4000 or 5000 kg./cm.^{2}; there is a gauge manufactured by Schaeffer and Budenberg of this rated capacity. I have used such a gauge to a maximum of 3000 for a number of years with entire satisfaction. It is possible to obtain in America commercial gauges rated to 50,000 lb./in.^{2}, but I have always had trouble with these because of leak past the soldered connections, as mentioned in the last chapter. The accuracy of the Bourdon gauge, like all elastic deformation gauges, is limited by hysteresis and similar effects in the steel of the spring. These effects become greater as the wall thickness of the spring becomes relatively greater, and are therefore greatest in those gauges for the highest pressures. As a rough indication, the accuracy of a spring gauge may be put at 1 per cent., although by special manipulations higher accuracy may be attained. Professor Richards used such a gauge up to 500 kg. for many years, which, by the exercise of unusual care remained accurate to considerably better than 1 per cent. There are now, however, so many other convenient methods of accurate pressure measurement, and in particular the construction of a free piston gauge has become so comparatively easy because of improved mechanical technique in grinding operations, that it would be unwise to trust to a gauge of this type in any accurate work. In my own experiments, this type of gauge plays little more than the rôle of a rough indicating instrument on the low-pressure part of the apparatus, where it is so convenient as to be wellnigh indispensable.

Other types of elastic deformation gauge have been proposed and used. In particular, the so-called 'Tait' gauge, invented by Tait [9] and extensively used by Barus,[10] may be mentioned. In this gauge the change of internal volume of a cylinder exposed to external pressure is indicated by the rise of mercury in a capillary. This gauge suffers from hysteresis. There is a real place for a convenient gauge capable of roughly indicating pressures up to 12,000 kg. or more. So far I have not been able to devise any of sufficient simplicity; in particular, any elastic deformation gauge capable of application over this range suffers from the delicacy of the multiplying mechanism needed to make visible the very small deformations in vessels capable of withstanding such stresses. The Tait gauge used to high pressures has large hysteresis, and also has inconveniently large temperature corrections.

A pressure effect which has several times been made the basis of a secondary gauge is the volume change in a fluid; a gas for pressures up to several hundred kg., or a liquid like water up to several thousand. In particular, the use of such a gauge by Lussana [11] may be mentioned. In Lussana's gauge a bulb filled with water was connected to a stem of glass tubing; pressure was transmitted to the interior through a column of mercury, which rose or fell in the glass tube, compressing the water behind it as pressure rose or fell. The glass bulb was, of course, subject to pressure on the outside. The height of the mercury in the tube was calculated from the resistance of a fine wire of platinum stretched along the axis of the tube, which was short-circuited to a greater or less extent as the mercury changed position. In reducing the readings to absolute pressure the results of Amagat for the compressibility of water were used, thus limiting the range of the gauge as used by Lussana to 3000 kg.

A source of inaccuracy in the gauge is the correction for the compressibility of the glass envelope, which may vary according to the compressibility of the glass by relatively large amounts, making possible an error of several per cent. in the calculated pressures. A further disadvantage arises from the irregular motion of the mercury meniscus; there is always a tendency for small drops of mercury to be left sticking to the wire, and the readings are likely to be irregular on this account.

The measurement of the compressibility of some liquid and its comparison with some value accepted as standard as a method of gauge calibration is not uncommon; it has, for example, been employed by Tammann, who calibrated his Bourdon gauges in this way.

Probably by far the simplest effect on which to base a secondary gauge is the change of electrical resistance of a metal wire. Once a method has been perfected for getting electrically insulated terminals into the pressure chamber, there is no physical measurement more easily made than one of resistance. The magnitude of the changes of resistance produced by pressure in metals is not large, so that the method is not particularly convenient for pressures of the order of hundreds of kilograms, but for pressures of thousands of kilograms the changes of resistance can be easily measured to 0·1 per cent. with conventional apparatus, and in this range the method has had extensive application. The suggestion that pressure should be measured in terms of change of resistance was perhaps first made by Lisell [12] in 1903. Lisell measured up to about 3000 kg. the resistance of a number of metals, among others the alloy manganin. This was found to increase in resistance under pressure, a rather unusual effect, and, furthermore, the rate of increase was linear, whereas all the other metallic substances examined were distinctly non-linear. This was so striking an effect that it was natural to propose its utilisation in the measurement of pressure. There was another great convenience in the use of manganin for this purpose, in that its temperature coefficient of resistance is so low that no special precautions are necessary in the way of keeping temperature constant, whereas the temperature coefficient of all the pure metals is so high that rather elaborate precautions are necessary to keep the temperature constant enough to allow a deduction of the pressure from the observed change of resistance. In 1909 La Fay [13] measured the effect of pressure on the resistance of a number of samples of manganin, and endorsed the suggestion of Lisell that it should be used as a secondary pressure gauge. La Fay found, as had also Lisell, that the pressure coefficient of different samples of manganin might differ by appreciable amounts, so that for accurate work the calibration of each sample of manganin is necessary, which is a disadvantage. The literature of this period contains several controversial articles by various authors, provoked by the contention of Lussana,[14] that the pressure coefficient of manganin is not positive, as found by Lisell and Lafay, but is negative. The solution of a situation, which at the time seemed inexplicable, was given by the discovery by Lussana [15] a number of years later, that the manufacturers had inaccurately labelled as manganin an alloy that was something quite different.

At the same time that I was working on the high-pressure form of the free piston gauge, I spent considerable time in the effort to find some form of resistance gauge that should be perfectly reproducible, and which therefore would not need calibration after the first measurements had once been made. The work of others made it seem unlikely that any ordinary metal, in which there are almost always internal strains, would be sufficiently reproducible, and the use of liquid mercury (B. 1, II), which is not subject to such effects, naturally suggested itself. I did succeed in getting measurements on mercury up to 7000 kg. of such reproducibility that anyone could now use these measurements as an accurate method of measuring pressure. There are, however, a number of precautions necessary, the most important of which concerns the glass capillary in which the mercury must be contained. The compressibility of the glass is so great compared with the pressure coefficient of resistance of the mercury that it is essential that the compressibility of the glass be controlled quantitatively. This demands either a measurement of the compressibility of the glass of the capillary, which at once sacrifices the hoped-for advantage of reproducibility, or else a definite variety of glass must be used, which is possible but not very convenient. The glass which I used was Jena 3660a. The use of this method proved to have other disadvantages; the glass must be carefully annealed, the temperature must be kept constant within slight variation, the capillaries are restricted to use in a vertical position, and finally, if there is an explosion by rupture of any part of the apparatus, which is not infrequent, the delicate gauge is almost inevitably shattered. I have, therefore, abandoned this method for measuring pressure; it is conceivable that situations might arise in which it would prove convenient, and in fact the method has been used by an investigator in Russia.

Later work has convinced me that it is not necessary to go to a liquid metal to secure freedom from the effects of internal strains, but that a soft metal ought to be suitable, and I have in fact found the pressure coefficient of two samples of lead (B. 32) the same within 0·01 per cent. The use of lead would still suffer from the disadvantage of demanding careful temperature regulation.

All these considerations of convenience led me to return to the manganin gauge of Lisell. By the development of an easy calibration procedure the disadvantage of the necessity for separate calibration is minimised, and this gauge is now fundamental to all my measurements. The gauge coil itself is of double silk-covered manganin wire 0·005 in. in diameter, and about 5 m. long, giving a resistance of about 120 ohms. The wire is doubled and wound non-inductively on itself into a coreless toroid of about 1 cm. diameter. The winding is done on a separate arbor, so mounted that the coil may be readily pushed off after the winding is completed, and the section of the toroid bound with silk thread to make it retain its shape. The grade of manganin used is important. I have used in almost all my work manganin of unknown German origin, purchased more than thirty years ago. An attempt at one time to use a manganin made in this country was not successful; the American manganin apparently was not so well seasoned, as shown by wandering of the zero and other irregularities. After winding, the coil is seasoned by exposure to a temperature of 140° C., or as high as possible without charring the silk covering, for from six to ten hours. It was my former custom to also season it by several applications of 12,000 kg./cm.^{2}, but this proved to be less necessary than I had supposed.

The gauge coil is now ready for calibration. This is done by determining with it the freezing pressure of mercury at 0° C. The gauge coil is attached to the insulating plug in the upper cylinder of fig. 9. In the lower cylinder is placed a long cylindrical container with approximately 200 gm. of mercury. The lower cylinder is then placed in an ice bath with suitable stirring arrangement, so that the temperature is maintained at 0°. The freezing pressure of mercury is now determined by the method described in greater detail in Chapter VII on melting under pressure. Briefly, the method consists in plotting the position of the piston by which pressure is generated against the resistance of the coil. Freezing is indicated by a discontinuity in the motion, the total amount of discontinuity corresponding to the difference of volume between the liquid and solid mercury. The freezing pressure of mercury at 0° is known by previous measurement with the free piston gauge to be 7640 kg./cm.^{2}, so that we have at once the constant of the coil. I had previously established, also by comparison with the free piston gauge, that the change of resistance of all the samples of manganin which I examined was linear with pressure up to 13,000 kg./cm.^{2}, thus extending the result found by Lisell to 3000. In making the calibration, the freezing-point of mercury should be approached from both directions, first raising pressure beyond equilibrium and allowing it to drop to the equilibrium value from above, and then, after a sufficient fraction of the mercury is frozen, releasing pressure below the equilibrium point and allowing it to rise to the equilibrium point from below. The equilibrium points reached in this way from above and below should be the same within a small difference. My measuring device has a sensitiveness corresponding to about 1/5000 at this pressure; when everything is running properly it is not possible to detect any difference of this amount in the pressure reached from above and below. In order to accomplish this it is desirable to use petroleum ether as the transmitting liquid; kerosene becomes so stiff that it would require a long time to shut the equilibrium pressure within limits as narrow as this.

The greatest difficulty that one is likely to encounter on first using the method is entirely overshooting the discontinuity—expecting it to be larger than it is. The change of volume of mercury on freezing at $0°$ is about 3.3 per cent. of its volume at atmospheric pressure. It would be well, before starting a calibration, to convert this figure into piston motion, in order to know what to expect. After the apparatus has once been constructed, the calibration itself need not occupy much time; I find two hours sufficient for the complete calibration.

All of my gauge coils have been made from wire from the same spool; it would be well for any one else embarking on a programme of pressure measurement to provide himself with a spool of manganin of sufficient size to supply all the gauge coils that he could possibly need. I find that different samples from the same spool differ at the outside by 1 per cent., so that once a calibration has been made, a new coil may be used without calibration to about 0.5 per cent. Samples of manganin from different sources differ considerably more. Values may be found in the literature for the pressure coefficient of manganin varying from 2.08 to 2.34×10^{-6} per kg./cm.^{2}.

With careful use a gauge coil should have a long life, almost indefinitely long if special effort is made. In practice the accident most likely to happen to it is the development of an internal short circuit, because of rubbing of one strand on another incident to rapid application and release of pressure. This may be partially avoided by careful winding, but the most important precaution is to use a transmitting liquid that does not become too viscous under pressure. It is possible to short circuit the coil after a very few applications of pressure transmitted with a heavy oil which becomes very viscous, or which may even freeze.

The temperature coefficient of manganin is so low compared with its pressure coefficient that only in very exceptional cases is it necessary to control its temperature. In fact I have never encountered a case in which a sufficiently accurate correction for the effect of change of temperature of the gauge coil could not be made by calculation from the observed temperature of the cylinder. Sometimes it may be necessary to make long-pressure runs extending over several days; during such a run one is likely to be made somewhat uncomfortable by possible shifting of the zero of the gauge. If the coil is in good condition, I have found that almost all of the possible wanderings of the zero, which correspond to a very small change of resistance, are due to the effect of variable room temperature on the other parts of the bridge. The effect may be minimised by comparing the manganin coil against another exactly like it, mounted close to the pressure cylinder in such a way as to have the same temperature as the cylinder and the gauge coil. Fluctuations of room temperature now have the same effect on both coils, and a large part of the wandering of the zero is eliminated.

The manganin gauge has been shown to be linear by direct measurement to 13,000 kg./cm.². I have used it by extrapolation up to 21,000 kg./cm.². This linear extrapolation is, of course, not perfectly safe, but in view of the smallness of the effect I think there can be little error in it. A possible method of check, which would not demand the extension of the use of the free piston gauge to this pressure, is to measure simultaneously the changes of resistance of a coil of manganin and of some other metal, and to calculate the exact pressure by extrapolation of formulæ known to be valid up to 13,000. If the pressure given by extrapolation with the formulæ for the two metals is the same, the probability is high that both are correct. I have done something that amounts to this with Bi and manganin (B. 73), and although the results have no high degree of accuracy, there can be no doubt that the error in making a linear extrapolation with manganin to 20,000 is small.

There are other transitions which may be used to give

fixed pressure points like the freezing-point of mercury at
0° C. Any transition that has been accurately measured
offers such a possibility, but it must satisfy certain other
requirements if it is to be convenient in practice. If the
transition is a melting, then the substance must be one that
can be easily obtained in a state of high purity. In general,
purity is a prerequisite to a sharp melting, that is, a melting
that takes place at a single definite pressure independent of
the amount of the substance that has melted. Purity is
not usually so important if the transition is between two
solid phases, however. It is only impurities which can be
dissolved in one or the other phase that are capable of affecting the sharpness of a transition. In fact, it is the exception
rather than the rule for a transition between solids to be
hazy because of this effect, whereas the majority of melting
points are not sharp for this reason. Mercury is rather an
exception in the ease with which it can be obtained pure,
and the sharpness of its melting. Organic liquids are particularly prone to impurities and hazy melting points. Another
requirement in a satisfactory transition is that the volume
change be large, so that it can be easily located; another
requirement is that the transition run rapidly. A transition
that satisfies all these requirements is the one between Ice I
and II at temperatures between −25° and −30°; at −30°
the equilibrium pressure is 2150 kg., the volume change of
the transition is 20 per cent., and it runs in this temperature
range with almost explosive rapidity, the latent heat being
very small. Further, water is easy to obtain in a state of
the requisite purity. A disadvantage in this transition is
the rather inconvenient temperature, and the fact that the
pressure is so low as not to give a high percentage accuracy,
particularly if the coil is to be used over a pressure range
considerably greater. The transitions of CCl₄ have also been
used as calibration points, but I believe that, all things
considered, the most convenient one is the freezing-point of
mercury at 0°.

[1] L. CAILLETET, *C.R.*, **84**, 82 (1877) ; **88**, 61 (1879).
[2] E. H. AMAGAT, *C.R.*, **87**, 432 (1878) ; **88**, 336 (1879).
[3] J. PERKINS, *Trans. Roy. Soc. Lond.*, p. 541 (1826).

4 PARROT et LENZ, *Mem. St. Pet. Acad.*, **2**, 595 (1833).
5 NATTERER, *Wien. Ber.*, **5**, 351 (1850) ; **6**, 557 (1851).
6 E. H. AMAGAT, *Ann. Chim. Phys.*, **29**, 68 (1893).
7 G. DIMMER, *ZS. Instrkd.*, **35**, 245 (1915).
   C. A. CROMMELIN and Miss E. J. SMID, *Proc. Amst.*, **18**, 472 (1915).
   H. F. WIEBE, *ZS. Instrkd.*, **30**, 205 (1910) ; *ZS. Kompr. u. Flus. Gase*, **13**, 83 (1910).
   G. KLEIN, *ZS. Ver. D. Ing.*, **54**, 791 (1910).
   W. MEISSNER, *ZS. Instrkd.*, **30**, 137 (1910).
   A. MARTENS, *ZS. Ver. D. Ing.*, **53**, 747 (1909).
   E. WAGNER, *Wied. Ann.*, **15**, 906 (1904).
   Report of the Reichsanstalt in ZS. Instrkd., **23**, 176 (1903).
   Report of Work at the National Physical Laboratory in Eng., **75**, 31 (1903).
   JACOBUS, *Eng.*, **64**, 464 (1897) ; *Trans. Amer. Soc. Mech. Eng.*, **18**, 1041 (1896–1897).
   C. BARUS, *Phil. Mag.*, **31**, 400 (1891).
8 A. MICHELS, *Ann. Phys.*, **72**, 285 (1923) ; **73**, 577 (1924).
9 P. G. TAIT, *Report of the Voyage of H.M.S. "Challenger,"* II, Appendix A (1881).
10 C. BARUS, *Proc. Amer. Acad.*, **25**, 93 (1889–1890).
11 S. LUSSANA, *Nuov. Cim.*, **7**, 1–22 (1904).
12 E. LISELL, *Upsala Universitets Arsskrift, Matematik och Naturvetenskap*, No. 1 (1903).
13 A. LAFAY, *C.R.*, **149**, 566 (1909).
14 S. LUSSANA, *Nuov. Cim.*, **5**, 1 (1903).
15 S. LUSSANA, *ibid.*, **15**, 149 (1918).

CHAPTER IV

SPECIAL SORTS OF RUPTURE PECULIAR TO
HIGH PRESSURES

THE body of this book will be occupied with the quantitative description of a large number of high-pressure phenomena, most of which may be expected to have some ultimate theoretical significance. There are, in addition, a number of other phenomena, equally characteristic, but difficult to describe in exact terms. Prominent among these are various phenomena of rupture; the steel containers and various parts of the high-pressure apparatus may fail in a variety of interesting ways, which as yet have not been satisfactorily explained, and which, therefore, are suggestive as indicating where modifications may be necessary in our present theories of rupture. It would perhaps be natural to collect these phenomena of rupture into a single chapter of miscellanies at the end of the book; my reason for giving them here is that some understanding of many of them is most important in the design of high-pressure apparatus, so that they have a close connection with the problems of technique just discussed.

An introductory word may make the status of these high-pressure rupture experiments plainer. The whole high-pressure field opened almost at once before me, like a vision of a promised land, with the discovery of the unsupported area principle of packing, by which the only limit to the pressures attainable was set by the strength of the metal parts of the apparatus. Immediately the question arose as to what the limit was. All that I had to guide me were various theories of rupture of a more or less engineering character, which indicated, among other things, that the maximum internal pressure any cylinder could stand, even if the walls were made infinitely thick, was not far numerically from the tensile strength as determined in ordinary tensile

78

tests. A few rough experiments sufficed to show, however, that this rough theory was very wide of the mark, so that obviously the first thing to be done, before the promised land could be entered, was to make an empirical survey of the general nature of the possibilities. During this survey, the following observations on rupture were collected; these are not to be taken in any way as systematic, but merely as incidental results. There is a rich field open here for further systematic investigation, and for theory. Except for getting a qualitative understanding of the situation for my own ulterior purposes, I have made little attempt to reduce these phenomena to the basis of a quantitative theory. One analysis of the collapse of hollow cylinders under external pressure may be mentioned (B. 7), and there is in addition a comparatively large amount of quantitative observation and some theorising which I have not yet had time to publish, done in connection with work at the Watertown Arsenal on the stretching of guns, and also some other work done for the U.S. Government during the war. Michels [1] has recently made the beginning of an attack on the problem of the rupture of thick cylinders under internal pressure, and doubtless more will be forthcoming from this quarter.

The problem which gave me most immediate concern was that of the piston. Elaborate devices were designed and partly constructed for decreasing the stresses in the piston by a sort of inverse differential free piston-gauge arrangement, but all complications of this kind, fortunately, proved unnecessary when it was discovered that glass-hard pistons of the proper grades of steel—such, for example, as file steels, or even better, ball-bearing steels—will stand forces in compression far beyond the limit set by the strength of the other parts of the apparatus. The piston is not, therefore, at present a problem. If in the future it proves possible to raise the limit set by the strength of the other parts to the limit of the piston, then the differential piston scheme allows the possibility of further progress, although it must be admitted that the packing problem for a piston of this character is not simple.

The bursting strength of the hollow cylinder in which

pressure is contained was the problem that demanded most attention in the preliminary investigation. It will pay to have before us the results of the theory of elasticity for the stresses and strains in a hollow cylinder exposed to internal pressure. These results are well known, and are given, for example, in Love's book. Two cases may be considered, according as the thrust expelling the end plugs is transmitted to the body of the cylinder by making the plugs integral with the cylinder, or as the end plugs are independently supported from outside. The first case is the one of interest to us. Here we have

$$\left.\begin{array}{l} R_r=\left(\dfrac{r_1}{r}\right)^2\dfrac{r^2-r_0{}^2}{r_0{}^2-r_1{}^2}P \\[2ex] \Theta_\theta=\left(\dfrac{r_1}{r}\right)^2\dfrac{r^2+r_0{}^2}{r_0{}^2-r_1{}^2}P \end{array}\right\}, \qquad \left.\begin{array}{l} e_{rr}=\dfrac{Pr_1{}^2}{r_0{}^2-r_1{}^2}\left[\dfrac{1}{3\lambda+2\mu}-\dfrac{r_0{}^2}{r^2}\cdot\dfrac{1}{2\mu}\right] \\[2ex] e_{\theta\theta}=\dfrac{Pr_1{}^2}{r_0{}^2-r_1{}^2}\left[\dfrac{1}{3\lambda+2\mu}+\dfrac{r_0{}^2}{r^2}\cdot\dfrac{1}{2\mu}\right] \end{array}\right\},$$

where r_0 is the external and r_1 the internal radius. There are a couple of important points about these solutions. In the first place, the maximum stress and strain are both to be found at the interior surface, where the values are

$$(\Theta_\theta)_{\max}=\dfrac{r_1{}^2+r_0{}^2}{r_0{}^2-r_1{}^2}P, \qquad (e_{\theta\theta})_{\max}=\dfrac{Pr_1{}^2}{r_0{}^2+r_1{}^2}\left[\dfrac{1}{3\lambda+2\mu}+\dfrac{r_0{}^2}{r_1{}^2}\cdot\dfrac{1}{2\mu}\right].$$

Guided by these formulas, which hold only in the range of elastic deformation, extensions were made by engineers and others to the region beyond the elastic limit in an attempt to get some information about the phenomena of rupture. There are various plausible suppositions that one can make about rupture. One natural point of view is that rupture will occur when the force tending to tear the fibres apart reaches a critical value. The critical value for this force is evidently set by the tensile strength T, as ordinarily measured. Since the greatest fibre stress at any point of the cylinders occurs at the inner surface, according to this view rupture would be expected to occur by separation of the circumferential fibres at the inner surface when the internal pressure equals $\dfrac{r_0{}^2-r_1{}^2}{r_1{}^2+r_0{}^2}T$, and the maximum value for this in a cylinder

PLATE II

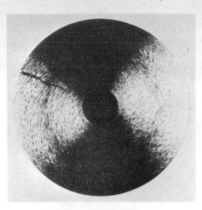

FIG. 19.—Cylinder of heat-treated steel, ruptured by internal pressure. The fracture started at the outside surface, travelled inward along a radial plane, and stopped before quite reaching the inside.

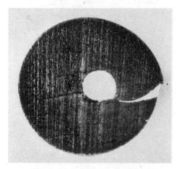

FIG. 20.—Cylinder of copper, ruptured by internal pressure. The rupture travelled from the outside in on a surface of shear, producing approximately an equiangular spiral.

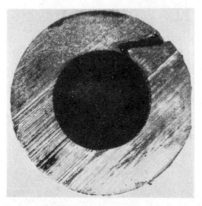

FIG. 21.—Cylinder of mild steel, ruptured by internal pressure. Rupture travels from the outside in, first on one shear plane and then on the other. This cylinder was originally 2 in. outside and $\frac{1}{2}$ in. inside diameter. The inner hole has been stretched to $1\frac{3}{8}$ in.

with infinitely thick walls is T.   Another plausible point of view is that rupture may be expected to occur when the elongation of any fibre exceeds a critical value, and this elongation is evidently the same as the elongation at rupture in ordinary breaking tests.   The maximum fibre elongation is the value of $e_{\theta\theta}$ at the inner surface, so that according to this point of view rupture should occur by separation of the circumferential fibres at the inner surface for an infinitely thick cylinder when $P = \dfrac{2\lambda + 2\mu}{3\lambda + 2\mu} T$, a result not very different from the result obtained above.   Other plausible criteria of rupture are that the maximum shearing stress is the determining factor, which in this case is the difference of R_r and Θ_θ, or that the maximum shearing strain $e_{\theta\theta} - e_{rr}$ is determinative.   Either of these criteria leads to a result much like the other criteria—namely, that rupture will occur at the inner surface at a pressure not far from the pressure numerically equal to the tensile strength.   It was, of course, realised that none of these criteria could be expected to be exact, because above the elastic limit there is flow, and the relations between stress and strain assumed in deriving the formulæ break down, but, nevertheless, there was a feeling that the results were not far from correct.

Experiment soon showed that expectations based on simple elastic theory were very far indeed from the mark, both with respect to the maximum pressure and the character of the rupture.   Cylinders stand a pressure very much higher than the tensile limit pressure, and rupture of cylinders made of ordinary grades of steel invariably starts at the outside surface and travels inward, rather than starting at the inside and running out.   When a cylinder bursts in this way from the outside under the action of internal pressure it will be found that the inside has received an amount of permanent stretch very much in excess of what would be expected from considerations like those above.   Figs. 19, 20, and 21 (Plate II) give some idea of the various forms that rupture may take. The rupture may progress inward for almost the entire distance on a radial plane, as in fig. 19, or it may progress along lines of shear, which gives the equiangular spiral effect shown in fig. 20 by a cylinder of copper, or it may travel first on one shear plane and then on another, as in fig. 21. It has always been my experience that at the internal surface the rupture is along a plane of shear; indeed, it is not uncommon to find the last stage of rupture at the inner surface taking place along the two planes of shear inclined to each other at nearly 90°, so that a triangular prism is detached at the inside, which may be projected through the opening crack by the internal pressure with considerable violence. The cylinder shown in fig. 22 (Plate III) represents about the maximum that I have observed in the way of internal pressure over the tensile value. This was a block of unhardened high carbon tool steel with a tensile strength of about 10,000 kg./cm.^{2}, and an elongation at the elastic limit of 25 per cent. The pressure required to burst it was 40,000 kg./cm.^{2}, and the maximum elongation at the inner surface was 125 per cent.

A little reflection will show the general nature of the reason why it is possible to reach pressures so much in excess of the tensile value. Most metals pass through a more or less plastic stage in the range of stress above the yield point, where the stress difference that the metal can support is nearly independent of the elongation. The inner part of the cylinder reaches this condition first; as pressure rises the fibre stresses in the inner circumferential fibres of the cylinder are unable to rise above this critical limit, but the radius of the part in which this critical maximum fibre stress prevails becomes greater, and the outer parts of the cylinder remain below the elastic limit. The inner layers are incapable of rupture under these conditions because of the support they receive from the outer layers; geometrical considerations will show that rupture at the inner surface would not result in a decrease of the total potential energy of strain in the metal, because the internal pressure acts in a direction to oppose the release of volume compression that would result from such a rupture. Any actual cylinder is not geometrically perfect, and after stretch and rupture as shown in fig. 22 the inner surface will usually be found marked with a great number of incipient slip lines, where rupture has started because of local conditions, but was checked before it could

PLATE III

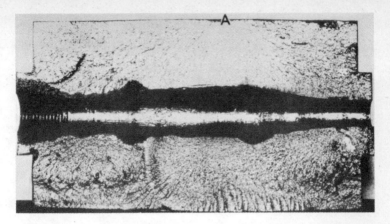

FIG. 22.—Cylinder of soft tool steel, ruptured by an internal pressure of 40,000 kg./cm.². Rupture started at the outside surface at the point marked A. The elongation at the inside was about 125 per cent.

FIG. 24.—Hollow quartz cylinder after exposure to external pressure. The original diameter of the cavity may be seen at A

FIG. 25.—Soft steel rod, ruptured by the 'pinching-off' effect.

progress to any marked extent because of the general geometrical conditions just mentioned.   There are other cases known in which support of the parts of the metal in which there is great stress prevents rupture from taking place under ordinary conditions; the simplest example is wire drawing, which should have been sufficient to suggest that maximum elongation could be expected to determine rupture only in a restricted range of conditions.

It is probable that only those substances which are capable of considerable plastic flow show characteristic rupture at the outside surface; if the substance is brittle, like glass, it is highly probable that rupture begins at the inside, more in the manner anticipated by the original theory.   It is very difficult to observe this in glass, but the phenomena can be shown very neatly in a cylinder of transparent gelatine, which when blown up on the inside may be seen to start to tear along a perfectly clean radial plane, beginning at the inner surface.   Glass cylinders, like steel cylinders, will stand pressures considerably in excess of that indicated by the simple theory.   The tensile strength of glass under ordinary conditions is seldom as much as 500 kg./cm.^{2}, but I have repeatedly reached pressures of 1000 kg./cm.^{2} in small glass capillaries, while Heydweiller[2] has reached the surprising figure of 3000 kg./cm.^{2}.   Glass under ordinary conditions does not have a point of yield appreciably below the ordinary rupture point, so that the stress and strain distribution given by the ordinary elastic theory must represent very closely the actual distribution right up to the point of rupture.   The fact that internal pressures are realisable up to six times the tensile limit shows how very far indeed from the truth are the maximum tensile stress and elongation criteria of rupture. A glass tube is a very treacherous thing; after it has been exposed to a high pressure it is very likely to break on the next application of a very much lower pressure.   The reason is doubtless that the glass does actually receive some internal strain from the first application of pressure, for glass is after all only a very viscous liquid, and this internal strain weakens it for the next application.

The problem inverse to the above—that is, the behaviour

of a thick hollow cylinder when exposed to high pressure over the external surface, was also subjected to an exploratory investigation (B. 7). This was done during my search for a suitable secondary pressure gauge. The Tait gauge, in which pressure is indicated by the change of internal volume of a cylinder when exposed to external pressure, is evidently one which by its form is capable of withstanding very high pressures without rupture, and for this reason it seemed promising.

The qualitative behaviour of a hollow cylinder exposed to external pressure depends in a very marked degree on the ratio of the wall thickness to the internal diameter. If the walls are relatively thin, the cylinder fails when pressure gets too high by folding in on itself, sometimes in a single fold, and sometimes in more, giving rise to star-shaped figures. This problem is more or less familiar, and may be treated by the usual methods of elastic theory for handling questions of geometric stability. The pressures required to produce collapse of this kind are not high, so that we need not trouble with further consideration of this problem. But if the walls are thick, so that the cylinder can withstand a high pressure without folding, then failure takes the form of simple flow of the metal toward the centre along radial lines, the geometric figure remaining that of concentric circles, the ratio of the outside to the inside diameter increasing as flow continues. Obviously if such a cylinder is to be used as a Tait gauge it can only be in the range of pressure below that required to produce flow. Two regions of experimental investigations are therefore to be distinguished, according as the pressure is above or below this limit. These two regions are not distinct, but the lower may be made to progressively overlap into the upper, because of an effect similar to one already found in cylinders exposed to internal pressure— namely, that the flow point is raised by the previous application of pressure to the highest pressure previously reached. The possibilities here in the way of raising the flow point are considerably greater than in raising the elastic limit by internal pressure because of the more favourable geometric relations, for here flow takes place in such a direction as to make the ratio of external to internal diameter greater, so that further flow is more effectively resisted. The ultimate

effect of a sufficiently high external pressure is to entirely close the interior cavity. I have observed cases in which the pressure required to produce complete closure was ten times as great as that required to start flow, so that an increase of tenfold in the elastic limit of cylinders exposed to external pressure is possible, against only three- or fourfold when the pressure is internal.

In the region below the current yield point there proved to be very great hysteresis effects. This is shown in fig. 23, in which is plotted the change of internal volume under successive cycles of pressure of continually increasing amplitude. The raising of the flow point to approximately the previous maximum is shown, and also hysteresis and other effects, such as hardening by resting after overstrain. These various effects are all known qualitatively for other sorts of test, such as tensile tests, for example; the rather unusual feature of these collapsing tests is the largeness of the effect. Cases have been observed in which the width of the hysteresis loop amounts to half the total deformation at the yield point. It was the existence of this very large hysteresis that convinced me

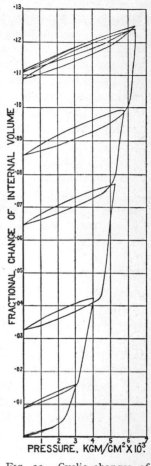

FIG. 23.—Cyclic changes of internal volume of a steel cylinder, subject to external hydrostatic pressure.

that it was not practical to expect a gauge of the Tait pattern to be satisfactory at pressures approaching the elastic limit of the metal, and indeed I do not believe that any type of deformation gauge will be satisfactory at pressures as high as this.

Sometimes, particularly when the ratio of the outside to the inside diameter is high and the composition of the metal is complicated, as in a tool steel of high carbon content, as opposed to a nearly pure iron, effects may be found more complicated than suggested in the figure. Thus I have found a case in which an initial increase of external pressure was accompanied by an initial *increase* of internal volume. I have also found the corresponding effect in cylinders exposed to internal pressure—that is, it is possible, after a thick cylinder has been strained severely beyond the elastic limit by internal pressure, that an initial application of internal pressure should produce a *decrease* of external diameter. Obviously such unusual effects must mean a highly complicated system of internal stresses.

The maximum external pressure that a hollow cylinder will stand without complete collapse is a function only of the material, and not of the initial dimensions. This might be expected, because a cylinder of any initial proportions must pass through all greater ratios of outside to inside diameter on its way to complete collapse; the fact that it is true means that hardening effects during flow toward complete collapse are not important. I have found the pressure for complete collapse for copper to be about 10,000 kg./cm.^{2}, and about 20,000 for "Bessemer" steel. After flow has once started the relation between internal diameter and pressure is approximately linear.

The fact that flow continues until complete collapse at a finite pressure is one that could not have been foreseen, and gives us a certain amount of information about the stress distribution during flow. The plausible assumption that has been made about flow in certain theoretical discussions is that, neglecting hardening effects, plastic flow takes place when the maximum shearing stress reaches a certain critical value. In this case the shearing stress is the difference between the principal stresses R_r and Θ_r. Let us see where this hypothesis leads us. According to it we would have during flow

$$R_r - \Theta_\theta = K,$$

where K is the stress difference at plastic flow. There is

also a condition of stress equilibrium that can be obtained from the equilibrium of a semicircular sector under the action of the stress Θ_θ acting across the diameter and the stress R_r on the curved surface, which is obviously

$$\int_{r_1}^{r} \Theta_\theta dr = rR_r,$$

r_1 being the internal radius and r the running value of the radius. Substitute in this equation the value of Θ_θ in terms of R_r and eliminate the integral sign by a differentiation with respect to r, obtaining the differential equation

$$\frac{dR_r}{dr} = -\frac{K}{r}.$$

The solution of this is

$$R_r = -K \log\frac{r}{r_1},$$

because at the inner surface $R_r = 0$. At the outside surface $R_r = -P$ and $r = r_0$, so that

$$P = K \log\frac{r_0}{r_1}.$$

This equation states that for a given external pressure P and a material with the constant of plasticity K, flow will take place until the outside and inside diameters satisfy the relation. In particular, r_1 will become zero only when the pressure P becomes infinite. But this is contrary to experiment, for we found the hole to entirely disappear at a finite pressure. That is, the material is actually weaker than we assumed, and the stress difference necessary to produce flow decreases after prolonged flow, instead of increasing, as some hardening phenomena under less extreme conditions might have led us to anticipate.

The increase of yield point produced by flow is an increase for that particular type of stress only. If a cylinder which has been collapsed by external pressure is exposed to internal pressure, it will be found to rupture at a pressure very much less than the pressure that would have ruptured it in the beginning before it was collapsed.

If the hollow cylinder is made of a non-plastic material, such as glass, the phenomena are quite different. Such a cylinder cannot flow, and if the walls are thick enough, it cannot fail unsymmetrically by geometrical instability, so that literally such a cylinder cannot break. I have exposed thick-walled glass capillaries to external pressures of 25,000 to 30,000 kg./cm.^{2} without any immediate effect. Sometimes such a cylinder will break spontaneously several days after the release of pressure, evidently because the glass had acquired some internal strain by viscous flow when exposed to the pressure, there still being at room temperature a remnant left of the plasticity of higher temperatures. If the glass cylinder is not geometrically stable—as, for example, when the inner hole is off centre—it may crush under the external pressure, disintegrating to an impalpable powder if the pressure is high.

The glass tube in the last paragraph was prevented from breaking by its perfect symmetry. If the hollow cylinder, instead of being made of amorphous material, is made of a crystalline material like quartz, then there is no longer symmetry, and rupture may occur. Fig. 24 (Plate III) is a photograph of a hollow cylinder of quartz after the application of a high external pressure (B. 30). The cylinder was originally in two pieces, which were cut from a single crystal of quartz, central holes drilled in them, and then put together in the original relative orientation on two optically flat planes, as shown. Some such method as this was necessary to get the hole into the inside of the crystal. A rubber tube was then snapped over the cylinder, and the assembly exposed to the action of hydrostatic pressures over the outside of the rubber tube. Failure takes two distinct forms. There is in the first place the development of minute cracks in various parts of the cylinder, which in the specimen of the photograph began at about 6000 kg./cm.^{2}. This effect is apparently not important; it never leads to complete rupture, and is apparently associated in some way with the failure of the two halves of the crystal to fit together in exactly the original orientation. The second effect is new, and, as far as I know, characteristic of crystals. At a pressure of 7500 kg./cm.^{2} excessively minute fragments begin to detach themselves from the interior wall. The frequency with which these particles are detached depends on the magnitude of the pressure; for the quartz cylinder shown in the figure an exposure of about ten minutes to 12,000 kg./cm.² was sufficient to completely fill the cavity with an impalpable sand, the diameter of the cavity being eroded in places to three times its original size. The process stops at any given pressure when the fragments exert sufficient pressure on the interior walls to support them against the external pressure. The fragments are projected with considerable violence when they are detached from the walls, with sufficient velocity, for example, to permanently embed themselves in a brass rod which on one occasion loosely occupied part of the cavity.

The quartz experienced no plastic flow during the erosion, there being no perceptible permanent change of the external diameter.

The elastic deformation experienced by a crystalline cylinder like quartz can be calculated from the equations of elasticity. The deformation is much more complicated than in an isotropic material; it can be regarded as a perturbation peculiar to the crystal superposed on the effect to be expected in an isotropic material of the mean elastic constants of the crystal. This crystalline perturbation effect consists, among other things, of a warping of the plane cross-sections, and a deformation of sections originally circular into a more or less trefoil-shaped outline. These additional strains involve additional stresses; the most important is a shearing stress tending to change the angle between the axis and the radius which does not occur at all in the isotropic case, and which may amount in quartz to 35 per cent. of the maximum isotropic stress. A surprising feature of the results was that the flaking-off seemed to have no particular connection with these additional stresses, but was apparently determined in large degree by the isotropic stress. The theory was roughly checked by measurements of the elastic deformation of the interior before rupture began; the accuracy was not great enough to show any of the pure crystalline effects, but did show that the mean radial displacement was consistent with the value to be expected from the mean constants.

The flaking-off of the internal surface is connected in some way with the orientation of the internal surface with respect to the crystal. Attempts to collapse the natural cavities which are sometimes found in quartz crystals, and which go by the name of 'negative crystals,' the faces of which consist of natural crystal faces, were entirely unsuccessful at pressures up to 18,000 kg./cm.^{2}.

Experiments were also made on the crushing of hollow cylinders of calcite, tourmaline, feldspar, barite, porphyry, and andesite. All of these were found to fail by flaking-off in the same way as the quartz, but the other phenomena were not always the same as in quartz. Thus a few cases were observed in which permanent flow without cracks accompanied the failure of the central part. The same phenomena were also observed in two minerals, granite and limestone, which of course consist of aggregates of microscopic crystals. The cavities in these minerals were found completely filled with the eroded fragments after exposure to 5000 kg./cm.^{2} for an hour.

These experiments have considerable geological interest in suggesting to what depths in the earth's crust open cavities may exist. The 'flaking-off' effect itself may be suspected to be very closely allied to phenomena in deep mines, where it is not unusual to find thick slabs of rock of considerable area separated from the supporting walls. The depths suggested by my experiments on granite are considerably less than those which had been previously deduced from experiments of Adams; his experiments were not performed with hydrostatic pressure, but by compressing cylinders of rock with hardened steel plungers into shrunk-on supporting hoops of mild steel. It appears from the above that the support afforded by these hoops must have played an important part in the final result.

Also suggested by the geological problem, I made experiments on the maximum density that could be produced in sands of various kinds by pressure up to 30,000 kg./cm.^{2}. No instances of the fusing together of the particles of sand were ever observed. After exposure to 30,000 the density of the sand might approach within 2·5 per cent. of the non-fractured original material; it is probable that the densities approached even more closely during the application of pressure. Reference must be made to the original paper for further discussion of these experiments.

Another type of rupture peculiar to high pressure has already been briefly referred to in the chapter on technique, the so-called 'pinching-off' effect. The conditions under which this type of rupture may be produced are indicated in fig. 25 (Plate III). A solid cylinder is exposed to pressure over the external curved surface only, the ends being unsupported. When pressure rises to a value approximating numerically to the breaking stress in pure tension, the cylinder parts somewhere in the median region, only by accident near either of the stuffing-boxes, as if it had been pulled apart by a tensile pull applied to the projecting ends. After rupture, the two severed ends are expelled by the pressure through the stuffing-boxes with considerable violence, and it is in this that the danger of this sort of rupture consists. If the rod is of brittle material, like glass, or glass-hard tool steel, the rupture takes place on a perfectly clean plane perpendicular to the axis, but if the rod is of a material that can yield before rupture, like soft steel, there is considerable contraction of area at the break, which looks very much like the break in an ordinary tensile test. The paradoxical thing about this rupture is that, neglecting friction in the stuffing-boxes, there is no force along the fibre across which rupture occurs. That the stress system is of this character should be evident to intuition after slight reflection, but a formal proof can be given, if required, by elastic theory. If the z axis is taken along the axis of the cylinder, and the other co-ordinates are chosen as r and θ in the conventional way, then, neglecting end effects, it can at once be seen that a stress system satisfying the boundary conditions is $R_r = -p$, $\Theta_\theta = -p$, $Z_z = R_\theta = R_z = \Theta_z = 0$, and, further, this must be the actual stress system by the uniqueness theorems of elastic theory. Whatever effect the stuffing-boxes have is to superpose a compressive stress Z_z, which makes the experimental result even more paradoxical, because rupture actually takes place against a small stress.

Although there is not longitudinal stress when the pinching off occurs, there is nevertheless longitudinal elongation, and its amount is $\dfrac{\lambda P}{\mu(3\lambda+2\mu)}$. One might be tempted to see in this elongation a sufficient explanation of the rupture, but this explanation proves to be, however, too easy, because there are other cases in which it is obviously not the determining factor in rupture.

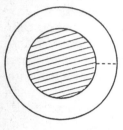

Fig. 26.—Indicates the manner of failure of a hard rubber ring when supported on the inside by a steel core (shaded), and exposed to hydrostatic pressure.

One of the simplest of such cases in which the elongation does not determine rupture is illustrated in fig. 26. This represents in section a cylindrical core of steel fitting closely inside a tube of hard rubber. The two together are placed in a pressure cylinder and exposed to hydrostatic pressure over the entire external surface. A pressure of a few thousand kilograms per square centimetre is sufficient to split the rubber tube, as indicated by the dotted line. The rupture here takes place exactly as if a cone had been driven into the tube, stretching it to the rupture point. In fact, reflection suggests that the analogy of the cone may have considerable truth in it, because it is evident that if it had not been for the steel core the rubber tube would have shrunk, because of its comparatively high compressibility, by much more than it was allowed to by the steel. From this point of view, therefore, the rubber tube was actually stretched beyond the size that it would naturally have assumed under the pressure, and so broke. If one works out the actual stress and strain in the rubber, however, it will be found that, allowing for the compressibility of the steel core, every strain and every stress in the rubber was compressive, but that, nevertheless, rupture occurred.

It is not difficult in any special case to invent considerations that appeal to our intuitions as adequate explanations of the observed rupture, as illustrated by the pinching-off effect and the split hard rubber tube; the difficulty comes in bringing all possible cases under one point of view.   It is possible to find among the examples discussed above cases which refute any one of the various criteria of rupture which have been seriously proposed.   Thus the maximum tensile stress criterion is inconsistent with the pinching-off effect, or the cored rubber tube rupture, and the rupture of cylinders under internal pressure;   the maximum tensile strain or maximum elongation criterion is inconsistent with the cored rubber tube or the cylinder burst by internal pressure.   Either the maximum shearing stress or maximum shearing strain criterion fails to apply to cylinders exposed to external hydrostatic pressure.   Furthermore, detailed considerations show that none of these criteria work when used as a criterion of yield instead of as a criterion of rupture.

For myself, I am exceedingly sceptical as to whether there is any such thing as a genuine criterion of rupture.   For example, it was obvious enough that the manner of failure of hollow cylinders exposed to external pressure was entirely different for amorphous and crystalline substances.   I believe we have no right to expect any general criterion of rupture in view of the extremely varied structure of different sorts of material.   It is sufficient in dealing with many phenomena to think of the molecules which constitute matter as being to a certain degree like small rigid bodies, in which are located, according to more or less complicated patterns, centres of electrical and magnetic force which hold the substance together and give it its properties.   When such a substance is subjected to stress, the molecules must readjust themselves to each other's irregularities in the most complicated ways, and according as one or another of the local centres of force are separated by more than the critical amount, we may have rupture under the most varied conditions.

These considerations, as well as the experimental facts summarised above, indicate therefore that we should attempt to establish criteria of rupture only as a matter of practical convenience, as for engineering purposes, and that we should expect any such criteria to be valid only in a narrow range of conditions, both of stress and material.

In addition to the types of rupture thus far discussed, which are more or less peculiar to high pressures, there are other types of rupture which are more or less incidental results of the high pressures—namely, rupture produced by the action of hydrogen and mercury, but which are most important to know about in the design of high-pressure apparatus.

Amagat [3] described in 1885 a shower of mercury driven by a pressure of 4000 kg./cm.[2] through the 8 cm. thick walls of a high-pressure cylinder made of a mild steel forging, in which no flaw could be detected. His explanation was that the mercury had been forced through the inter-molecular pores of the solid steel. This effect would evidently impose important limitations on high-pressure technique, and I therefore made a systematic investigation of the subject (B. 3). I was not able to repeat Amagat's observation, and I am convinced that his effect must have been due to microscopic mechanical flaws in the forging, in spite of his failure to find such flaws. This is a rather natural explanation, because only recently have the manufacturers succeeded in producing uniformly sound steel ingots. The fact that Amagat avoided the difficulty by making another apparatus with thicker walls suggests the same explanation. I did, however, find an effect entirely different from that of Amagat at a pressure which was usually in the neighbourhood of 6000 kg./cm.[2], but once was as low as 3000. A hardened steel cylinder, filled with mercury, in which pressure is produced, will split along a radial crack. Cylinders of the same steel similarly heat-treated were able to stand without rupture pressures up to 24,000 kg./cm.[2] when the transmitting liquid was water and glycerine, or ether, or carbon disulphide. The rupture with mercury takes place a few minutes after the application of pressure. If the cylinder were of untreated steel, so that its yield point was reached at a pressure lower than that usually required to produce rupture in the hardened steel, then the results when pressure was transmitted to the walls by mercury were not different from the results when the transmitting liquid was water, in those experiments where the action of the pressure continued only for a short time. But in one case a cylinder of mild steel was exposed to a pressure of 4000 kg./cm.^{2} exerted by mercury for three weeks continuously, and then was exposed to a pressure of 12,000 kg./cm.^{2}, where it broke at once with no preliminary yield, although normally such a cylinder could have withstood a considerably higher pressure and then would have broken only after considerable yield.

The explanation of the effect involves the chemical affinity between mercury and iron.   Under ordinary conditions mercury will not wet iron, and it is difficult to amalgamate an iron surface.   The reason for this is not that there is no chemical affinity between iron and mercury, but that the surface of iron is usually protected with a coating of dirt or oxide.   This may be proved by breaking a piece of iron under the surface of mercury, so that the mercury comes immediately into contact with the freshly broken surface.   Such a surface will be found to be brightly amalgamated.   Furthermore, the amalgamation can be made to spread from such a surface into the interior by warming a piece of iron with an amalgamated surface under mercury to a temperature of 200° C. for several hours.   The presence of the mercury in the interior of such a piece can best be shown by fracturing it; the amount of mercury is so small that it cannot be found by microscopic analysis, but it gives to the fracture a coarse granular structure and a characteristic silvery appearance.   Examination of the cylinders which had been broken by internal mercury pressure, by again breaking them across so as to give a section perpendicular to the axis, always showed an amalgamated band on both sides of the pressure fracture, running from the inside out.   The explanation obviously is that the mercury was forced by the internal pressure into the pores of the solid steel at the inner surface, where the pores are distended by the action of the pressure; that once inside the metal it combined chemically, thereby weakening the metal to a certain extent; that the weakening of the metal was greatest in the region where, because of some accident, the penetration was greatest; and that because of this weakening the stresses accumulated in the weakened part, opening up the pores to still more rapid penetration, so that a state of instability was reached by which the amalgamated region travelled rapidly to the outside, when rupture took place. The fact that cylinders of soft steel are not so rapidly attacked is due to two facts. In the first place, soft steel is less liable to attack by mercury than hardened steel; this may be proved by direct experiment, and is doubtless connected with the fact that the density of hardened steel is less than that of soft steel. The second effect is that the plastic stage in soft steel is reached at a pressure below that required to force mercury into the pores of the steel, and above the plastic limit the inner layers of the soft steel cylinder are compressed instead of dilated by the pressure, so that mercury cannot get between the pores. Mild steel is, however, subject to slow attack by mercury under pressure, as shown by the experiment with the cylinder exposed to pressure for three weeks.

It appears, then, that rupture is not produced by mercury, unless the stress is of such a character as to dilate the metal and to force the mercury into the pores. A consequence of this is that steel exposed to hydrostatic pressure all over is not attacked by mercury, and this gives the suggestion as to the technique of employing mercury in high-pressure experiments. If the mercury is held in a separate container, which is itself exposed to pressure on the outside as well as on the inside, then the mercury is as inert as any other liquid. Whenever mercury is used in high-pressure apparatus, the greatest precautions must be used to prevent it from escaping from the container and coming into direct contact with the walls of the vessel, where the strain is a dilation. A cylinder exposed to the action of even a very small amount of mercury is very likely to rupture eventually, it may be a long while after exposure.

An effect in some respects similar to that of mercury is shown when pressure is transmitted to the walls of the cylinder with hydrogen, in that the hydrogen escapes through the cylinder walls with rupture (B. 47). The pressure is somewhat higher than with mercury, about 9000 kg./cm.[2]

being required at room temperature, and there are various differences in the details of the appearance. On the first application of pressure by means of hydrogen, escape of the hydrogen takes place at about 9000 kg./cm.² with explosive violence, but the cylinder is not obviously ruptured, and in fact I could not tell by the most careful scrutiny where the escape had taken place. This performance may be repeated a number of times, a crack of visible size gradually developing with repetition. It is more difficult to prevent action of hydrogen on the steel than action of mercury. Hydrogen may be forced by pressure into the pores of a block of steel subjected to hydrostatic pressure all over, thus compressing all the pores of the metal, conditions which are effective in preventing any action by mercury. A piece of steel exposed to hydrogen in this way gradually becomes permeated with hydrogen with great loss of strength. The fracture of a piece of steel which has been rotted by the action of hydrogen is characteristically coarse and granular in appearance, in distinction to the fine granular fracture of heat-treated steel. This action of hydrogen imposes a very serious limitation to any investigation of its properties at high pressures, because hydrogen goes into solution to some extent in any transmitting medium, and the steel is weakened and breaks after use a few times, with considerable incident danger.

It is not impossible that other gases may exert a similar action. Thus I have observed a rupture after long exposure to high pressures exerted by air, which suggests something similar to the effect of hydrogen, with the difference that the effect is very much slower. Experiments now in progress indicate that helium may be somewhat similar in its action to mercury; at any rate, ruptures are produced with helium as the transmitting fluid which would not be produced by ordinary liquids. I have not found any effects with ordinary liquids indicating that they can be driven into the steel to the slightest extent by any pressures at my command.

[1] A. Michels, *Proc. Roy. Acad. Amst.*, **31**, 55 (1928).
[2] A. Heydweiller, *Wied. Ann.*, **64**, 725 (1898).
[3] E. H. Amagat, *C.R.*, **100**, 633 (1885).

CHAPTER V

P-V-T RELATIONS IN FLUIDS

Introduction.—The p-v-t relations of a substance cover a variety of phenomena which are often treated separately. If one's chief interest is in the pressure aspect of phenomena, it is most natural to measure the volume as a function of pressure at some constant temperature; such measurements are usually listed as compressibility measurements. If compressibility measurements are carried out at several different temperatures, the results may be described as giving the temperature coefficient of compressibility. But if instead of varying pressure at several different constant temperatures one had varied temperature at several different constant pressures, then the results would have been described as a study of the way in which the thermal expansion varies with pressure. Going a step farther, if the measurements of volume are accurate enough to give the second temperature derivatives, then one may employ the relation

$$\left(\frac{\partial C_p}{\partial p}\right)_\tau = -\tau\left(\frac{\partial^2 v}{\partial \tau^2}\right)_p$$

to study the variation of C_p with pressure, and by an integration to obtain C_p as a function of pressure. Various other effects which have been the object of direct measurement are also implicitly involved in a complete knowledge of the p-v-t relations. For example, the adiabatic compressibility as distinguished from the isothermal compressibility, and also the temperature rise accompanying adiabatic compression, involve nothing not determined in terms of the complete p-v-t relation and a knowledge of C_p as a function of temperature at atmospheric pressure, which we may assume to be known. Or there is another p-v-t phenomenon which has been the object of direct experiments, namely, the way in which the temperature of the maximum density of water is affected by pressure.

It is a well-known result of thermodynamics that a substance is completely characterised thermodynamically when the p-v-t relation is given, and in addition C_p along some line not an isothermal. If we may assume that C_p is already known as a function of temperature at atmospheric pressure, then all we need do to completely characterise the substance thermodynamically at all pressures and temperatures is to determine with sufficient precision the p-v-t relation. We shall in this chapter confine ourselves to this aspect of the subject. The attempts that have been made to get equivalent information by other methods—for example, the adiabatic rise of temperature accompanying compression—have already been mentioned in the historical introduction, and have not led to results of sufficient importance to justify further attention here.

In this chapter there will be discussed all the information that can be obtained from the p-v-t relations of fluids, that is, liquids and gases. We regard the various critical phenomena between liquids and gases as below the pressure range of interest here, and in order to limit this discussion to reasonable compass, we will somewhat arbitrarily consider only those experiments on gases which have been made at pressures greater than 1000 kg./cm.². We shall consider in this chapter only phenomena connected with a substance in a single phase; in later chapters certain aspects of the p-v-t relations in two-phase systems, that is, melting and transition phenomena, will be considered. In this chapter the discussion will be first of systems of one component, and then very briefly of the few cases of two components, solutions or mixtures, which have been examined experimentally.

The methods usually employed in the measurement of the compressibility of fluids have been simple modifications of the original method of Canton,[1] in which the fluid is enclosed in a large bulb provided with a capillary by which any volume changes of the fluid may be appropriately magnified. Usually pressure is applied to the fluid by means of a mercury column in the capillary, so that the volume of the fluid is determined in terms of the position of the mercury column. The bulb and capillary are usually exposed to an external pressure equal to the internal pressure, but in some early experiments at low pressures the pressure was exerted only on the inside. The mercury may be replaced by other fluids; in the original experiments of Canton, the pressure transmitting fluid was air under the receiver of an air-pump. If the pressure is low, the position of the mercury in the capillary may be determined optically by enclosing the entire piezometer in a heavy glass vessel provided with glass windows for observation. If the pressure is too high for convenient optical observation, various devices may be used to determine the position of the mercury column, such as a hair index floating on the mercury column, which is left in the extreme position where it may be observed after release of pressure (Tait [2]). Or the interior of the capillary may be gilded, and the extreme height reached by the mercury determined by the height to which the gilding has been dissolved away by the mercury (Tait,[2] Cailletet [3]). A disadvantage of schemes like these is that only one reading can be obtained with a single set-up of the apparatus. The method may be converted into a continuous reading one, as was done by Carnazzi,[4] by stretching a fine wire along the bore of the capillary and measuring its electrical resistance. As the mercury column rises, the wire is short-circuited and its resistance decreases. A difficulty with this method is irregular capillary action. A compromise between the single and continuous reading methods may be made by fusing into the capillary a number of platinum contacts, and determining the pressure at which the mercury column reaches in succession the various contacts. This method was devised by Tait, and used extensively by Amagat.[5] Amagat used it for both liquids and gases; when applied to gases at high pressures it is necessary to provide the piezometer with a sort of antechamber in which the compression of the gas to the first few hundred kg. can be accomplished before the mercury enters the tube with the contacts. It is obviously necessary to provide several tubes of different dimensions to cover different pressure ranges.

If the pressure range is high, the volume change becomes so great that it is not necessary to magnify it by the use of a capillary, and other methods become available. Of course the simplest of all methods is to use the liquid by itself in a steel cylinder, subjecting it to pressure by pushing a plunger into the cylinder, and measuring the motion of the plunger. This demands a plunger without leak, and such have not been easily made. The method was used by Parsons and Cook,[6] and was used by me in an extended series of measurements to be described in great detail later. A modification of this is the method used by Perkins.[7] The liquid fills a cylinder, which is provided with a plunger moving through a stuffing-box as free from leak as possible. The whole arrangement is placed in a larger pressure chamber, and subjected to external hydrostatic pressure. The plunger moves in until pressure inside and outside are equalised; the compression is calculated from the displacement of the piston, which may be determined in any convenient way, as by a ring sliding on it which remains in the extreme position. Essentially the same method was used by me in measuring the compressibility of mercury. A disadvantage of the method is that it gives only a single reading for one set-up. Another method is that of Aimé.[8] The liquid is enclosed in a vessel provided with an inward opening trap-door or its equivalent. The vessel is subjected to external hydrostatic pressure in any convenient way, by Aimé by lowering into the sea, and mercury is forced in through the trap-door until pressure inside and outside are equal. The mercury is permanently trapped by falling to the bottom of the vessel, and its total amount may be determined by weighing after release of pressure. This also gives only a single reading per set-up, and has the added disadvantage that it is an integrating method, recording the sum of all the increasing pressure steps, so that pressure must increase or decrease monotonically. The results obtained by Aimé with this method were very bad, but it may be made to give good results under properly controlled laboratory conditions, and I have successfully used it in a double application to get the compressibility of water and mercury simultaneously.

It is obvious that the compressibility of a fluid determined in any of the ways outlined above is only an apparent compressibility, and that to determine the true compressibility a correction must be made for the effect of the distortion of the containing vessel.   The correct method of doing this was not for a long time understood, and positively incorrect methods were used by physicists of eminence.   The correction is comparatively unimportant if the fluid is very compressible, as a gas, but is more important in the case of ordinary organic liquids, and in the case of mercury in glass the correction rises to more than half the total measured effect.   The early attempts to obtain the correction were indirect, using the theory of elasticity, attempting to infer the distortion of the vessel under hydrostatic pressure from the observed deformation of the material of the piezometer under other types of stress.   There was not at first sufficient understanding of the theory of elasticity to enable this to be done;   for example, there was much discussion of the deformation of a strictly homogeneous and isotropic piezometer when exposed to uniform hydrostatic pressure inside and out.   Elaborate pieces of apparatus were made in the endeavour to answer this question, as in the accurately turned spherical piezometer of Regnault.[9]   A direct method by which the correction could be obtained was indicated by Buchanan,[10] although his numerical results were far from the truth, and Amagat [11] also tried his hand at it, but not with satisfactory results.   The question will be discussed later in connection with the compressibility of solids.   The necessity for knowing the correction may be avoided if one is satisfied to determine the difference of compressibility of different materials, for if the apparent compressibilities of two different liquids are determined in the same apparatus, the compressibility of the piezometer approximately cancels when one compressibility is subtracted from the other.   For many years the results of Richards were given in the form of differential compressibilities.

In order to see in detail that differential compressibilities may be obtained as just suggested, we write out the equations which govern the measurement of compressibility in an ordinary piezometer, the experiment being so arranged that the piezometer is exposed to equal hydrostatic pressure all over. Let V_0 be the original internal volume of the bulb of the piezometer, l_0 the length of the capillary originally filled with liquid, S_0 its original cross-section, and l the length of the liquid-filled capillary after pressure has been applied. We will suppose that the position of the liquid in the capillary is measured by means of a scale etched into the glass of the capillary, which therefore is distorted along with the capillary itself. Then we have

$$\text{original volume of liquid} = V_{0l} = V_0 + l_0 S_0,$$
$$\text{final volume of liquid} = V_{pl} = V_0' + l S_0',$$

where V_0' and S_0' are the volumes of the bulb and the cross-section as distorted by pressure. Now, if the solid of which the piezometer is made is homogeneous and isotropic, it is equally deformed in all directions, and to a sufficient approximation for low pressures:

$$V_0' = V_0(1 - kp), \quad \text{and} \quad S_0' = S_0(1 - \tfrac{2}{3}kp),$$

where k is the cubic compressibility, and therefore $k/3$ the linear compressibility of the material of the piezometer. Hence

$$V_{0l} - V_{pl} = V_0 kp + l_0 S_0 - l S_0(1 - \tfrac{2}{3}kp).$$

We must now distinguish between the actual positions of the liquid in the capillary and the positions as read on the distorted scale. If we call l_R the position as read while under pressure, then

$$l = l_R(1 - \tfrac{1}{3}kp).$$

Furthermore $l_0 - l_R = \Delta l_R$, where Δl_R is the displacement of the liquid column as read on the distorted scale. We now have

$$V_{0l} - V_{pl} = V_0 kp + S_0[l_0 - l_R(1 - \tfrac{1}{3}kp)(1 - \tfrac{2}{3}kp)]$$
$$= V_0 kp + S_0[\Delta l_R + l_R kp].$$

Put

$$V_{pl} = V_{0l}(1 - k_l p),$$

where k_l is the compressibility of the liquid. Then

$$V_{0l}k_lp = (V_{0l} - l_0S_0)kp + S_0[\Delta l_R + l_Rkp]$$
$$= V_{0l}kp + \Delta l_RS_0(1 - kp),$$

and

$$V_{0l}(k_l - k)p = \Delta l_RS_0(1 - kp).$$

Hence, except for the factor $1 - kp$, measurement of the change of position of the liquid with respect to the capillary gives the difference of compressibility between the liquid and the solid material of the piezometer. The factor $1 - kp$ is so nearly unity for pressures of the order of a hundred kilograms that its effect is negligible. For example, if we are measuring the compressibility of water, which is approximately 5×10^{-5}, in a glass piezometer, the compressibility of which may be taken as $2 \cdot 5 \times 10^{-6}$, it is evident that we are making a mistake of 5 per cent. in neglecting the difference between $k_l - k$ and k_l, whereas in neglecting kp in comparison to unity, we are making an error of only 0·025 per cent., if the pressure range is 100 kg.

Compressibility of Gases. Returning now to the discussion of the actual data, it will be necessary to consider nothing before the time of Amagat in dealing with measurements on fluids which are gaseous under ordinary conditions. Amagat [4] measured the compressibility of H_2, O_2, N_2 and air up to 3000 kg./cm.² The method was that of a number of platinum contacts in a glass tube, and has already been sufficiently indicated. Amagat displayed his results by plotting the product pv against p. If the gas satisfies Boyle's law this plot should be a straight line parallel to the pressure axis. Actually it is well known that at small volumes a gas becomes less compressible than the perfect gas law demands, because of the finite size of the molecules and their interference with each other, so that pv increases with increasing pressure. Amagat found that after various initial abnormalities in the first few hundred kilograms are wiped out, the increase with pressure is nearly linear. The increase in pv is least rapid for H_2, as might be expected. At 3000 kg./cm.² the departures from the perfect gas law for O_2 have become so great that its density is approximately equal to that of water at atmospheric pressure, whereas, if Boyle's law had continued to hold, its density would have been three times as great.

Kohnstamm and Waldstra [12] in 1914 have checked the work of Amagat on H_2 up to 2200 kg./cm.[2] Below 2000 the agreement is fairly good, the pressures at equal volumes differing by about 8 kg./cm.[2] at 2000. Above 2000, however, the discrepancy becomes greater and amounts to 20 kg./cm.[2] at 2200.

In 1929 Bartlett, Cupples, and Treamearne [13] studied the compressibility of H_2, N_2, and a 3 : 1 mixture of them, up to 1000 kg./cm.[2] between 0° and 400° C. This work was done at the Fixed Nitrogen Laboratory, and the principal object of the work was to obtain data for the reaction to NH_3. Extensive tables are given for the volumes, the accuracy of which is stated to be 0·2 per cent. It is found that the volumes of the mixtures cannot be calculated by the rule of mixtures from the volumes of the components until a temperature of 200° is reached; below this temperature the actual volume may differ by a maximum of 1·7 per cent. from that calculated by the rule.

There is considerable experimental activity in this field at present, and it is probable that at some time in the near future there may be papers from Michels in Leyden and Keyes at the Massachusetts Institute of Technology.

My own measurements of the compressibility of the so-called permanent gases were made in 1923 (B. 42, 44, 47). Five gases were used—H_2, He, N_2, A, and NH_3. The design of a proper method proved to be unexpectedly difficult. The simplest method was tried first, in which a cylinder containing a manganin pressure gauge is charged with the gas through a by-pass at the top (like that shown in fig. 9) to a preliminary pressure of perhaps 2000 kg./cm.[2], and the compressibility above 2000 determined by measuring the displacement of the piston as a function of pressure. Because of the high compressibility of the gases, the corrections for the distortion of the cylinder are unusually low in this method, which would have been all that could be desired except for the unfortunate fact, already mentioned, that H_2

ruptures the steel by actually passing through the pores of the metal at pressures of 9000 or greater.   It is also probable that air and O_2 produce a similar effect, although the action is much slower.   Another method was then tried in which the gas was compressed by a column of mercury into a thin

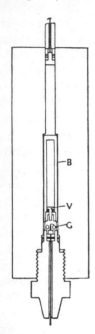

FIG. 27.—Apparatus for determining the compressibility of gases by the piston-displacement method.

steel capillary, and the position of the mercury determined by measurements of the electrical resistance.   The difficulty with this method was bad contact between the mercury and steel and irregular capillary action.   Finally a method was adopted much like that for the compressibility of liquids.   The apparatus is shown in fig. 27. The gas under an initial pressure of 2000 kg./cm.² is introduced into the pressure cylinder in a cylindrical bomb B, closed at the bottom by an inward opening valve V of my conventional type.   The lower end of the cylinder is closed with a plug bearing a manganin resistance gauge G, and the remainder of the cylinder is filled with a weighed amount of kerosene, air having been removed by several applications of vacuum during filling.   The upper part of the cylinder was closed by a conventional moving plug, and pressure produced by pushing in this plug with a hardened piston actuated by a hydraulic press.   The whole apparatus, including the press, which was specially constructed, was somewhat smaller than that used for most of my high-pressure work.   The method of functioning is plain: the plug is pushed in, increasing the pressure; when pressure reaches that of the gas in the bomb, the valve opens inwardly, and from here on the compressibility of gas and kerosene together are measured, whereas below this pressure the compressibility of the kerosene alone determines the displacement of the plug.   The compressibility of the kerosene may be eliminated by measurements made by other methods, or special calibrating runs may be made in which the gas is replaced by a core of steel, the compressibility of which is known.   There is a small correction, not more than 1 per cent. at the maximum pressure of 15,000 kg./cm.^{2}, for the stretch of the steel cylinder under internal pressure; this correction was calculated by elasticity theory.   It might be feared that there would be an error in this method due to the solution of the gas in the kerosene; if such an error exists, it was too small to detect, the volume always being a single valued function of the pressure, independent of the time available for the gas to go into solution or to pass by diffusion from one part of the apparatus to another.   There was doubtless a certain amount of solution of gas into the kerosene, as shown by the fact that when used to measure the compressibility of H_2 the H_2 acted on the steel from solution, for the bomb was rotted after several exposures and ruptured, the valve stem at the bottom of the bomb was also ruptured, as well as the stem of the moving plug at the top, and eventually the outer cylinder itself was ruptured.   For this reason the results on H_2 are not as satisfactory as those on other gases.   It would, however, be a complicated matter to devise a method capable of measuring the volumes of H_2 to the highest pressures with an accuracy equal to that attainable with more inert gases, and I have not yet attempted it. The method also suffers from the restriction that the use of kerosene to transmit pressure to the gas makes it impossible to measure the compressibility of O_2 or air because of the danger of explosion.

The bomb was filled to its initial pressure of 2000 by a special arrangement which need not be shown in detail. The amount of gas was determined by weighing the bomb before and after filling.   The source of the gas was the conventional gas bottle, in which there was a pressure of 100 kg./cm.^{2} or so.   The gas was raised from this to 2000 by a couple of intensifiers.   The gas in the bottles was not perfectly pure; the amount of impurity was found by a chemical analysis, and a correction applied, assuming the law of mixtures for low concentrations.   The most impure gas was He, in which there was an impurity of 4 per cent. of N_2.

The experiments were made to a maximum pressure of from 15,000 to 16,000 kg./cm.², and at three temperatures— 30°, 65°, and 95° C.   The data for the effect of temperature

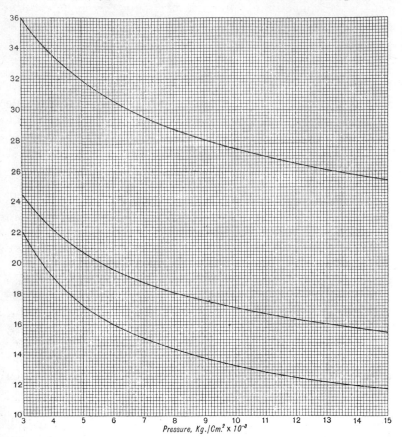

Fig. 28.—The volume in c.c. per mol. as a function of pressure at 65° of nitrogen, hydrogen, and helium, reading from the top down.

were not particularly good, however, and there is room for further work here.   It is unfortunate that the method gives only the changes of volume beyond 2000 kg./cm.², so that to get absolute volumes the data of other observers must be used, which exist only in the case of H_2 and N_2.   To obtain the fiducial volumes at 2000 kg./cm.² for the other gases, A, He, and NH_3, some extrapolation formula must be used which has proved applicable over the previous low-pressure range. There is a fairly satisfactory such formula for He, but there is nothing of the sort for A and NH_3, so that for these gases my measurements can give only the changes of volume of a known mass beyond 2000 kg./cm.[2].

The results are shown in the appended figures and tables.

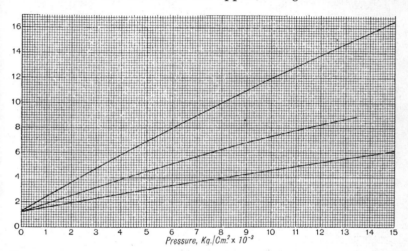

FIG. 29.—The product pv as a function of pressure for nitrogen, hydrogen, and helium, reading from the top down.

In fig. 28 the volume in cm.[3] per molecule is shown at pressures above 3000 for N_2, H_2, and He. If the perfect gas law were satisfied these curves would be identical for the three gases, and would be part of an equilateral hyperbola. If the curve for N_2, for example, were an equilateral hyperbola, its ordinate would drop from 36 at 3000 to 7·2 at 15,000, whereas the ordinate at 15,000 is actually 25·4. The great departure from the perfect gas law is exhibited in another way, as was done by Amagat, in fig. 29, where pv is plotted against pressure. At 15,000 the volume of N_2 is more than 16 times greater than it would be if the perfect gas law were valid, that of H_2 more than 9 times greater, and that of He 6 times greater. In fact, at these high pressures the compressibility of these gases approaches very closely to that of ordinary liquids. In fig. 30 the 'instantaneous' compressibility is plotted as a function of pressure for these three gases and also for water and CS_2. The extreme variation of the compressibility of these five substances at 15,000 is by a factor less than 2.

To a not bad first approximation pv is linear in pressure,

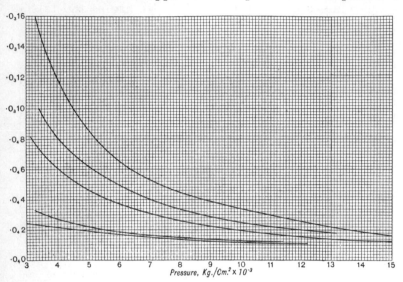

FIG. 30.—The compressibility $\frac{1}{v}\left(\frac{\delta v}{\delta p}\right)_\tau$ as a function of pressure of helium at 55°, hydrogen at 65°, nitrogen at 68°, CS_2 at 65°, and water at 65°, reading from the top down.

which obviously means that the relation between pressure and volume can be approximately written $p(v-a)=$ const. If this equation were rigorously true, 'a' would have the significance of the volume under infinite pressure. To a second approximation, however, pv against pressure is distinctly not linear, but is concave downward. If we assume that pv is of the second degree in pressure, then to a second approximation the equation of an isothermal can be written in the form

$$p[v-(a-bp)]=\text{const.},$$

which exhibits the 'co-volume' $(a-bp)$, as decreasing with increasing pressure.

Passing now to the finer details in the behaviour at high pressure, in fig. 31 is plotted the change of volume in cm.[3]

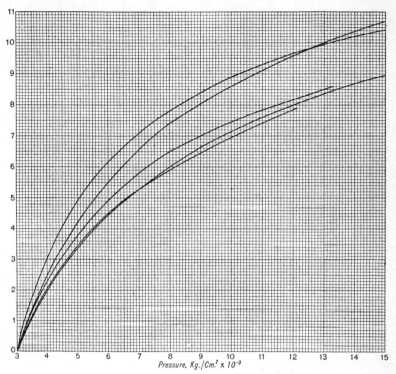

Pressure, Kg./Cm.² x 10⁻³

FIG. 31.—The change of volume in c.c. per mol. as a function of pressure reckoned from 3000 kg. as the fiducial pressure. At 10,000 kg. the order of the curves, reading from the top down, is: helium at 55°, nitrogen at 68°, hydrogen at 65°, argon at 55°, and ammonia at 30°.

per molecule of the five gases reckoned from 3000 kg./cm.² as the fiducial pressure. A significant feature of the diagram is the crossing of some of the curves: that of A crosses that of NH_3 at about 6800 kg./cm.², and that of N_2 crosses that of He at 12,500, and it seems probable that there will be other crossings beyond the pressures of the diagram. The meaning of a crossing is a change of relative compressibility. Thus at first, at pressures below 3000, He is more compressible than N_2, as would be expected because it is monatomic, and its atom is considerably smaller than the molecule of N_2. But at higher pressures there is a reversal, and the gas with the complicated molecule becomes more compressible than that with the simple atom. It is natural to associate this reversal with the same compressibility of the atom or molecule which manifests itself as a downward curvature of the pv curve. The compressibility at high pressures is naturally to be ascribed in most part to the loss of volume of the atoms or molecules themselves, the vacant spaces between atoms or molecules having been mostly squeezed out of existence already at lower pressures. The reason, then, that at high pressures a system of N_2 molecules becomes more compressible than a system containing an equal number of He atoms is that the complicated N_2 molecule is larger and has in it greater possibilities of further compression than the simple He atom. The determining factor apparently is the electronic structure of the fundamental unit. Consider the case of A and H_2; at pressures beyond those of the diagram it seems highly probable that the curve for A will rise and cross that for H_2. At low pressures, diatomic H_2 is more compressible than monatomic A, because the molecule of H_2 is smaller than the atom of A, and its co-volume smaller, but the initial superiority of the compressibility of H_2 disappears at moderate pressures when the empty spaces have been wiped out, and above this A becomes more compressible, because there is more room for compression inside the complicated structure of the atom itself, which has 18 extra-nuclear electrons against 2 for H_2. The crossing of the curves for A and NH_3 is to be explained in the same way; the NH_3 molecule has 10 extra-nuclear electrons against 18, and is therefore less compressible at extreme pressures.

Our conclusion that an important part of the compressibility at high pressures is contributed by the compressibility of the atoms or molecules themselves is confirmed by the behaviour of the various equations of state. Practically all of these equations have been built around the data of Amagat up to pressures of 3000 kg./cm.², or on other data limited to even lower pressures. In this range of pressure the compression of the molecules is a relatively unimportant factor, and since it is one that is not usually considered in elementary discussions, we are to expect that these equations of state do not properly emphasise this factor, with the result that when the equations are used by extrapolation to pressures of 15,000 kg./cm.², the volumes which they predict are too high and the compressibilities too small. This is, as a matter of fact, the way in which nearly all the equations which have been seriously proposed act; in my detailed paper the figures are given for some of these equations. It is to be said, however, that an equation proposed by Becker [14] for dealing with the data of Amagat, and specially designed by Becker to be used by extrapolation to considerably higher pressures, does reproduce the experimental results for N_2 up to 15,000 kg./cm.² with very considerable success.

The measurements on gases were a disappointment in that the values which they gave for the thermal expansion of gases at high pressures were not entirely satisfactory. It is possible, however, to draw the general conclusion that the thermal expansion, that is, the increase in volume in cm.³ of 1 gm./mol. for 1° rise of temperature, decreases very little over the entire range of pressure between 3000 and 15,000, in spite of the tenfold or even greater decrease of compressibility in the same range. From one point of view this is not surprising. It is known that the magnitude of the thermal expansion of a substance involves the failure of the restoring forces on the atoms or molecules to be linear functions of the displacement, the thermal expansion of a substance in which the force is linear being zero. As the molecules are squeezed into tighter contact at higher pressures, the restoring force may be expected to depart from linearity by an increasing amount, so that the thermal expansion will stay large, or even increase, as the data indicate for H_2.

It is interesting to compare the volumes of the gases at high pressures with the volumes which would be calculated, assuming the molecules to be of invariable dimensions and packed so as to be in contact. In Table I are shown the

8

TABLE I

MOLECULAR DISTANCES AT HIGH PRESSURES COMPARED WITH
MOLECULAR DIAMETERS

Substance	Density at 15,000 at 65°	Distance between Molecular Centres		Molecular Diameter from Kinetic Theory	
		Spheres Close Packed	Spheres Normal Cubic Piling	Boyle's Law	Free Path
Hydrogen	0·1301	$2·97 \times 10^{-8}$	$2·65 \times 10^{-8}$	$2·54 \times 10^{-8}$	$2·72 \times 10^{-8}$
Helium .	0·340	2·87	2·56	1·98	2·20
Nitrogen .	1·102	3·89	3·47	3·56	3·80

distances between molecular centres which would be computed from the observed densities of the gas, assuming, first, closely packed spheres and, secondly, normal cubic piling. The molecular diameters calculated by kinetic theory from the departures from Boyle's law and from such phenomena as viscosity are also included. These diameters are all of the same order of magnitude; the packing in the actual compressed gas is closest for N_2, and least close for He. In view of the fact that the volume curves show no break in smoothness up to the maximum pressure, it is difficult to see how the volumes can approach so closely to the volumes of rigid spheres in contact unless there is some actual compression of the molecules themselves.

The volumes at the highest pressures may also be made to give a lower limit for the compressibility of the solid phases of these gases. Consider solid H_2 at atmospheric pressure at −259·9° C. Its density here is 0·076. Imagine now this solid compressed at −259·9° C. to 15,000 kg./cm.^{2}, and then warmed at 15,000 from −259·9° to +65° C. Its density almost certainly decreases during this latter process, both because of the rise of temperature and the discontinuous increase of volume when the solid melts to the amorphous fluid phase. Hence the density at $-259 \cdot 9°$ is greater than $0 \cdot 1301$, the density at 15,000 at $+65°$ C. Hence the mean compressibility of the solid at $-259 \cdot 9°$ between atmospheric pressure and 15,000 is greater than that which would produce an increase of density from $0 \cdot 076$ to $0 \cdot 1301$. This corresponds to a decrease of volume to less than $0 \cdot 58$ of the initial volume, which means a higher average compressibility than that of any other solid hitherto measured, or any liquid either, for that matter, the relative volume of the most compressible solid measured, Cs, being $0 \cdot 67$ at this pressure, and that of the most compressible liquid, pentane, about the same. In the same way it is possible to compute that the volume of liquid H_2 at $-252°$ C. is decreased by a pressure of 15,000 to less than $0 \cdot 54$ of its initial volume. Similarly, liquid He at $-271 \cdot 6°$ is compressed by 15,000 to less than $0 \cdot 43$ of its atmospheric volume. It is, therefore, not unlikely that the solid phases of the noble gases are the solids of highest compressibility, and if one may reason by analogy with the alkali metals, solid xenon would be expected to be the most compressible solid element.

Interesting suggestions concerning the deformation with pressure of the atoms or molecules of these gases may also be obtained from a consideration of the change of internal energy with pressure. This is given by the thermodynamic relation $\left(\dfrac{\partial E}{\partial p}\right)_\tau = -\tau\left(\dfrac{\partial v}{\partial \tau}\right)_p - p\left(\dfrac{\partial v}{\partial p}\right)_\tau$, where E is the internal energy. It is obvious that at low pressures the sign of $\left(\dfrac{\partial E}{\partial p}\right)_\tau$ is negative, the physical interpretation of which is that at low pressure the attractive forces are greater than the repulsive forces, so that, on the whole, work is done by the internal forces when volume decreases at low pressures, and the internal energy decreases. At sufficiently high pressures, however, the repulsive forces must preponderate, and the sign of $\left(\dfrac{\partial E}{\partial p}\right)_\tau$ must reverse. We shall see later in discussing ordinary liquids and solids that this reversal does take place for them at pressures in the experimental range. The volume at which the reversal takes place marks the volume at which attractive and repulsive forces are in equilibrium. But at 0° abs. at atmospheric pressure $\left(\dfrac{\partial E}{\partial p}\right)_\tau = 0$, and the attractive and repulsive forces are in equilibrium. Hence, to that degree of approximation to which the average molecular forces are functions of volume only, and this is a fair approximation if the molecules act like rigid structures, the volume at room temperature at which $\left(\dfrac{\partial E}{\partial p}\right)_\tau = 0$ may be expected to be equal to the volume at 0° abs. at atmospheric pressure. This expectation turns out to be roughly verified for ordinary liquids and solids. But it is far from verified for these gases; the point here is that $\left(\dfrac{\partial v}{\partial \tau}\right)_p$ stays so large over the entire pressure range and $\left(\dfrac{\partial v}{\partial p}\right)_\tau$ becomes small so rapidly that $\left(\dfrac{\partial E}{\partial p}\right)_\tau$ remains negative up to the extreme pressure of 15,000 kg./cm.² at room temperature, and the mean forces continue to be attractive. But at 15,000 kg./cm.² at room temperature the volume is approximately only half that at 0° abs. at atmospheric pressure, where attractive and repulsive forces balance. It is evident, therefore, that the whole atom or molecule is so compressed by a pressure of 15,000 kg./cm.² that its effective boundary —that is, the region in which the repulsive forces rapidly assume the ascendancy—is compressed into less than half its volume at 0° abs. This is about as striking a demonstration as can be imagined of the necessity of taking account of the deformability of the molecule, or, expressed in another way, of the impossibility of assigning to the independent molecule a 'law of force,' which shall be independent of the interaction of neighbouring molecules.

Compressibility of Liquids. — Little further discussion is necessary of the methods that have been used for the measurement of the compressibility of fluids liquid under ordinary conditions. The method of Richards,[15] however, has received such wide application that it should be specifically mentioned. The piezometer is of the form shown in fig. 32. The liquid to be measured is enclosed in a large glass bulb, closed at the upper end with a carefully ground stopper. At the lower end a capillary is attached, bent around and enlarged as shown. At the junction of the capillary and the larger vertical part there is a platinum contact. The lower part of the bulb and the capillary to a little above the contact are filled with mercury. The whole arrangement is placed in a pressure chamber, and after temperature equilibrium has been reached, pressure raised until contact between the platinum and mercury is broken in consequence of the compression of the liquid in the bulb and the mercury. Pressure is now lowered slightly until contact is just restored, and this pressure noted. The apparatus is now opened and an additional weighed quantity of mercury added, and the procedure repeated, noting now the new higher pressure at which contact is made and broken. This may be repeated at as many pressures as desired. The operation may be repeated with other liquids replacing the original liquid. In particular, the whole piezometer may be filled with mercury. In the end one has information of this kind: a known number of grams of liquid A or B or C plus a known number of grams of mercury occupies the same

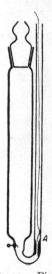

FIG. 32.—Piezometer of Richards for differential compressibility.

volume at a definite pressure as a known number of grams of mercury. This information is available for a number of different pressures. From this we may at once find the difference of volume under pressure of the various liquids and an amount of mercury equal in volume at atmospheric pressure, except for a correction due to the compression of that part of the mercury which varies in amount with the pressure. For a pressure of 100 kg./cm.^{2} this correction is only 0·04 per cent., so that up to several hundred kilograms measurements by the method of Richards give directly, within a small fraction of 1 per cent., the difference between the compressibility of any liquid and mercury, without demanding at all a knowledge of the compressibility of the material of the piezometer.   If the compressibility of mercury is approximately known, the correction may be approximately determined, and the difference of compressibilities found even more accurately.   The method was used by Richards for the compressibility of a great many liquids up to 500 kg./cm.², all at 25° C.   The quantities of mercury added were chosen as closely as possible so as to make contact at 100, 300, and 500 kg./cm.².   From the difference of the differences for these pressure ranges the change of compressibility with pressure could be found for those substances in which the effect is large enough to be detectible over this small pressure range.   Various precautions have to be employed.   The glass of the piezometer shows hysteresis effects analogous to the well-known temperature hysteresis effects in thermometers.   To eliminate or minimise effects of this kind, it is necessary that the routine of manipulation always be exactly the same.   On one occasion Richards replaced the glass piezometer with one of steel in order to get rid of hysteresis effects, but the advantage did not prove to compensate for the greater experimental difficulty.

Richards and his pupils have measured by this method the compressibility of more than seventy-five liquids up to 500 kg./cm.² at 25°.   The method can also be used to obtain the compressibility of solids by merely replacing part of the liquid in the piezometer by the solid; in this way Richards measured the compressibility of a large number of the elements.   It is evident that when applied to solids the accuracy is less, because of the much smaller compressibility of solids, and also because the compressibility of mercury may now be much larger than the effect sought, which must be obtained by difference.   In addition to Richards, the method has been used by a number of other experimenters,[15] chiefly in measuring the compressibility of solutions.   The pressure range of this other work has extended up to 1500 kg./cm.².   At much higher pressures the method will encounter difficulties of various sorts; for instance, it is not possible to seal ordinary platinum contacts through glass successfully at high pressures, because the glass breaks from differential compressibility; hysteresis and elastic after-effects become more serious at high pressures, and the reduction of the observations becomes more complicated because the correction for the compressibility of the mercury increases in importance.

The experiments of Amagat on the compressibility of liquids cover the highest range of any previous work. The methods were essentially the same as for the compressibility of gases, except that the large antechamber for the preliminary compression of the gases could be dispensed with. An optical method was used up to 1000, and an electrical contact method between 1000 and 3000. His measurements were on water and twelve other liquids. These same twelve liquids have been measured in my experiments to 12,000, and the numerical results will be discussed later.

In my own measurements to higher pressures it was not possible to use many of the methods successful in lower ranges, so that new methods had to be devised or else new adaptations made of old methods. These methods will first be discussed and then the results. One of my first attempts was to measure the compressibility of mercury by a method modified from that of Perkins, already sufficiently described on pp. 2 and 101. A prime essential is that the plunger shall slide freely and without leak; this was evidently not attained in Perkins's form of the apparatus, because his compressibility, even for as compressible a liquid as water, was 250 per cent. too high. The modified piezometer used for mercury is shown in fig. 33. The plunger P is a piece of drill rod, $\frac{1}{16}$ in. in diameter, fitting its hole within about 0·0002 in. To ensure as much freedom from leak as possible, it slides at the upper end E through a stuffing-box containing a viscous mixture of molasses and glycerine, covered with mercury to prevent mixing with the pressure-transmitting liquid. The displacement of the piston was measured by the movable brass ring D, the whole apparatus being placed in the high-pressure vessel, and the position of D observed after each experiment. With this apparatus the compressibility of mercury was measured to 6200 kg./cm.[2]. The agreement over this range with values determined by a later more accurate method was

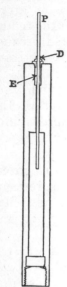

fair; the accuracy was sufficient to settle the main point at issue at the time of its application, namely, whether the compressibility of mercury decreases by a large amount with increase of pressure. The method suffers from the inconvenience in application that only a single reading can be obtained for each set-up, but this is not prohibitively serious, as the manipulations can be made rapidly. A more serious disadvantage is that there seems to be no way of correcting for the slight leak past the piston; with the apparatus actually used this error was of the order of 2 or 3 per cent.; it would probably be possible to decrease this error by more careful construction.

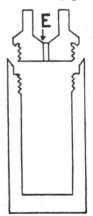

FIG. 33. — Piezometer for determining the compressibility of mercury by the piston displacement method.

The next method tried made possible the measurement of the compressibility of water and mercury up to 12,000 (B. 56). This was a modification of the method of Aimé,[8] described on pp. 3 and 101.

Aimé's results, as well as those of Perkins, were affected by large error. Probably a large part of the error of Aimé arose from the large drops of mercury clinging to the mouth of the orifice through which mercury was forced into the piezometer. This source of error can be minimised by making the orifice very small; this was done by using for the orifice a fine longitudinal scratch made in a steel pin E, driven into

FIG. 34. — Piezometer by which the compressibility of water and mercury was determined by a modified Aimé method.

a hole in the upper part of the piezometer, which is shown in fig. 34. In adapting the method to high pressures, there are two important points to be considered. In the first place, it was desirable to replace the glass of which the piezometer had usually been made by some more isotropic material less likely to show hysteresis and other irregular effects, since I had already found by direct measurement that the elastic deformation of glass at pressures no higher than 6000 may have irregularities of as much as 4 per cent. because of these effects. Steel suggested itself naturally as the material with which to replace glass; its compressibility is about one-fourth that of glass, it may be made free from internal strains by proper annealing, and direct experiment had shown that even a piece of boiler plate is equally compressible in all directions. Secondly, the correction for the compressibility of the mercury forced in at the high pressures is comparatively large, so that accurate results demanded the determination of this correction. This correction is roughly proportional to the pressure; thus for measurements up to only 500, the correction on the compressibility of water because of this effect is only 0·2 per cent., whereas at 12,000 it is 3·5 per cent. This situation was met by making two sorts of determinations, one in which the piezometer was nearly filled with water and mercury was forced in, giving the compressibility of water except for the correction for the compressibility of mercury, and a second in which the piezometer was nearly filled with mercury and water was forced in while the piezometer was in an inverted position, giving the compressibility of mercury except for the correction for the compression of the water forced in. By combining the two sets of measurements, the compressibility of water and mercury could both be found from two sets of equations. The distortion of the steel piezometer evidently enters the result, and this demands a knowledge of the compressibility of the steel. This was determined by direct experiment, and will be described later.

With this method, the compressibility of water and mercury was determined to 7000 kg./cm.^{2} at 0° C. and to 11,000 at 22°. The pressure limitations were set by the freezing under pressure of both water and mercury. The fact that water will freeze at high pressure at temperatures above 0° C. was not known before this work was started, and it was first discovered from the irregularity of the results for compressibility at high pressures. Freezing is to be avoided because the strains introduced into the steel of the piezometer by the sudden change of volume incident to freezing produce rather large irregularities in the subsequent behaviour, doubtless because of permanent distortions in the steel. There is no reason why this method should not be extended to other liquids, and I believe it can give as accurate results and as conveniently as the method of Richards, over which it has manifest advantages for use at high pressures. It has a distinct field of usefulness whenever a few measurements of compressibility with simple apparatus at a temperature close to room temperature are desired. But for an extended series of measurements it suffers from the disadvantage of so many other methods of giving only one reading for a single set-up, and at temperatures greatly different from that of the room, waiting for temperature equilibrium between successive set-ups is very time consuming.

The next development (B. 11) was with a continuous reading method, which has already been described in some detail as applied to gases. In this the compressibility is calculated from the displacement of the piston by which pressure is produced. The pressure cylinder contains a pressure-measuring coil of manganin, and the liquid to be measured in some sort of a container, in which it may be separated from the pressure-transmitting liquid by mercury. The motion of the piston involves the compressibility of the liquid, of its container, of the mercury, and of the transmitting liquid, and also the distortion of the whole cylinder. The various extraneous effects can be eliminated by auxiliary measurements, or by replacing the liquid by pieces of steel, the compressibility of which has been determined by other methods. A correction which cannot be determined in this way is for the change of cross-section of the cylinder under pressure; this can be calculated from the theory of elasticity and is at most, at 12,000 kg./cm.^{2}, of the order of only 1 per cent. Hysteresis effects in the steel cylinder play quite an appreciable part, more important than in the case of gases; to eliminate these it is necessary to make several seasoning applications of pressure before the measurements, and then to use the mean of readings with increasing and decreasing pressure. With proper precautions it is possible to obtain measurements with a single set-up of the apparatus consistent to nearly the limit of the accuracy with which the position of the piston can be measured with an ordinary micrometer, that is, to about 0·0002 in. on a stroke of 2 or 3 in. All measurements of compressibility by this method were made with two independent fillings of the apparatus, separated by an interval of a month or more of time. The compressibilities determined from these independent fillings agreed within 0·1 or 0·2 per cent., and the thermal expansions, determined from measurements at 20° intervals of temperature, were consistent to 2 per cent.

The liquids measured by this method were the same twelve as those measured by Amagat; this choice was made in order to afford a check on the readings over the common range up to 3000, and also because the method of piston displacement is not particularly sensitive at the lower pressures, where it was desirable to supplement my measurements with the more sensitive measurements of Amagat. In order to increase the sensitiveness at low pressures, an independent set of readings was made over the pressure range between atmospheric and 2500 kg./cm.², making the pressure steps shorter than in the wider range. It must be recognised, however, that the method of piston displacement is not very well adapted with the particular apparatus used to give good results below 500 kg./cm.². There is no reason why the method should not be good at these lower pressures if the apparatus were specially constructed for this range.

The method of piston displacement I believe to be the simplest and most rapid that had been used up to that time for mapping out the p-v-t relations, but there were nevertheless certain features capable of improvement. The corrections for the transmitting liquid were an undesirably large fraction of the total effect, and the discovery of the best procedure to avoid error from hysteresis demanded troublesome experimenting by the method of trial and error. But the most important difficulty with the method has to do with the temperature effects. Thermal expansion is determined from

measurements of pressure as a function of temperature at
constant piston displacement, or from the difference of two
isothermals.   In either case, the apparatus is so constructed
that the pressure-measuring coil must be subjected to the
same changes of temperature as the liquid being measured.
It is easy to see that at the high pressure end of the range the
accuracy with which the difference of pressure at two different
temperatures may be inferred from resistance measurements
may be seriously affected by slight variations in the pressure
coefficient of resistance with temperature.   This source of
error was eliminated as well as possible by special calibrations
of the coil at different temperatures, but at best the accuracy
of the pressure calibration is limited, so that the accuracy
with which a thermal expansion may be measured is very
materially less than that possible in a compressibility measurement.   The measurements of the thermal expansions of
liquids made by this method may, therefore, all be affected
by the same error.   The relative thermal expansions of different liquids, however, should be correct, since the differential
expansions depend only on the *sensitiveness* of the resistance
measurements, which is high, as opposed to the absolute
accuracy, which is lower.   Since from certain points of view
the behaviour of thermal expansion at high pressures seemed
to be of particular importance, I have attempted to devise
other methods permitting a more accurate determination.

The first of these was unsuccessful for the desired purpose,
but it is nevertheless of sufficient usefulness to justify mention
(B. 64).   This is another modification of the plunger method
of Perkins.   There are two principal features—in the first
place the plunger is fitted with extreme accuracy to avoid
error from leak, and in the second place the method is made
continuous reading, so that the cylinder need not be opened
for each reading after every application of pressure.   The
piezometer is shown in fig. 35.   The piston is of the same
diameter as the body of the cylinder, about $\frac{3}{8}$ in., and is
ground so as to fit the hole with an accuracy of 0·00001 in.,
a fit which has only recently become possible because of
improvements in the technique of grinding.   The position
of the piston at any pressure is measured with an electrical

device, which has been useful in other places in my high-pressure work. A high-resistance wire is attached to the plunger, and slides over a contact rigidly fixed to the cylinder, but electrically isolated from it. The potential difference between the sliding contact and another terminal fixed to the wire, when a known current flows in the wire, is measured on a potentiometer. From this the length of wire between the fixed and sliding electrodes can be found, and so the position of the plunger. The use of the potentiometer evidently eliminates error from contact resistance. The method is astonishingly sensitive, and permits readings to better than a wave-length of light, so that it is of the same order of sensitiveness as an interference method. Because of the sensitiveness of measurements, the cylinder may be made comparatively small, so that it is possible to mount it in a separate cylinder apart from the manganin gauge, as is possible in most of my other high-pressure methods, but which was not possible in the method of compressibility by piston displacement. As a consequence, the cylinder containing the gauge coil may be kept at a constant temperature while the liquid varies in temperature, and in this way the error from changing the temperature of the gauge coil is avoided. The apparatus has given very satisfactory results

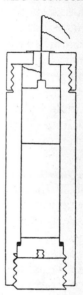

Fig. 35.—Sliding piston piezometer.

in measuring isothermal compressibilities, and in measuring the changes of volume accompanying certain polymorphic transitions at constant temperature, and has a useful field here, but unfortunately it does not work for the precise purpose for which it was designed, that of thermal expansion measurements at high pressures. The reason for this is that during the changes of temperature incident to measuring thermal expansion the cylinder of the piezometer must experience the rise of temperature before the piston, and during this period of temperature inequality there is a relative expansion of cylinder with respect to the plunger, and consequent leak.   An obvious method of meeting the difficulty is to make cylinder and plunger of invar, but unfortunately invar is not hard enough to permit application of present technical methods of high-precision grinding; this difficulty may perhaps be surmounted in the future.

The second method is the method of the sylphon (B., yet unpublished).   The sylphon is a flexible metal bellows that has a number of interesting applications in industry.   It is made commercially by spinning, from a single piece of metal. Unfortunately the commercial sizes are too large and too stiff for my purpose, and also, unfortunately, it was a matter of prohibitive expense ($10,000) to make the jigs and tools required for the production by the commercial method of a single sylphon of more suitable dimensions.   Accordingly, sylphons for this purpose were made by soldering together stampings of thin sheet brass.   The method of using the sylphon in a compressibility measurement is very simple; the sylphon is filled with the liquid to be measured and sealed; it is then exposed to external hydrostatic pressure, under which it shortens until the internal pressure is equal, except for a small fraction of an atmosphere depending on the stiffness of the sylphon, to the external pressure.   The change of length of the sylphon is measured by the same electrical device as used for the determination of the position of the plunger in the piston piezometer, and so the change of internal volume may be found.   The effective cross-section of the sylphon is practically constant, independent of its length, as may be checked by independent calibration, so that the volume change may be calculated easily from the change of length.   There are, of course, various corrections to apply: for the cubic compressibility of the material of the sylphon, and for the effect of pressure on the resistance of the measuring wire, etc., but these corrections are all small, and may be determined easily by direct experiment. It is obvious that the sylphon may be mounted in a separate cylinder, so that the pressure-measuring gauge is not exposed to temperature changes.   Furthermore, there are no hysteresis or other time effects to be guarded against.   The sylphon is maintained in rectilinear motion during the shortening by an internal plunger and cylinder which act as a guide, and which are shown in the detailed view in fig. 36. The sylphon is capable of giving rapid p-v-t measurements, and has proved very convenient. In use, rather more care must be taken than usual to avoid sudden changes of pressure, which may easily ruin the sylphon by causing a permanent set in the flexible walls by an inertia effect, before the length can respond to the changed external pressure. Care must also be taken not to exceed the freezing pressure of the liquids.

Experimental Results for Liquids. The following liquids have been measured: by the small plunger method, mercury; by the modified Aimé method, water and mercury; by the piston displacement method, water, kerosene, methyl alcohol, ethyl alcohol, n-propyl alcohol, i-butyl alcohol, i-amyl alcohol, ether, acetone, CS_2, PCl_3, C_2H_5Cl, C_2H_5Br, C_2H_5I; by the piston piezometer method (rough measurements only), glycerine; and by the sylphon method, pentane, i-pentane, normal hexane, and its four isomers, n-heptane, n-octane, n-decane, C_6H_6, C_6H_5Cl, C_6H_5Br, CCl_4, bromoform, i-propyl alcohol, n-butyl alcohol, and water. The volumes of these liquids are given as

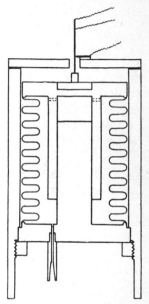

FIG. 36. — Cross-section of the sylphon, and attachments for determining the compressibility of liquids.

a function of pressure and temperature in Table II; in the original papers the volumes will be found tabulated at closer intervals.

In the paper on twelve liquids measured by the piston displacement method an elaborate discussion will be found of the effect of pressure on thermal expansion, compressibility, pressure coefficient, work of compression, heat of compression, change of internal energy, C_p and C_v. The discussion in the paper on liquids measured by the sylphon method is less detailed; only a hasty summary can be attempted here.

TABLE II

Relative Volumes of Liquids

Pressure. kg./cm.²	Methyl alcohol		Ethyl alcohol		n-Propyl alcohol		i-Butyl alcohol		i-Amyl alcohol		Ether	
	20°	80°	20°	80°	20°	80°	20°	80°	20°	80°	20°	80°
0	1·0238	1·0005	1·0212	1·0934	1·0173	1·0865	1·0195	1·0880	1·0181	1·0814	1·0315	0·9906
1,000	0·9530	1·0023	0·9506	0·9944	0·9498	0·9934	0·9486	0·9918	0·9526	0·9936	0·9363	0·9223
2,000	0·9087	0·9456	0·9081	0·9407	0·9142	0·9448	0·9097	0·9410	0·9158	0·9452	0·8871	0·8912
3,000	0·8792	0·9095	0·8786	0·9055	0·8897	0·9145	0·8822	0·9080	0·8892	0·9136	0·8530	0·8515
4,000	0·8557	0·8814	0·8545	0·8787	0·8700	0·8919	0·8601	0·8830	0·8682	0·8903	0·8275	0·8289
5,000	0·8354	0·8588	0·8443	0·8568	0·8529	0·8732	0·8409	0·8619	0·8508	0·8713	0·8071	0·8112
6,000	0·8192	0·8412	0·8178	0·8387	0·8390	0·8579	0·8269	0·8463	0·8373	0·8560	0·7916	0·7953
7,000	0·8053	0·8262	0·8038	0·8229	0·8266	0·8442	0·8130	0·8317	0·8251	0·8418	0·7773	0·7813
8,000	0·7936	0·8134	0·7917	0·8094	0·8163	0·8328	0·8028	0·8210	0·8148	0·8302	0·7645	0·7687
9,000	0·7827	0·8013	0·7807	0·7973	0·8069	0·8230	0·7927	0·8105	0·8044	0·8190	0·7525	0·7574
10,000	0·7725	0·7905	0·7703	0·7863	0·7984	0·8142	0·7832	0·8007	0·7949	0·8091	0·7418	0·7469
11,000	0·7634	0·7813	0·7606	0·7765	0·7909	0·8061	0·7742	0·7913	0·7860	0·8001	0·7312	0·7365
12,000	0·7559	0·7738	0·7521	0·7682	0·7840	0·7982	0·7662	0·7827	0·7782	0·7926	0·7216	

Pressure. kg./cm.²	Acetone		CS₂		PCl₃		C₂H₅Cl		C₂H₅Br		C₂H₅I	
	20°	80°	20°	80°	20°	80°	20°	80°	20°	80°	20°	80°
0	1·0279	··	1·0235	1·1092	1·0234	1·1039	··	··	1·0275	··	1·0214	1·0935
1,000	0·9553	1·0107	0·9586	1·0083	0·9593	1·0065	0·9276	0·9827	0·9478	1·0018	0·9509	0·9969
2,000	0·9108	0·9497	0·9173	0·9552	0·9205	0·9557	0·8774	0·9264	0·9044	0·9407	0·9092	0·9425
3,000	0·8775	0·9105	0·8877	0·9185	0·8926	0·9220	0·8442	0·8749	0·8776	0·9090	0·8802	0·9065
4,000	0·8532	0·8819	0·8647	0·8902	0·8705	0·8962	0·8200	0·8462	0·8505	0·8772	0·8583	0·8790
5,000	0·8334	0·8586	0·8453	0·8676	0·8521	0·8757	0·7994	0·8230	0·8317	0·8546	0·8394	0·8581
6,000	0·8175	0·8403	0·8295	0·8501	0·8375	0·8596	0·7821	0·8040	0·8163	0·8371	0·8236	0·8418
7,000	0·8028	0·8243	0·8147	0·8347	0·8245	0·8450	0·7680	0·7887	0·8020	0·8220	0·8093	0·8264
8,000	0·7898	0·8101	0·8022	0·8220	0·8133	0·8323	0·7561	0·7762	0·7900	0·8091	0·7968	0·8134
9,000	··	0·7974	0·7911	0·8107	0·8029	0·8210	0·7454	0·7644	0·7787	0·7968	0·7856	0·8013
10,000	··	0·7858	0·7805	0·7997	0·7929	0·8109	0·7352	0·7533	0·7686	0·7858	0·7755	0·7909
11,000	··	0·7750	0·7715	0·7894	0·7838	0·8014	0·7259	0·7432	0·7598	0·7762	0·7665	0·7817
12,000	··	0·7651	0·7638	0·7795	0·7761	0·7928	0·7176	0·7336	0·7524	0·7677	0·7588	0·7737

TABLE II—continued

Pressure kg./cm.²	n-Pentane 0°	n-Pentane 50°	n-Pentane 95°	i-Pentane 0°	i-Pentane 50°	i-Pentane 95°	n-Hexane 0°	n-Hexane 50°	n-Hexane 95°	2-Methyl pentane 0°	2-Methyl pentane 50°	2-Methyl pentane 95°	3-Methyl pentane 0°	3-Methyl pentane 50°	3-Methyl pentane 95°
0	1·0000	1·0837	1·1869	1·0000	1·0843	1·1741	1·0000	1·0712	1·1535	1·0000	1·0736	1·1549	1·0000	1·0727	1·1530
1,000	0·9021	0·9395	0·9768	0·9028	0·9415	0·9809	0·9191	0·9567	··	0·9131	0·9496	0·9851	0·9120	0·9471	0·9815
2,000	0·8546	0·8820	0·9078	0·8571	0·8845	0·9119	0·8763	0·9048	0·9304	0·8711	0·8976	0·9227	0·8689	0·8971	0·9205
3,000	0·8229	0·8454	0·8671	0·8264	0·8490	0·8712	0·8490	0·8720	0·8914	0·8419	0·8638	0·8826	0·8408	0·8630	0·8824
4,000	0·7997	0·8193	0·8371	0·8025	0·8222	0·8403	0·8259	0·8472	0·8629	0·8191	0·8382	0·8552	0·8200	0·8379	0·8553
5,000	0·7811	0·7985	0·8125	0·7830	0·8014	0·8178	0·8068	0·8262	0·8404	0·8009	0·8184	0·8336	0·8022	0·8189	0·8342
6,000	0·7647	0·7807	0·7933	0·7670	0·7850	0·8002	0·7905	0·8091	0·8225	0·7844	0·8023	0·8179	0·7866	0·8026	0·8158
7,000	0·7506	0·7657	0·7775	0·7534	0·7729	0·7863	··	0·7943	0·8071	··	0·7878	0·8022	0·7723	0·7888	0·8006
8,000	0·7381	0·7520	0·7641	··	0·7592	0·7706	··	0·7817	0·7942	··	0·7754	0·7886	0·7592	0·7761	0·7876
9,000	0·7281	0·7409	0·7527	··	0·7474	··	··	0·7707	0·7824	··	0·7644	0·7763	0·7476	0·7647	0·7753
10,000	0·7192	0·7316	0·7433	··	··	··	··	0·7615	0·7723	··	0·7538	0·7658	0·7372	0·7544	0·7649
11,000	··	··	··	··	··	··	··	0·7529	0·7632	··	0·7450	0·7564	0·7276	0·7455	0·7550
12,000	··	··	··	··	··	··	··	··	··	··	··	0·7484	··	0·7373	0·7452

Pressure kg./cm.²	n-Octane 0°	n-Octane 50°	n-Octane 95°	n-Decane 0°	n-Decane 50°	n-Decane 95°
0	1·0000	1·0595	1·1230	1·0000	1·0530	1·1083
1,000	0·9311	0·9654	0·9943	0·9383	0·9683	0·9952
2,000	0·8924	0·9200	0·9422	··	0·9263	0·9466
3,000	0·8640	0·8882	0·9068	··	0·8952	0·9146
4,000	··	0·8639	0·8802	··	0·8714	0·8880
5,000	··	0·8428	0·8592	··	··	0·8675
6,000	··	0·8251	0·8416	··	··	0·8481
7,000	··	0·8103	0·8267	··	··	0·8341
8,000	··	··	0·8134	··	··	0·8215
9,000	··	··	0·8014	··	··	··
10,000	··	··	0·7915	··	··	··

Pressure kg./cm.²	2-2 Dimethyl butane 0°	2-2 Dimethyl butane 50°	2-2 Dimethyl butane 95°	2-3 Dimethyl butane 0°	2-3 Dimethyl butane 50°	2-3 Dimethyl butane 95°	n-Heptane 0°	n-Heptane 50°	n-Heptane 95°
0	1·0000	··	··	1·0000	1·0722	1·1496	1·0000	1·0633	1·1350
1,000	0·9154	0·9517	0·9824	0·9147	0·9503	0·9841	0·9223	0·9684	0·9919
2,000	0·8737	0·9016	0·9218	0·8695	0·8960	0·9198	0·8813	0·9083	0·9337
3,000	0·8473	0·8703	0·8855	0·8395	0·8633	0·8836	0·8531	0·8753	0·8954
4,000	0·8260	0·8464	0·8594	0·8180	0·8383	0·8562	0·8315	0·8508	0·8676
5,000	0·8083	0·8269	0·8392	0·8004	0·8186	0·8340	0·8145	0·8312	0·8457
6,000	··	0·8106	0·8223	0·7855	0·8028	0·8162	··	0·8146	0·8276
7,000	··	··	0·8079	··	0·7881	0·8008	··	0·8001	0·8125
8,000	··	··	0·7958	··	0·7747	0·7874	··	0·7875	0·7992
9,000	··	··	0·7850	··	0·7624	0·7757	··	0·7763	0·7870
10,000	··	··	0·7756	··	0·7509	0·7645	··	0·7659	0·7762
11,000	··	··	··	··	··	0·7546	··	··	0·7660

TABLE II—continued

Pressure. kg./cm.²	Benzene			Chloro-benzene			Bromo-benzene			Carbon tetrachloride		Bromoform	
	0°	50°	95°	0°	50°	95°	0°	50°	95°	50°	95°	50°	95°
0	1·0000	1·0630	1·1295	1·0000	1·0502	1·1013	1·0000	1·0467	1·0940	1·0000	…	1·0000	1·0427
1,000	…	0·9841	1·0201	0·9541	0·9882	1·0215	0·9570	0·9891	1·0169	0·9192	0·9540	0·9369	0·9662
2,000	…	…	0·9684	0·9244	0·9511	0·9755	…	0·9527	0·9765	…	0·9049	…	0·9225
3,000	…	…	0·9325	…	0·9268	0·9463	…	0·9261	0·9460	…	0·8726	…	0·8915
4,000	…	…	…	…	0·9068	0·9247	…	0·9047	0·9229	…	…	…	…
5,000	…	…	…	…	0·8903	0·9072	…	…	0·9033	…	…	…	…
6,000	…	…	…	…	0·8762	0·8924	…	…	0·8868	…	…	…	…
7,000	…	…	…	…	0·8652	0·8794	…	…	0·8726	…	…	…	…
8,000	…	…	…	…	…	0·8675	…	…	0·8604	…	…	…	…
9,000	…	…	…	…	…	0·8580	…	…	0·8501	…	…	…	…
10,000	…	…	…	…	…	0·8487	…	…	…	…	…	…	…
11,000	…	…	…	…	…	0·8406	…	…	…	…	…	…	…
12,000	…	…	…	…	…	…	…	…	…	…	…	…	…

Pressure. kg./cm.²	i-Propyl alcohol			n-Butyl alcohol			n-Hexyl alcohol		
	0°	50°	95°	0°	50°	95°	0°	50°	95°
0	1·0000	1·0540	1·1097	1·0000	1·0455	1·0907	1·0000	…	…
1,000	0·9387	0·9725	1·0112	0·9459	0·9779	…	0·9488	0·9791	1·0052
2,000	0·9040	0·9296	0·9562	0·9111	0·9372	0·9595	0·9183	0·9420	0·9619
3,000	0·8769	0·8995	0·8961	0·8874	0·9087	0·9276	…	0·9159	0·9337
4,000	0·8571	0·8765	0·8750	0·8681	0·8867	0·9033	…	0·8960	0·9118
5,000	0·8398	0·8579	0·8572	0·8512	0·8684	0·8834	…	0·8787	0·8931
6,000	0·8244	0·8415	0·8419	…	0·8530	0·8667	…	…	0·8773
7,000	0·8116	0·8279	0·8289	…	0·8390	0·8519	…	…	0·8637
8,000	0·8002	0·8159	0·8168	…	0·8269	0·8388	…	…	…
9,000	0·7901	0·8052	0·8062	…	0·8153	0·8277	…	…	…
10,000	0·7813	0·7948	0·7971	…	0·8076	0·8174	…	…	…
11,000	…	0·7861	0·7894	…	…	0·8082	…	…	…
12,000	…	0·7784	…	…	…	0·7999	…	…	…

Pressure. kg./cm.²	Water			Mercury		Glycerine
	0°	50°	95°	0°	20°	30°
0	1·0000	1·0119	1·0395	1·00000	1·00362	1·000
1,000	0·9567	0·9741	0·9984	0·99626	0·99972	0·958
2,000	0·9248	0·9439	0·9661	0·99261	0·99593	0·932
3,000	0·8996	0·9201	0·9409	0·98905	0·99232	0·911
4,000	0·8795	0·8997	0·9194	0·98561	0·98877	0·893
5,000	0·8626	0·8824	0·9009	0·98231	0·98540	0·879
6,000	…	0·8668	0·8849	0·97914	0·98216	0·866
7,000	…	0·8530	0·8705	0·97607	0·97914	…
8,000	…	0·8407	0·8577	…	0·97608	…
9,000	…	0·8296	0·8461	…	0·97327	…
10,000	…	0·8192	0·8352	…	0·97059	…
11,000	…	…	0·8256	…	0·96806	…
12,000	…	…	…	…	0·96567	…

With regard to the magnitude of the decrease of volume produced by pressure, all the non-metallic liquids listed above —that is, all except mercury—fall roughly into three classes. The first contains only glycerine, the second group contains water, C_6H_5Cl, and C_6H_5Br, and the third all the others. Under a pressure of 12,000 kg./cm.^{2} at room temperature glycerine loses 13·4 per cent. of its original volume; the three substances in the second group lose roughly 20 per cent., and those in the third group something of the order of 30 per cent. It will be understood that there is considerable scattering among the liquids in the 30 per cent. class; of these the most compressible is pentane, and the least the higher alcohols. Compared with these liquids, mercury loses somewhat less than 4 per cent. The reason for the unusually low compressibility of glycerine, C_6H_5Cl and C_6H_5Br, is not apparent; particularly striking is the very much smaller compressibility of the two latter than C_6H_6. An interesting subject for further investigation would be to find other organic liquids of abnormally small compressibility. The curve of volume against pressure of at least most of the liquids examined by the piston displacement method is so nearly the same over the entire pressure range that in making the calculations it proved advantageous to consider an "average" liquid, the volume of which has the following equation for pressures over 500 kg.:

$$\Delta V/V_{500} = a(p-500)^{0\cdot8} + \beta(p-500)^{0\cdot6} + \gamma(p-500)^{0\cdot4} + \delta(p-500)^{0\cdot2},$$

where

$$a = 0\cdot0029, \quad \beta = -0\cdot0546, \quad \gamma = +0\cdot2969, \quad \text{and} \quad \delta = -0\cdot1804.$$

The curve for any one of the actual liquids could be approximately obtained from the curve for the average liquid by multiplying the four constants just given by a single characteristic constant, which was greatest, 1·104, for ether, and least, 0·8726, for n-propyl alcohol. This gives an idea of the actual variation in the 30 per cent. class of liquids.

It is instructive to compare the volumes at high pressures of various isomers. If the most important part of the compression at high pressures comes from the compression of the molecules themselves, then one would expect that at high pressures there would be a tendency for the effect of structural differences to be wiped out, and for the volumes to become more nearly equal than at low pressures. This is very definitely the case with the two isomers, ether and n-butyl alcohol ($C_4H_{10}O$), the ratio of the volumes of equal weights being 1·096 at atmospheric pressure and 1·037 at 12,000. The tendency is not nearly so definite, however, in the other isomers investigated. Thus the ratio of equal weights of normal and iso-propyl alcohol is 1·0250 at atmospheric pressure and 1·0049 at 12,000, a change in the expected direction. The corresponding ratios for n- and i-butyl alcohols are 1·0051 and 0·9901, not in the expected direction, for n- and i-pentane 1·0080 and 1·0163, again not in the expected direction, and in the series of the five hexanes the ratio more often than not departs from unity as pressure increases. However, in all those cases in which the variation is not in the expected direction, the absolute value of the ratio is much nearer unity than for ether and butyl alcohol or for the two propyl alcohols, so that we may still think that the main tendency is for isomers to approach equality of volume at high pressures, but in those cases where the volume differences are not large, there may be disturbing secondary effects superposed on the main effect.

The compressibility, defined as $\dfrac{1}{v_0}\left(\dfrac{\partial v}{\partial p}\right)_\tau$, may be at once obtained from the shapes of the isotherms which may be plotted from the tables of volumes. The compressibility will be found to decrease greatly with increasing pressure, as of course is necessary if the volume is not to become negative. In Table III are shown the relative compressibilities at various pressures of a number of liquids. On the average the compressibility at 12,000 is only $\frac{1}{15}$th of the compressibility at atmospheric pressure. By far the most rapid decrease of compressibility occurs at the low pressures; on the average the compressibility has decreased to about one-half its initial value in the first 1000 kilograms, whereas the decrease in the last 6000—that is, in the interval between

TABLE III

CHANGES OF COMPRESSIBILITY AND THERMAL EXPANSION
PRODUCED BY PRESSURE

Liquid	Compressibility, $\varkappa$				Dilatation, δ			
	$\dfrac{\varkappa_1}{\varkappa_{12000}}$	$\dfrac{\varkappa_{1000}}{\varkappa_{12000}}$	$\dfrac{\varkappa_{6000}}{\varkappa_{12000}}$	$\varkappa_{12000}$	$\dfrac{\delta_1}{\delta_{12000}}$	$\dfrac{\delta_{1000}}{\delta_{12000}}$	$\dfrac{\delta_{6000}}{\delta_{12000}}$	δ_{12000}
Methyl alcohol .	18·4	8·2	2·20	$0 \cdot 0_5 74$	4·29	2·76	1·23	$0 \cdot 0_3 298$
Ethyl alcohol .	13·7	7·4	2·02	81	4·50	2·73	1·30	268
n-Propyl alcohol	15·8	7·8	1·94	70	4·80	3·06	1·33	237
i-Butyl . .	16·6	6·3	1·68	86	4·15	2·62	1·17	275
i-Amyl alcohol .	14·4	7·1	1·88	74	4·40	2·84	1·30	240
Ether . . .	..	7·7	1·62	96	..	3·65	1·32	248
Acetone . .	..	7·3	1·85	87	..	3·27	1·35	282
Carbon bi-sulphide	13·8	6·3	1·82	87	5·47	3·16	1·31	262
Phosphorus tri-chloride .	14·2	7·1	1·81	80	4·84	2·83	1·31	278
Ethyl chloride .	..	8·4	1·78	90	..	3·44	1·37	267
Ethyl bromide .	14·9	8·3	1·87	82	..	3·46	1·33	260
Ethyl iodide .	14·9	7·2	1·89	81	4·86	3·15	1·22	248
Water . .	4·9	3·7	1·64	89	1·00	1·00	1·00	400
Kerosene . .	..	..	1·82	87	..	..	1·14	280

6000 and 12,000—is only by another factor of 2. The
natural explanation of this is that at low pressures the
molecules fit loosely together with considerable free space
between, and the major part of the compressibility at low
pressures arises from the occupancy of this free space; at
high pressures, where the free space has become more or less
squeezed out of existence, this easy sort of compressibility
disappears, and the compressibility that remains is that
furnished by the molecules themselves, which persists with
smaller change over comparatively wide ranges of pressure.

The variations in compressibility from substance to substance are much greater at low pressures than at high. The
compressibilities at atmospheric pressure vary among the
organic liquids examined by a factor of 10, whereas at 12,000
the extreme variation is by a factor of about 1·8, or about

1·4, omitting glycerine but including water. The great difference between the behaviour of glycerine and other liquids is confined to the low-pressure end of the range; at atmospheric pressure the compressibility of glycerine is very much less than that of most organic liquids, but its compressibility drops much less rapidly with pressure, and at 12,000 it has risen relatively to about three-fourths of that of i-amyl alcohol, for example.

In spite of the fact that compressibility drops rapidly with increasing pressure, it does not drop as rapidly as might be inferred from its very rapid drop between atmospheric pressure and 1000, for example. This point is well brought out by the various equations of state that have been proposed on the basis of the work of Amagat to 3000 kg., and the situation is much like that with regard to gases, already discussed. Nearly all the equations proposed assign to the liquid a limiting volume at infinite pressure, and in practically every case this limiting volume is too high, higher in fact than can be reached by pressures in the experimental range. A clean-cut example is afforded by ether. For this substance Tumlirz [16] has proposed an equation of state with a limiting volume at infinite pressures of 0·7274, and Tammann [17] has proposed another equation with a limiting volume of 0·7246, whereas at a pressure of 12,000 at 20° C. the measured volume is less than this, namely, 0·7216. The explanation is evidently along the lines already suggested: at low pressures so large a part of the phenomenon of compressibility is concerned with wiping out the free spaces between the molecules, that the part played by the compression of the molecules, which persists with little change over wide ranges of pressure, is obscured, and the limiting volume is set too high.

The work of compression at constant temperature can be obtained at once by an integration, and is $W = -\int p \left(\frac{\partial v}{\partial p} \right)_\tau dp$.

If W is calculated by this equation and plotted against pressure, it will be found, of course, that the curve starts out as a second degree parabola tangent to the pressure axis in the range in which pressure is low and $\left(\frac{\partial v}{\partial p} \right)_\tau$ approximately constant. But the drop of $\left(\dfrac{\partial v}{\partial p}\right)_\tau$ is so rapid that the parabolic range is very low, and at pressures above 2000 the work of compression can be represented with fair approximation by a linear relation, $W = a + bp$ (a negative, b positive). This equation gives a fair means of extrapolating the volume to pressures far beyond the reach of experiment. Differentiation gives $\left(\dfrac{\partial v}{\partial p}\right)_\tau = -b/p$, and integration gives for the volume

$$\Delta V = -b \log p + C.$$

This does not hold for low pressures nor for excessively high pressures, but in an extensive range above 500 kg. it is a fairly good approximation. Average values for the twelve liquids give the following:

$$\Delta V = -0.1 \log_\epsilon p + 0.602 \ . \ (p \text{ in kg./cm.}^2).$$

The pressure at which this formula must of necessity cease to hold is that at which $\Delta V = -1$. With the values of the coefficients given, this occurs at a pressure of 9,000,000 kg./cm.^{2}, which is so very much beyond the experimental range that the equation may be used for extrapolation with fair assurance over a range several fold greater than the present range.

The numerical values of the work of compression are of interest. At the higher pressures, this work does not vary much from liquid to liquid, the maximum variations being only about 25 per cent. The work of compression of ether is one of the largest; the work of compressing 1 gm. of ether at 40° C. to a pressure of 12,000 kg./cm.^{2} is about 15 kg./cm., which would raise it to a height of 15,000 m., or give it a velocity of about 550 m./sec. This amount of energy is not so extreme as many persons are prepared to expect, and constitutes one reason why high-pressure experiments with liquids are not more dangerous; the steel cylinders in which the liquids are confined are usually so much more massive than the liquid that any velocity which may be imparted to the metal parts is comparatively low.

The thermal expansion decreases with increasing pressure, as might be expected, but by an amount materially less than the decrease of compressibility, the factor by which the expansion at 12,000 is less than its value at atmospheric pressure varying from 4 to 8, against an average factor of 15 for the compressibility. The effect of pressure on the average expansion between 20° and 80° has been shown in Table III. A mathematical consequence of the decrease of expansion with increasing pressure is an increase of compressibility with increasing temperature. There are two factors concerned in the behaviour of expansion at high pressures which work in opposite directions. On the one hand there is the fact that the natural frequency of the molecules increases at high pressures, so that an effect is to be expected analogous to that of lowering temperature towards 0° abs., where the thermal expansion drops to zero. On the other hand, as the molecules are forced into closer proximity by higher pressures, their mutual forces of inter-action must depart by increasingly large amounts from linearity, and it is known that a high thermal expansion accompanies a large departure from linearity. It is evident that this latter consideration is by far the most important, for in many cases the decrease of volume produced by 12,000 at room temperature is sufficient to reduce the volume to considerably less than its value at 0° abs. at atmospheric pressure.

At low pressure the thermal expansion of all liquids increases with rising temperature, that is, $\left(\dfrac{\partial^2 v}{\partial \tau^2}\right)_p > 0$. This is what would be expected, because as temperature is raised at low pressures the liquid approaches the condition of a gas, the thermal expansion of which is greater than that of any liquid. At high pressures, however, this relation is reversed, and $\left(\dfrac{\partial^2 v}{\partial \tau^2}\right)_p < 0$. The first trace of this effect was found by Amagat near the end of his pressure range, 3000, but such an effect seemed to him so unlikely that he ascribed it to experimental error. Measurements over a wider range prove the undoubted reality of the effect, however, and in fact this reversal is shown by all the liquids of my investigation for which measurements could be made over a wide enough temperature interval to establish the sign of $\left(\dfrac{\partial^2 v}{\partial \tau^2}\right)_p$. The pressure of reversal is in almost all cases less than 4000, and is usually near 3000.

This reversal in the sign of $\left(\dfrac{\partial^2 v}{\partial \tau^2}\right)_p$ is obviously a most important point for any theory of liquids. Doubtless, the explanation is essentially connected with the effect of non-linearity in the intermolecular forces. At high pressures, where these effects are important, the non-linearity becomes increasingly important at small volumes. If the liquid is cooled at constant pressure, it is carried into the region of smaller volumes, where the non-linear effects are increasingly important, with a consequent increase of thermal expansion, that is, $\left(\dfrac{\partial v}{\partial \tau}\right)_p$ increases as

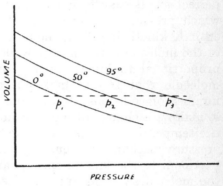

FIG. 37.—The ratio $\dfrac{p_3 - p_2}{p_2 - p_1}$ should be equal to 0·9 if $\left(\dfrac{\delta p}{\delta \tau}\right)_v$ is a volume function only.

temperature is lowered at constant pressure, or $\left(\dfrac{\partial^2 v}{\partial \tau^2}\right)_p$ is negative.

The ratio of thermal expansion to compressibility, or $\left(\dfrac{\partial v}{\partial \tau}\right)_p \bigg/ \left(\dfrac{\partial v}{\partial p}\right)_\tau$, is mathematically identical to $-\left(\dfrac{\partial p}{\partial \tau}\right)_v$. This derivative has received the special name 'pressure coefficient,' and has played an important rôle in many theoretical speculations. The graphical significance of the derivative is evident from fig. 37. If the isotherms are drawn for several successive temperatures, an approximate value for the derivative may be found from the intersections of lines parallel to the pressure axis, as indicated, by forming the ratios $\frac{p_2-p_1}{t_2-t_1}$, etc.   If the substance is a perfect gas, or if it obeys van der Waals's equation, then $\left(\frac{\partial p}{\partial \tau}\right)_v$ is a function of volume only, which in the graph means that $\frac{p_2-p_1}{t_2-t_1}=\frac{p_3-p_2}{t_3-t_2}$, etc.   The physical basis of this relation in the case of a perfect gas is easy to understand.   Here the mechanism by which pressure is exerted on the walls of the container is entirely kinetic in origin, and is given by the total momentum of the molecules striking the walls in unit time.   But this is proportional to kinetic energy or temperature, since each molecule that strikes makes a contribution mv, and the number of collisions at constant volume is proportional to v, making a change of momentum proportional to mv^2 or to the temperature.   The same argument applies whenever the pressure mechanism is entirely kinetic; temperature is proportional to the kinetic energy, and the effective volume of the molecules is constant.   In fact, the relation holds under somewhat more general conditions.   If one puts $\left(\frac{\partial p}{\partial \tau}\right)_v=f_1(v)$, integration gives $p=\tau f_1(v)+f_2(v)$, a type of equation which includes van der Waals's as a special case.   The physical meaning of this result is that pressure can be regarded as exerted by two mechanisms acting independently and additively, one part, given by $f_2(v)$, is at constant volume the same at all temperatures as at $0°$ abs., and is evidently the part arising from the attractive and repulsive forces of the molecules, and a second part, $\tau f_1(v)$, arising from the kinetic mechanism just discussed.   Van der Waals's equation in particular satisfies these general assumptions, and is of this type.   Many theoretical speculations on liquids have assumed conditions such as these, and the expectation has been strong in some quarters that liquids must satisfy the relation $\left(\frac{\partial p}{\partial \tau}\right)_v=f(v)$.

The experimental evidence for this relation on the basis of low-pressure measurements has not been unfavourable. However, at high pressures the relation definitely breaks down. This is shown by my measurements, using the sylphon method, for which the temperature intervals were wider and small-scale irregularities less confusing than by the piston displacement method. Turning to fig. 37, it is evident that if the isotherms are drawn for 0°, 50°, and 95°, and if $\left(\dfrac{\partial p}{\partial \tau}\right)_v$ is a function of volume only, then $\dfrac{p_3-p_2}{p_2-p_1}$ should equal $\dfrac{95-50}{50-0}=0 \cdot 9$. In Table IV are shown the

TABLE IV

CHECK OF SUPPOSED RELATION $\left(\dfrac{\partial p}{\partial \tau}\right)_v = f(v)$ BY EVALUATION OF $\dfrac{p_3-p_2}{p_2-p_1}$

Liquid	Relative Volume	$\dfrac{p_3-p_2}{p_2-p_1}$	Relative Volume	$\dfrac{p_3-p_2}{p_2-p_1}$	Relative Volume	$\dfrac{p_3-p_2}{p_2-p_1}$
n-Pentane	0·92	0·84 ±	0·75	0·98		
i-Pentane	0·94	0·905	0·80	0·972		
n Hexane	0·92	0·95 ±	0·79	0·774		
2-Methyl pentane	0·94	0·897	0·80	0·910		
3-Methyl pentane	0·94	0·865	0·80	0·817	0·75	0·597
2·2-Dimethyl butane	0·92	0·743	0·82	0·68		
2·3-Dimethyl butane	0·95	0·895	0·80	0·747		
n-Heptane	0·99	0·914	0·82	0·832		
n Octane	0·99	0·806	0·87	0·767		
n-Decane	0·94	0·837				
C_6H_5Cl	0·95	0·893				
C_6H_5Br	0·95	0·870				
i-propyl alcohol	0·96	0·947	0·86	0·925	0·79	0·900
n-butyl alcohol	0·96	0·837	0·86	0·850		
n-Hexyl alcohol	0·80	0·845				

values of this ratio for a number of liquids at a number of different volumes. This ratio is certainly not always equal to 0·9, and often departs from it by an amount far beyond possible experimental error. In general, the tendency is for the ratio to become less than 0·9 at small volumes (high pressures).

An explanation of the failure of $\left(\dfrac{\partial p}{\partial \tau}\right)_v$ to be a function of volume only lies at hand in the compressibility that we have had to ascribe to the molecules. If the molecule is deformable, it is obvious that the pressure cannot be regarded as exerted by two mechanisms acting independently and additively, but there must be interaction between the two mechanisms. In particular, as temperature increases at constant volume, the molecules become compressed, and the part of the pressure exerted by the attractive and repulsive forces changes. In the detailed paper on the liquids, measured by the sylphon method, is given a working-out of the equation of state of the simplest conceivable substance consisting of a single molecule, but a molecule with elastic deformability. It is there shown that $\left(\dfrac{\partial p}{\partial \tau}\right)_v$ is not a function of volume only, but must involve the temperature also.

A topic closely connected with that just discussed is the change of internal energy with pressure. For this we have the thermodynamic relation $\left(\dfrac{\partial \mathrm{E}}{\partial p}\right)_\tau = -\tau\left(\dfrac{\partial v}{\partial \tau}\right)_p -p\left(\dfrac{\partial v}{\partial p}\right)_\tau$, already quoted in connection with gases. At low pressures the second term is negligibly small, the first term preponderates, and since $\left(\dfrac{\partial v}{\partial \tau}\right)_p$ is positive for most substances, the normal behaviour is that internal energy decreases as pressure increases at constant temperature. That is, energy flows out faster in the form of heat to maintain the substance isothermal than it is put in in the form of mechanical work. This result is likely to strike one at first as paradoxical; the obvious explanation is that on the average the forces between molecules are attractive under ordinary conditions, and as volume decreases the potential energy of the mutual forces decreases. This process cannot go on indefinitely, however, but when volume has decreased sufficiently the repulsive forces preponderate, so that $\left(\dfrac{\partial E}{\partial p}\right)_\tau$ must ultimately reverse sign and become positive. The pressure at which reversal takes place is evidently given by

$$p = -\tau\left(\frac{\partial v}{\partial \tau}\right)_p \bigg/ \left(\frac{\partial v}{\partial p}\right)_\tau \left[= \tau\left(\frac{\partial p}{\partial \tau}\right)_v \right].$$

The pressure of reversal for the liquids investigated above is in the experimental range, and, in fact, is usually at 7000 or higher. If the molecules can be treated as undeformable, the volume at which this reversal takes place would be expected to be equal to the volume at 0° abs. at atmospheric pressure. In the case of ether Hildebrand has found that the two volumes are about the same; data are lacking for most other liquids, but, as will be shown in the next chapter, the volume at reversal is usually less than at 0° abs. in the case of solids.

The "heat of compression," that is, the heat which flows out of the liquid as it is compressed isothermally, is at once determined when the change of internal energy and the work of compression are known. The numerical value is of interest. It turns out that the heat of compressing an average liquid to 12,000 kg./cm.² would be sufficient to raise its temperature to about 70° C. if applied adiabatically. In actual practice the temperature changes during compression are much less than this, because of the comparatively large mass of the containing vessels, but the changes of temperature are nevertheless so large that the greater part of the time in making high-pressure measurements is consumed in waiting for equalisation of temperature after changes of pressure.

The variation of specific heat with pressure can be found, when the p-v-t relation is accurately enough determined, by means of the thermo-dynamic relations

$$\left(\frac{\partial C_p}{\partial p}\right)_\tau = -\tau\left(\frac{\partial^2 v}{\partial \tau^2}\right)_p, \quad \text{and} \quad C_v = C_p + \tau\left(\frac{\partial v}{\partial \tau}\right)_p^2 \bigg/ \left(\frac{\partial v}{\partial p}\right)_\tau.$$

The effect of pressure on specific heat has been calculated for the twelve liquids measured by the piston displacement method, and a detailed discussion will be found in that paper. The results are in general complicated. As a rule C_p at first decreases as pressure is increased, passes through a minimum at 2000 or 3000 kg. at roughly 90 per cent. of its initial value, and from here on increases somewhat irregularly, usually not recovering its initial value under a pressure of 12,000. The behaviour of C_v is somewhat similar, except that the minimum tends to come at lower pressures and at a higher relative value of C_v, and the subsequent increase usually carries the value of C_v at 12,000 to something more than its atmospheric value. The simple picture of the mechanism of a liquid which led to the expectation that $\left(\dfrac{\partial p}{\partial \tau}\right)_v$ would be a function of volume only also leads to the expectation that C_v as well will be a volume function only. That this cannot be true will be evident from an examination of the detailed curves in the original paper; it is not worth while to emphasise the point further here.

The features just discussed are broad general features common to the behaviour of all liquids, except highly abnormal ones like water. But superposed on these broad effects there are small-scale effects of a high degree of irregularity, and specific to the different liquids. The original papers must be consulted for the details of these small-scale effects. Examples of nearly every conceivable type of behaviour may be found; the compressibility and thermal expansion and specific heats may increase or decrease with increasing temperature or pressure, and the curves for different temperatures may cross and recross in the most bewildering way. The reason doubtless is that the molecule is a very complicated structure when the details of its force field are taken into account, and the nature of the complications is not the same for any two kinds of molecule. The effect of these local complications in the molecular field may be expected to become important when the molecules are forced into close proximity by high pressure, but is comparatively unimportant when the molecules are free to move as they please and so act with approximate spherical symmetry at low pressures. One conclusion that I draw from these exceedingly complicated small scale phenomena is that the hope of ever deducing for liquids 'an' equation of state is futile. No general considerations can possibly be broad enough to give a single type of equation which shall be elastic enough to reproduce the fine scale differences of behaviour of all possible complicated molecules. All that can be expected is to find an equation capable of reproducing the broad general types of behaviour discussed above; the fine scale differences of individual substances must be treated by special considerations.

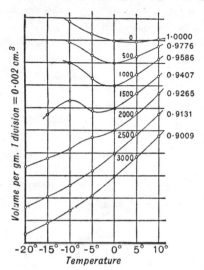

FIG. 38.—Relative volume of water as a function of temperature at several constant pressures. The pressures, in kg./cm.², are given by the figures on the curves, running from 0 to 3000. The figures on the right give the volumes of the liquid at the pressure in question and a temperature of 0° C. The curves are crowded together to save space : on a correct scale they would be separated by spaces about ten times as great as those shown.

Water is known to be a highly abnormal substance at atmospheric pressure, so that we would expect it also to be abnormal under high pressure. The most striking manifestation of abnormality at high pressure is the existence of several varieties of ice, to be discussed later, but there are also definite irregularities in the p-v-t relations of the liquid. In the neighbourhood of 0° C. at atmospheric pressure the most striking abnormality is the minimum in the curve of volume against temperature at 4°. On the curves of volume against temperature at constant pressures higher than atmospheric, the temperature of this minimum is found to be displaced. In fig. 38 are shown curves of volume against temperature for a number of successively increasing pressures. In order to save space in the diagram, the different curves are crowded together, but the scale of the ordinates gives the correct change of volume on any one curve. The curves show that as pressure increases the minimum temperature is shifted toward lower values, and at the same time the change of volume beyond the minimum temperature becomes less pronounced, until at 1500 kg./cm.² a maximum volume as well as a minimum comes into the region of realisation. At 2000 the minimum and maximum coalesce to a point of inflection, and beyond that the curves become normal. This apparently complicated behaviour can be completely explained qualitatively by the usual assumption of two molecular species in liquid water of different degrees of association and different volumes. As water is cooled at atmospheric pressure, the relative number of the associated molecules increases. Water in the form of associated molecules occupies more volume than in the form of simple molecules. The increase of volume due to the increased association with decreasing temperature eventually becomes so great as to more than compensate for the normal decrease of volume due to falling temperature, with the result that there is a minimum volume. At higher pressures, the associated molecules may be thought to be abnormally compressible, so that the volume difference between the two kinds of molecules becomes smaller, and the whole effect less pronounced, eventually disappearing at high enough pressures. There are, however, certain difficulties in this explanation by association; perhaps the chief is that we would expect the compressibility of liquid water to be abnormally high, whereas it is abnormally low.

As might be expected from the p-v-t relations shown in fig. 38, the compressibility and thermal expansion of water must behave abnormally. In fig. 39 is shown the thermal expansion of water as a function of pressure at several different temperatures. In the low pressure part of the range the behaviour is highly abnormal, but above 5000 or 6000 the normal behaviour for all liquids is assumed, the thermal expansion being less at the higher temperatures, $\left(\dfrac{\partial^2 v}{\partial \tau^2}\right)_p$ negative. In general, water tends to become completely normal at high pressures. The abnormalities of water which play so important a part in biological phenomena are more or less local abnormalities, confined to the first few thousand atmospheres. We shall see in a later chapter that water possesses a number of different solid forms. Tammann [18] has made a detailed study in which various parallelisms are shown between the abnormalities of the liquid and the existence of the solid forms. Tammann's idea is apparently that corresponding more or less to each solid form there is a particular kind of associated molecule in the liquid. I do not believe that the parallelism can be made quite as close as this, or that the concept of association can be made quite as

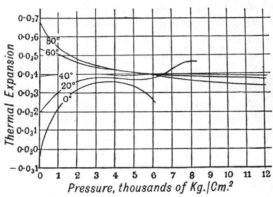

Fig. 39.—Thermal expansion of water at several constant temperatures as a function of pressure.

clean cut as many chemists have supposed. In this connection should be mentioned direct measurements of the adiabatic temperature effect of compression $\left(\frac{\partial \tau}{\partial p}\right)_s$, by Pushin, and Grebenshchikov [19] up to pressures of 4000 kg./cm.². According to the thermodynamic formula, $\left(\frac{\partial \tau}{\partial p}\right)_s$ should vanish at the maximum density point. According to their observations $\left(\frac{\partial \tau}{\partial p}\right)_s$ vanishes at 0° between 500 and 1000 kg./cm.²; at 0° my directly observed maximum density point also occurs between 500 and 1000. However, my calculated values of $\left(\frac{\partial \tau}{\partial p}\right)_s$ at 0° differ by an almost constant amount from their observed values, so that the presumption is that there is some error in my calculated values. At higher temperatures the discrepancy disappears.

Two Component Systems. As regards the compressibility of two component liquid systems—that is, of solutions—a great deal of experimental material has been collected in this field, but none of it is particularly systematic, and practically none of it is to pressures higher than 1000 kg. An exhaustive account will be found in the *Piezo Chemie* of Cohen and Schut; little has been done since the publication of that book. The phenomena are exceedingly complicated, but there are a few generalisations that may be made. The first effect of adding a solute to water is to depress the compressibility of the solution to less than the compressibility of pure water, irrespective of whether the compressibility of the added substance is greater than that of water, as when alcohol is added, or less, as in the case of electrolytes. The same is also true irrespective of whether the volume of the solution is greater or less than that of its constituents. In those cases where the pure solute has a greater compressibility than pure water, the compressibility passes through a minimum as concentration increases.

These facts have been made the basis of a theory of solutions by Tammann; the leading idea of this theory is that the first and most important effect of a small quantity of solute is to increase the internal pressure in the water by chemical affinity, so that the physical properties of the solution at atmospheric pressure, including its compressibility, are the same as the properties of pure water under an external pressure higher than atmospheric by the pressure of chemical affinity. Tammann was able to correlate a number of effects by this theory, but by no means all, and the theory definitely breaks down when applied to non-aqueous solutions. Here the phenomena may be the exact opposite of those in aqueous solutions; for example, the compressibility of solutions of benzene and methyl alcohol passes through a maximum greater than that of either pure component.

Tammann and Schwarzkopf [20] have also shown that according to their views the behaviour of aqueous solutions is inconsistent with my observations of the disappearance of the minimum in the volume of water at high pressure, but they use this as an argument against my observations rather than as an argument against their theory.

Theoretical considerations have been advanced by Richards and Chadwell [21] which contain the possibility of explaining, at least qualitatively, the variety of effects. According to this view, the first effect of adding a solute to a solvent in which the molecules exist in different states of association is to change the equilibrium distribution between the different sorts of molecules. If, furthermore, the molecules in different states of association are supposed to have different compressibilities, then a change in the relative number will be accompanied by a change in compressibility, which may be either an increase or a decrease, depending on the relative compressibility of the different kinds. Superposed on this effect is the specific effect of the added molecules, which becomes important at high concentrations.

The compressibility of a few amalgams has been studied by Lussana. The results are irregular, but in general the compressibility of tin amalgams was found to pass through a minimum with increasing tin content.

[1] John Canton, *Phil. Trans. Roy. Soc.*, p. 640 (1762) ; p. 261 (1764).

[2] P. G. Tait, *Report of the Voyage of H.M.S. "Challenger,"* II, Appendix A (1881).

[3] L. Cailletet, *Ann. Chim. Phys.*, **19**, 386 (1880).

[4] Carnazzi, *Nuov. Cim.*, **5**, 180 (1903).

[5] E. H. Amagat, *Ann. Chim. Phys.*, **29**, 68 (1893).

[6] C. A. Parsons and S. S. Cook, *Proc. Roy. Soc.*, **85**, 332 (1911).

[7] J. Perkins, *Trans. Roy. Soc.*, **72**, 324 (1819–20).

[8] G. Aimé, *Ann. Chim. Phys.*, **8**, 257 (1843).

[9] V. Regnault, *Mem. Inst. France*, **21**, 329 (1847).

[10] J. Y. Buchanan, *Trans. Roy. Soc. Edin.*, **29**, 589 (1880) ; *Proc. Roy. Soc. Edin.*, **73**, 296 (1904).

[11] E. H. Amagat, *C.R.*, **108**, 727 (1889).

[12] Ph. Kohnstamm and K. W. Walstra, *Proc. Amst. Acad.*, **17**, 203 (1914).

[13] E. P. Bartlett, H. L. Cupples, and T. H. Treamearne, *Jour. Amer. Chem. Soc.*, **50**, 1275 (1928).

[14] R. Becker, *ZS. f. Phys.*, **4**, 393 (1921).

[15] T. W. RICHARDS, *Pub. Carnegie Inst. Wash.*, No. 7 (1903).
  C. DRUCKE, *ZS. f. Phys. Chem.*, **52**, 662 (1905).
  A. RITZEL, *ibid.*, **60**, 319 (1907).
  W. WATSON, *Proc. Roy. Soc. Edin.*, **33**, 282 (1912–1913).
  E. COHEN and R. B. DE BOER, *ZS. f. Phys. Chem.*, **84**, 41 (1913).
[16] O. TUMLIRZ, *Sitzber. K. Akad. Wiss. Wien*, **118**, 1 (1909).
[17] G. TAMMANN. *Ann. Phys.*, **37**, 975 (1912).
[18] G. TAMMANN, *Gött. Nachr.*, p. 1 (1913).
[19] N. A. PUSHIN and E. GREBENSHCHIKOV, *Trans. Chem. Soc.*, **123**, 2717 (1923).
[20] G. TAMMANN and E. SCHWARZKOPF, *ZS. anorg. und allgem. Chem.*, **174**, 215 (1928).
[21] T. W. RICHARDS and H. M. CHADWELL, *Jour. Amer. Chem. Soc.*, **47**, 2283 (1925).

CHAPTER VI

COMPRESSIBILITY OF SOLIDS

No direct method for measuring the volume compressibility of a solid has been devised. The reason for this is not difficult to see if one reflects that the only direct method of measuring the volume of a solid is by some immersion method in a liquid, which must, in the ideal case, remain of constant volume during the experiment and must be confined in a container of invariable volume. But the volume of no fluid or container is invariable under stress, so that corrections have to be applied, and the method becomes indirect. The situation is complicated because any method of measuring the compressibility of a fluid involves a correction for the distortion of the vessel in which it is contained, which in turn demands, in the simplest case, a knowledge of the compressibility of the solid metal of the container. The situation may be dealt with in two ways: one is to determine indirectly the compressibility of the solid by observation of its deformation under stresses other than a hydrostatic pressure; the second is to determine the linear changes of dimensions of the solid, that is, the linear compressibility, in some way, and from this to calculate the change of volume. This latter is in principle perfectly satisfactory, since it is just as satisfactory to determine the volume of a rectangular block by measuring its three linear dimensions as by using a method of immersion in a fluid. There have, however, been various difficulties in the way of applying the second more straightforward method which will be discussed in detail presently.

Practically all the early measurements of the compressibility of solids were by indirect methods, involving the theory of elasticity. For example, in the case of an isotropic substance, which has only two independent elastic constants, the cubic compressibility may be calculated from Young's

modulus and the shearing modulus. The first attempts to
apply this method were obscured by uncertainties in the
theory of elasticity itself; the controversy between the
'multi-' and the 'rari-' constant schools lasted for some
time, and was not really settled until Voigt, in the 1880's,
began his famous experiments on the elastic constants of
crystals. In applying the results of elasticity theory, very
widely divergent results were often obtained; the explanation of these early discrepancies is now felt to be almost
certainly insufficient isotropy; ordinary drawn rods of the
metals of commerce are likely to have internal strains which
can only be removed by very careful annealing, and in the
early days the materials themselves were intrinsically in-
homogeneous because of impurities, as shown, for example,
by the inclusions of slag in all early samples of wrought iron.
All of this unfortunate early experience led to considerable
mistrust of the theory of elasticity itself, so that in some
quarters there has been an unwillingness to accept any compressibility results obtained by such indirect means. This
distrust does not seem to be justified, however, provided the
material satisfies the proper requirements of homogeneity.
There are, nevertheless, still certain objections that apply
to any determination of compressibility by such indirect
methods, and now that other methods have been developed
capable of yielding results of high accuracy, there is usually
little excuse for having recourse to an indirect method
utilising stresses other than hydrostatic pressures. Slight
departures from perfect homogeneity may be very serious
in those methods using non-hydrostatic stresses; furthermore, no such indirect method at present proposed is capable
at all of showing how compressibility changes with pressure,
which is one of the significant data to be obtained by high-
pressure experiments.

If an indirect method utilising non-hydrostatic stresses
must be applied, one of the simplest, as far as computations
are concerned, is that of Mallock.[1]   In this method the change
of length of a tube when exposed to internal pressure is
measured; it may be shown by elasticity theory that this
change of length involves only the cubic compressibility, so

that a determination of the other elastic constants is not necessary. This method has been applied by Grüneisen [2] to the determination of the compressibility of a number of metals at low temperatures. Here a direct determination would be difficult if not impossible, so that the indirect method may have a real field.

The first attempt at a direct measurement of the linear compressibility of a solid by Buchanan [3] has already been mentioned in Chapters I and V. His earliest experiment was in 1880, when he obtained for the cubic compressibility of glass $2 \cdot 92 \times 10^{-6}$ per atmosphere. This is a reasonable enough value, but there is no way of checking it, because the compressibility of glass may vary through wide limits. A later paper in 1904 applied the method to a number of solids up to 300 kg./cm.[2]. These values were not at all good, and were particularly bad for the metals, leading in the worst case, that of platinum, to the value $5 \cdot 5 \times 10^{-7}$ for cubic compressibility against $3 \cdot 6 \times 10^{-7}$, the correct value. There must have been some more or less constant error in Buchanan's measurements, perhaps a refraction effect in the glass tube through which he made his observations. Amagat [4] in 1889 published results obtained by a somewhat similar method to 2000 atmospheres. The pressure range was too high to allow optical observations, and the changes of length were determined by an electrical contact device. He found for the compressibility of glass a value 2 or 3 per cent. greater than by indirect methods, and stated, without giving the data, that the agreement was equally good for a number of metals. Amagat never published the details of the method, however, and apparently was not satisfied with it. His results for the metals were not particularly good; for example, his value for iron was $6 \cdot 8 \times 10^{-7}$, against the probably correct value of $5 \cdot 9 \times 10^{-7}$. Lussana [5] in 1904 attempted a complete set of determinations, measuring the compressibility of a number of solids in a glass piezometer. He determined the absolute linear compressibility of the glass by measuring electrically the relative displacement of one end of a glass rod and a platinum wire attached to the pressure cylinder and dipping into a mercury cup attached to the rod. But his results were very bad, giving, for example, compressi-bilities of Sn, Pb, and Cu 50 per cent. too high. Richards [6] in 1907 published an attack on the problem by a method much like that of Amagat. An iron rod 2·7 meters long was mounted vertically in a pressure cylinder with a mercury cup on the upper end into which a platinum point dipped attached to a pressure-tight screw passing through the top of the cylinder, the height of which could be measured with a cathetometer. The difference of position of the platinum point when making contact with and without pressure gave at once the linear compressibility. Richards's pressure range was about 500 kg./cm.². The results again were not good, his value for the compressibility of iron being $3·85 \times 10^{-7}$ against $5·9 \times 10^{-7}$. The reason for the discrepancy is not entirely clear; it is probably connected with a change of shape under pressure of the mercury drop.

The very considerable discrepancies in the various values for the compressibility of the metals when I took up high-pressure experimenting made it important that the matter should be cleared up as a preliminary to the measurement of any compressibility over an extended pressure range. My first attack on the problem was published in 1907 (B. 2) up to 6200 kg./cm.². A rod of iron, enclosed in a pressure cylinder, was kept tightly pressed against a shoulder at one end with a spring, and at the other end carried a brass sleeve, which pressed against another shoulder, and which could slide on the rod when pressure was applied, making the rod shorten relatively to the cylinder. After each application of pressure the apparatus was opened and the displacement of the sleeve measured. Such measurements give directly the relative change of length of rod and cylinder; the cylinder lengthens under pressure, and this lengthening must be determined to obtain the absolute change of dimensions of the rod. The lengthening is only about 5 per cent. of the total effect, and was determined from observations with a micro-scope on the outside of the cylinder, neglecting any warping of the cross-section, which was justifiable because the ends of the cylinder were some distance away.

A disadvantage of the method was that the continual

opening of the apparatus made work very slow; furthermore, there were errors from particles of dirt getting into the apparatus every time it was opened.

These first measurements were not accurate enough to show any change of compressibility with pressure; the average compressibility of iron between 0 and 6200 kg./cm.^{2} was $5 \cdot 59 \times 10^{-7}$ at 20° C. An important result obtained with this method was to establish by direct measurement that the compressibility of a piece of boiler plate was the same within at least $0 \cdot 1$ per cent. in directions parallel and perpendicular to the direction of rolling, thus justifying the calculation of the cubic compressibility from the measured linear compressibility, assuming equal compression in all directions. Later experiments (B. 41, p. 191) established the same result for rolled copper. In fact, there is no reason why the relation should not apply if the metal is cubic and well annealed; if the metal crystallises in some system not cubic, large departures may be expected.

The next attempt was in 1911 (B. 5), in connection with measurements of the compressibility of mercury and water by the modified Aimé method already discussed in Chapter V. These measurements were made to 10,000 kg./cm.^{2} at 10° and 50°; the method was practically the same as that used before, except that the containing cylinder was made of a different grade of steel, which had a higher elastic limit, there having been some evidence that the elastic limit of the steel had been exceeded in the previous measurements. The change of length of the outer cylinder was measured with a Maarten's mirror device. Again the results were not accurate enough to show a change of compressibility with pressure, but a considerable change was found with temperature, the mean compressibility between 0 and 10,000 kg./cm.^{2} being $5 \cdot 83 \times 10^{-7}$ at 10°, and $6 \cdot 01 \times 10^{-7}$ at 50°.

The third and last attempt (B. 41) was in connection with measurements of the compressibility of thirty metals in 1923. The broad features of this method were the same as the two others, in that the change of length of an iron rod exposed to hydrostatic pressure was measured relative to the containing cylinder, and a correction was determined for the change of length of the cylinder by external measurements. The method was greatly improved, however, in that it was changed from a single reading method, requiring that the apparatus be taken apart for each reading, into a continuous reading method, by the use of a sliding electrical contact of the kind already described in some detail in connection with the piston piezometer on page 125. The sensitiveness of the method permits the detection of relative motion of $1 \cdot 5 \times 10^{-6}$ cm.; the regularity and consistency of the readings was such that the average departure of a single reading from a smooth curve was only $0 \cdot 16$ per cent., corresponding to about 10^{-4} cm. With this method departures of compressibility from linearity could be easily measured. The final result for the volume compressibility of pure iron, depending on something over fifty readings at different pressures, and assuming, as was justified by direct experiment, that the compressibility is equal in all directions, was—

at 30°, $-\Delta V/V_0 = 10^{-7}(5 \cdot 87 - 2 \cdot 1 \times 10^{-5}p)p$,
at 75°, $-\Delta V/V_0 = 10^{-7}(5 \cdot 93 - 2 \cdot 1 \times 10^{-5}p)p$,

pressure in kg./cm.[2].

These results have been taken as fundamental in all my later calculations, and have also apparently been accepted by other high-pressure experimenters, such as Adams and his co-workers at the Geophysical Laboratory in Washington.

The absolute compressibility of some one metal having been established, the way is open for the application of any of the methods of measuring relative compressibility. The first method is to determine the difference of the volume compressibility of the solid and some liquid, by replacing part of the liquid in the piezometer by the solid, and measuring the net compressibility of solid and remaining liquid. This method was applied as long ago as 1881 by Roentgen and Schneider,[7] in measuring the compressibility of NaCl. The most extensive use of this method was by Richards, whose measurements give immediately only the difference of compressibility between the substance in question and mercury; in fact, Richards [8] for many years emphasised that his results were to be corrected when a more accurate value for the compressibility of mercury was found, which in turn demanded a more accurate value for some fiducial solid. Richards's mistrust of the accuracy of some of his absolute values was justified, the fundamental value which he used for iron in many of his calculations being 0.2×10^{-7}, or 35 per cent., too low.

The use of a differential method of measuring the volume compressibility of solids has also received extensive application at the Geophysical Laboratory,[9] where my method of determining the compressibility of liquids by piston displacement has been applied with very little modification to the measurement of the compressibility up to 12,000 kg./cm.^{2} of a large number of rocks and minerals, substances of geophysical interest. There is a disadvantage in measurements by this method in that the compressibility of liquids is manyfold greater than that of the ordinary solid, so that the correction for the transmitting liquid is an important part of the whole measured effect, with a resulting loss of accuracy in the final result. The method as used at the Geophysical Laboratory is capable of giving the change of compressibility with pressure of only the most compressible rocks or minerals. Furthermore, the method has been applied only at room temperature; there would probably have been further complications in attempting to extend the method to a determination of the temperature coefficient of compressibility.

These differential methods determine volume compressibility; the method which I have exploited measures differential linear compressibility. Such a method is particularly adapted to studying the behaviour, under pressure, of single crystals, especially non-cubic crystals, the compressibility of which is not the same in different directions. For such substances the partial information given by a determination of the volume compressibility is of small value compared with the more complete information obtainable from a measurement of the linear compressibility in all the crystallographically independent directions. On the other hand, a measurement of volume compressibility is perhaps significant enough for non-isotropic substances which are full of internal strains, like many rocks. The

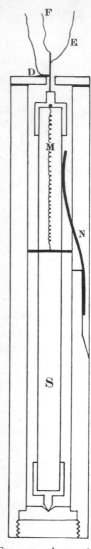

FIG. 40.—Apparatus for determining the relative linear compressibility of iron and substances which may be obtained in the form of long rods.

linear compressibility of these is not the same in different directions (B. 48), but the variations are so complicated as to be of no immediate interest.

I have used two somewhat different methods for measuring differential linear compressibility, depending on whether the compressibility is large or small. If the compressibility is relatively large, the arrangement shown in fig. 40 is used. The specimen, in the form of a long rod S, is kept pressed against the bottom of the holder of iron by a spring M. Attached to the upper end of the rod is a high-resistance wire sliding over a contact D, attached to the holder but insulated from it. The spring N has the function of keeping the wire pressed against its contact. The relative position of holder and wire is determined by a potentiometer measurement of the difference of potential between the sliding contact D and a terminal E fixed to the wire, exactly as in the method of the piston piezometer. One current terminal is F; the other is grounded to the apparatus. The whole arrangement is placed in a high-pressure vessel, and exposed to hydrostatic pressure. If the solid S is differently compressible from the iron of the holder, there is a relative change of length when pressure is applied, which is measured on the potentiometer. The effect of various distortions in various parts of the holder—as, for example, in the mica washers by which insulation is secured—may be eliminated by replacing S by a bar of pure iron, the compressibility of which has already been measured by the absolute method described above.

If the substance to be measured is relatively incompressible, its motion relative to the holder may be conveniently magnified by a lever arrangement as shown in fig. 41; the magnification of the lever shown is about sevenfold. The motion of the upper end of the lever is measured with a sliding-contact device similar to the other, which will be sufficiently obvious from the diagram without detailed description. In the construction of this apparatus the pivots on which the lever turns must be made with extreme care in order that there shall be no irregularity in the very small relative motions, which are usually a small fraction of a millimetre, even for a pressure range of 12,000 kg./cm.². Except for this, the compressibility apparatus is very easily constructed; the apparatus of fig. 40 in particular may be constructed by anyone. The lever apparatus may be adapted to specimens of a wide variety of dimensions; a simple modification not shown makes it possible to measure the linear compressibility of wires, these being held in tension instead of in compression. The accuracy and self-consistency of measurements made with this apparatus is surprisingly high. It is not unusual for the average departure from a smooth curve of a single reading out of fourteen or fifteen to be 0·2 per cent., or even 0·1 per cent., which means a proportionally greater accuracy in the mean. The uncertainty in the absolute value of the pressure itself may be 0·1 per cent.

There are evidently a number of corrections to be applied, such as for the

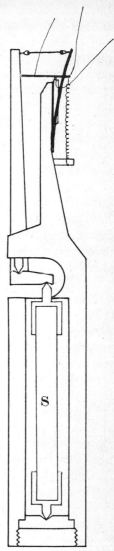

FIG. 41.—Lever piezometer, for determining the relative linear compressibility of iron and substances too short for the use of fig. 40.

effect of pressure on the resistance of the measuring wire, and for differential compressibility between the wire and the holder, but it is a great advantage of the method that all these corrections can be easily determined, and are all very small, the total correction being only 2 or 3 per cent., whereas in the differential volume methods with a liquid, the corrections are often materially larger than the effect itself.

The experimental details of the actual measurements are the same whether the material is isotropic or in the single crystal condition. The chief difficulty with single crystals is in obtaining the material. The most favourable length for specimens to be measured in the apparatus of fig. 41 is about 2·5 cm., and the diameter may vary from 1 mm. up to 6 mm. It is easy to obtain single crystals of the low melting metals of these dimensions by methods which I have developed (B. 53); a detailed description of the method here would be out of place. Not many natural minerals occur in perfect crystals of the optimum size, and in most such cases smaller crystals must be used. By piling several pieces end to end it is possible to make good measurements with single pieces only a few mm. long.

Organic crystals usually require a special technique. These are usually so much more compressible than metallic or mineral crystals that the use of the piezometer with multiplying lever is not necessary, but it is sufficient to transmit the displacement of the specimen directly to the measuring wire in a modified form of the apparatus of fig. 40. The compressibility is often so great that sufficiently accurate measurements can be made on specimens no longer than 5 mm. The chief modification in technique necessary in handling organic crystals arises from the fact that they are nearly all soluble in the pressure-transmitting liquid, so that the piezometer must be so designed that the crystal is protected from the transmitting medium by immersion in mercury. The details of the modified apparatus are described in B. 53.

Not many organic crystals are easily grown even to such dimensions as 5 mm., and those which I have measured have been produced with considerable difficulty. In the endeavour to obtain suitable inorganic crystals I have searched the most important museums of the United States. Any reader of this book who is able to supply single crystals of the requisite dimensions of any substances not listed in the following Table V would confer a real favour by permitting their use for compressibility determinations.

With the apparatus above the compressibility of a large number of solid materials has been determined. The data for the most important of these are collected in Table V; in order to economise space the numerical values for the more complicated substances are not reproduced, but a list of these substances is given and the references where the numerical values may be found. Included in the tables are data of Slater, obtained with my apparatus for a number of alkali halides. The values obtained at the Geophysical Laboratory for a number of rocks and minerals are not given, but the original sources must be consulted for these. Most of the measurements can be represented within experimental error by a second-degree expression in the pressure, and for these it is sufficient to tabulate only the coefficients of the two terms of the expansion. There are some materials, however, in which the variation of compressibility with pressure is high, which cannot be represented within experimental error by a second-degree expression, or even by a three- or four-constant expression, such as the alkali metals, and for these substances the results are given in tabular form at various intervals of pressure.

Discussion of Compressibility of Solids.—The simplest materials, and from many points of view the most significant, are the elements. It was first pointed out by Richards that the compressibility of the elements is a strongly periodic function of atomic weight. In fig. 42 this periodic relation is shown. In order to get a manageable scale, 7 plus the logarithm to the base 10 of the compressibility at room temperature is plotted, instead of compressibility itself, against atomic number. This figure shows all the elements whose compressibility is at present known; most of them have been taken from my work, but there are a few from other sources: C (diamond) was measured by Adams,[10] and Ga, As, Se, In, and I by Richards.[11] The experimental methods

TABLE V

COMPRESSIBILITY OF SOLIDS

$$-\frac{\triangle V}{V_0} = a \times 10^{-7} p - b \times 10^{-12} p^2 \quad (p \text{ in kg.|cm.}^2)$$

Element and reference		a	b	Element and reference		a	b
Li (B. 41)	30° / 75°	86·92 / 89·72	97·5 / 107·3	Ni (B. 41)	30° / 75°	5·29 / 5·35	2·1 / 2·1
Be (B. 62)	30° / 75°	8·55 / 8·55	3·88 / 3·88	Cu (B. 41)	30° / 75°	7·19 / 7·34	2·6 / 2·7
B (B. 72)	30°	5·51	2·2	Zn $\frac{\triangle l}{l_0}$ ∥ hex. axis	30° / 75°	12·98 / 13·55	5·32 / 7·82
C (1)	20°	1·8		(B. 53) ⊥ hex. axis	30° / 75°	1·946 / 2·025	1·11 / 1·47
Na		See special table.		Ga (3) Solid 20° / liquid 30°		$20\begin{Bmatrix} \text{to} \\ 500 \\ \text{kg.} \end{Bmatrix}$ 40	
Mg $\frac{\triangle l}{l_0}$ ∥ hex. axis	30° / 75°	9·84 / 10·15	6·51 / 7·78				
(B. unpublished) ⊥ hex. axis	30° / 75°	9·84 / 9·66	9·19 / 6·95	Ge (B. 41)	30° / 75°	13·78 / 13·64	6·8 / 6·8
Al (B. 41)	30° / 75°	13·43 / 13·76	5·0 / 5·1	As (hexagonal)	20°	44 (2) (to 500 kg.) There is a direction in which the linear compressibility is at least 7 times smaller than the average (B. 72).	
Si (2)	20°	3·1 (to 500 kg.)					
P		See special table.					
S		See special table.					
K		See special table.		Se non-cubic (2)	20°	11·8 (to 500 kg.)	
Ca (B. 41)	30° / 75°	56·97 / 58·50	47·2 / 52·7	Rb		See special table.	
Ti (B. 72)	30° / 75°	7·97 / 8·68	−0·12 / +4·5	Sr (B. 41)	30° / 75°	81·87 / 82·68	72·5 / 71·7
Va (B. 62)	30° / 75°	6·09 / 6·12	2·58 / 2·55	Zr (B. 66)	30° / 75°	10·97 / 11·06	7·44 / 7·80
Cr (B. 62)	30° / 75°	5·19 / 5·31	2·19 / 2·19	Cb (B. unpublished)	30° / 75°	5·70 / 5·78	2·2 / 2·2
Mn (B. 72)	30° / 75°	7·91 / 8·08	5·3 / 4·8	Mo (B. 41)	30° / 75°	3·61 / 3·62	1·0 / 1·0
Fe (B. 41)	30° / 75°	5·87 / 5·93	2·1 / 2·1	Rh (B. 45)	30° / 75°	3·72 / 3·81	2·67 / 2·67
Co (B. 41)	30° / 75°	5·39 / 5·47	2·1 / 2·1	Pd (B. 41)	30° / 75°	5·28 / 5·31	2·1 / 2·1

TABLE V—*continued*

Element and reference		a	b	Element and reference		a	b
Ag (B. 41)	30° 75°	9·87 10·04	4·4 4·5	Ce low pressure modification, below 4000. (B. 62)	30° 75°	45·63 45·03	−161·4 −151·5
Cd, $\dfrac{\Delta l}{l_0}$ { ‖ hex. axis		18·3		Pr (B. 62)	30° 75°	33·8 34·6	13 13
indirect, from elastic constants, at 20°.				Hf (B. 66)	30° 75°	9·01 8·81	2·37 2·37
(B. 53) { ⊥ hex. axis		2·1		Ta (B. 41)	30° 75°	4·79 4·92	0·25 0·25
In (4)	25°	25·0 (to	500 kg.)	W (B. 53, 41)	30° 75°	3·18 3·18	1·4 1·5
Sn, ‖ $\dfrac{\Delta l}{l_0}$ tetrag. { axis	30° 75°	6·719 6·956	4·07 3·91	Ir (B. 45)	30° 75°	2·68 2·81	1·3 2·2
⊥ (B. 53) tetrag. { axis	30° 75°	6·022 6·144	4·20 4·26	Pt (B. 41)	30° 75°	3·60 3·64	1·8 1·8
Sb, ‖ $\dfrac{\Delta l}{l_0}$ trig. { axis	30° 75°	16·48 16·37	20·5 18·0	Au (B. 41)	30° 75°	5·77 5·70	3·1 2·1
⊥ (B. 53) trig. { axis	30° 75°	5·256 5·091	4·56 3·04	Hg		See special table.	
Te, ‖ $\dfrac{\Delta l}{l_0}$ trig. { axis	30° 75°	−4·137 −5·132	−9·6 −13·2	Tl, non-cubic (4)	25°	2·77 (to	500 kg.)
⊥ (B. 53) trig. { axis	30° 75°	27·48 27·77	52·7 53·6	Pb (B. 41)	30° 75°	23·73 24·33	17·25 17·7
I (2)	20°	127 (to	500 kg.)	Bi, $\dfrac{\Delta l}{l_0}$ ‖ hex. { axis	30° 75°	15·92 15·80	11·1 11·6
Cs		See special table.		(B, 53, 41) ⊥ hex. { axis	30° 75°	6·624 7·044	4·39 8·40
Ba (B. 62)	30° 75°	101·9 106·3	129 149	$\Delta V/V_0$ {	30° 75°	29·17 29·89	22·43 31·13
La (B. 62)	30° 75°	35·13 35·01	14·7 17·1	Th (B. 62)	30° 75°	18·18 18·46	12·78 13·29
				Ur (B. 41)	30° 75°	9·66 9·55	2·5 2·2

TABLE V—*continued*

SPECIAL TABLES

Element and reference			4000 kg.	8000 kg.	12,000 kg.	15,000 kg.
Na, (B. 41)	$-\dfrac{\Delta V}{V_0}$	$\begin{cases}30° \\ 75°\end{cases}$	0·0570 0·0602	0·1050 0·1093	0·1465 0·1528	
P, $-\dfrac{\Delta V}{V_0}$	red	$\begin{cases}30° \\ 75°\end{cases}$	0·0190 0·0189	0·0342 0·0344	0·0469 0·0476	
(B. 62)	black	$\begin{cases}30° \\ 75°\end{cases}$	0·0095 0·0095	0·0158 0·0158	0·0205 0·0209	
S, $-\dfrac{\Delta l}{l_0}$	'a' direction	$\begin{cases}30° \\ 75°\end{cases}$	0·0180 0·0190	0·0304 0·0323	0·0412 0·0427	
(B. 62)	'b' direction	$\begin{cases}30° \\ 75°\end{cases}$	0·0165 0·0201	0·0277 0·0334	0·0370 0·0433	
	'c' direction	$\begin{cases}30° \\ 75°\end{cases}$	0·0079 0·0082	0·0143 0·0148	0·0198 0·0203	
	$-\dfrac{\Delta V}{V_0}$	$\begin{cases}30° \\ 75°\end{cases}$	0·0419 0·0466	0·0707 0·0784	0·0949 0·1027	
K, $-\dfrac{\Delta V}{V_0}$ (B. 41)		45°	0·1095	0·1877	0·2544	
Rb, $-\dfrac{\Delta V}{V_0}$ (B. 54)		50°	0·143	0·211	0·254	0·279
Cs, $-\dfrac{\Delta V}{V_0}$ (B. 54)	solid liquid	50° 75°	0·161 0·063 (1000)	0·240 0·107 (2000)	0·294	0·328
Hg, $-\dfrac{\Delta V}{V_0}$ (B. 5)	liquid	20°	0·01485	0·02754	0·03795	

TABLE V—*continued*

COMPRESSIBILITY OF A FEW SIMPLE CUBIC COMPOUNDS

Compound and reference		*a*	*b*	Compound and reference		*a*	*b*
LiF (5)	30° 75°	11·5 11·6	8·6 8·6	KI (5)	30° 75°	83·7 86·0	150·7 150·7
LiCl (5)	30° 75°	33·4 34·4	32·4 32·4	RbBr (5)	30° 75°	77·8 78·4	133·4 133·4
LiBr (5)	30° 75°	42·2 43·8	50·2 50·2	RbI (5)	30° 75°	9·38 9·67	197·7 197·7
NaF (B. unpublished)	30° 75°	20·7 20·8	17·7 18·1	BaF_2 (B. unpublished)	30° 75°	19·33 19·65	14·8 14·6
NaCl (6)	30° 75°	41·82 43·44	50·4 51·9	SrF_2 (B. unpublished)	30° 75°	15·78 16·12	10·3 10·8
NaBr (5)	30° 75°	49·8 51·5	62·1 62·1	CdF_2 (B. unpublished)	30° 75°	11·02 10·96	8·5 8·4
KF (5)	30° 75°	32·4 32·6	32·0 32·0	TiN (B. unpublished)	30° 75°	3·32 3·51	2·1 2·1
KCl (5)	30° 75°	55·2 56·4	73·5 73·5	TiC (B. unpublished)	30° 75°	4·72 4·78	2·2 2·2
KBr (5)	30° 75°	65·7 67·5	102·5 102·5				

[1] L. H. ADAMS and E. D. WILLIAMSON, *Jour. Frank. Inst.*, **195**, 493 (1923).
[2] T. W. RICHARDS, *Jour. Amer. Chem. Soc.*, **37**, 1646 (1915).
[3] T. W. RICHARDS and S. BOYER, *ibid.*, **43**, 274 (1921).
[4] T. W. RICHARDS and J. D. WHITE, *ibid.*, **50**, 3290 (1928).
[5] J. C. SLATER, *Phys. Rev.*, 23, 488 (1924) ; *Proc. Amer. Acad.*, **61**, 135 (1926).
[6] P. W. BRIDGMAN, *ibid.*, **64**, 33 (1929).

TABLE V—*continued*

REFERENCE TABLE FOR COMPRESSIBILITIES OF MORE COMPLICATED SOLID COMPOUNDS WHICH HAVE BEEN MEASURED TO HIGH PRESSURES

Alum, ammonium	B. 72	Jeffersonite		B. 63
,, chrome	B. 72	Limestone, Solenhofen		B. 48
,, potassium	B. 72	Magnetite		B. 56
Ammonium tartrate	B. 72	Nichrome		B. 41
Andradite	B. 63	Ni 37·5, Fe 62·5		B. 65
Apatite	B. 63	$(NH_3C_2H_5)_2SnCl_6$		B. 72
Argentite	B. 56	Obsidian		B. 52
Bakelite	B. 72	Orthoclase		B. 63
Barite	B. 63	Pitchstone		B. 52
Basalt	B. 48	Pyrite		B. 56
Beryl	B. 63	Quartz	B. 56, 63	
Calcite	B. 56	Rochelle salt		B. 72
Catlinite (pipestone)	B. 48	Rutile	B. 56, 63	
Celestite	B. 56	Sodium bromate		B. 72
Cobaltite	B. 56	,, chlorate		B. 72
Crocoite	B. 56	,, nitrate		B. 72
Diphenylamine	B. 72	Sphalerite		B. 56
Fluorite	B. 56	Spodumene	B. 56, 63	
Galena	B. 56	Tachylite, Kilauea		B. 52
Garnet	B. 63	,, Torvaig		B. 52
Glass, pyrex	B. 48, 52	Talc		B. 48
,, 9 miscellaneous	B. 52, 68	Tartaric acid		B. 72
,, SiO_2	B. 52	Topaz		B. 63
Hanksite	B. 63	Tourmaline	B. 56, 63	

have already been sufficiently indicated. All my own determinations were made by the method of linear compressibility, except those for Rb and Cs, which are so compressible that the piston displacement method was used for them.

The periodic character of the relation is so striking as to need no comment. There are certain minor irregularities in the curves, such, for example, as in the Zn, Ge, As, Se sequence, and in the Cd, In, Sn, Sb, Te sequence. These irregularities are without doubt connected in some way with the crystal structure, since all these elements are non-cubic. It is rather surprising that the factor of crystal system is not more disturbing to the periodic relationship; Zn, for example, is seven times more compressible in one direction than another, but nevertheless the average compressibility does not fall far from the main sequence.

The most important gap in the results is in the compressibilities of the rare gases, none of which are known in solid form. The direct experimental determination of these would be difficult, because of the necessity for making the measurements at low temperatures; we have seen, however, that

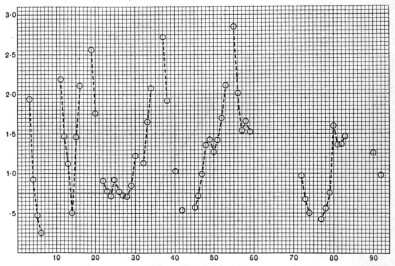

FIG. 42.—The logarithm to the base 10, plus 7, of the cubic compressibility of the elements plotted as ordinate against the atomic number as abscissa.

measurements on gaseous H_2 and He enable lower limits to be set to the compressibilities of the corresponding solids, and that the probable compressibilities are very high, in fact much higher than that of any of the elements shown in the figure. It seems almost certain that the positions of greatest compressibility in the completed diagram will be occupied by the solid rare gases, instead of by the alkali metals as at present. Other gaps in the figure include the other permanent gases. There is indirect evidence that the compressibility of solid N is higher than that of most liquids, and that it is probably of the same order as that of solid Cs. By far the largest gap among the elements ordinarily solid

is among the rare earth metals, few of which have been isolated in sufficiently large pieces to permit measurement.

The great range of numerical values of compressibility is
striking, the range from C to Cs being by a factor of 240;
the variation would be much greater if the compressibility
of the solidified rare gases were known. This range of
variation is much greater than the range in the compressibility of ordinary liquids; if, however, the temperature
range were extended sufficiently to include liquid metals and
liquified rare gases, there is no reason to think that the
range of compressibilities characteristic of the solid phase
would be any greater than that for the liquid.

If the changes of volume that can be produced by pressure
are compared with those that can be attained by changing
temperature, it is significant that for many elements the
decrease of volume produced by applying 12,000 kg./cm.^{2} at
room temperature is materially greater than the decrease of
volume produced by cooling at atmospheric pressure from
room temperature down to 0° abs. That is, more extensive
information about the way in which atomic forces vary when
the distance apart of the atoms varies can be obtained by
changing the pressure within limits experimentally attainable than by changing the temperature.

It is probably of considerable significance that the compressibility of the metallic elements and of ionic lattices like
NaCl are not of different orders of magnitude, indicating that
the mechanisms are not essentially different. The compressibility of ionic lattices like NaCl has, of course, been the
subject of much theoretical discussion, first by Born from
more classical points of view, and later from the point of
view of the wave mechanics, and it is now possible to calculate
rather satisfactorily the compressibility of a material like
NaCl. In the ionic lattice the attractive forces are determined merely by the electronic charges at the lattice points.
At equilibrium the repulsive forces are of the same general
magnitude as the attractive forces. Now it is significant
that the compressibility has the same dimensions as $e^{-2}d^{-4}$,
where d is the mean distance between atomic centres and e
the electronic charge. If the compressibility is determined

essentially by the interaction of ionic charges, then it is to be expected that the compressibility will be of the order of magnitude of $e^{-2}d^{-4}$, to which it is dimensionally equivalent. An equivalent statement, assuming simple cubic arrangement, and expressing d in terms of the atomic weight, density, and mass of the hydrogen atom, is found to be, after substituting numerical values, that the compressibility is of the order of

$$8 \cdot 6 \times 10^{-14}\left(\frac{\text{At. Wt.}}{\text{Density}}\right)^{\frac{1}{3}}, \quad \text{or} \quad \frac{\text{Compressibility}}{8 \cdot 6 \times 10^{-14}\left(\dfrac{\text{At. Wt.}}{\text{Density}}\right)^{\frac{1}{3}}}$$

should be of the order of unity.   Table VI contains results for a number of metals and alkali halides.   It seems highly

TABLE VI

Substance	Compressibility $8 \cdot 6 \times 10^{-14}\left(\dfrac{\text{At. Wt.}}{\text{Density}}\right)^{\frac{1}{3}}$	Substance	Compressibility $8 \cdot 6 \times 10^{-14}\left(\dfrac{\text{At. Wt.}}{\text{Density}}\right)^{\frac{1}{3}}$
Li	3·20	U	0·38
Na	2·67	Fe	0·51
K	2·56	Ni	0·50
Ca	0·86	Co	0·48
Sr	0·85	Pd	0·31
Mg	1·02	Pt	0·22
Zn	0·91	LiF	0·70
Cd	0·78	LiCl	0·71
Cu	0·60	LiBr	0·68
Ag	0·51	NaCl	0·60
Au	0·30	NaBr	0·58
Ge	0·51	KF	0·57
Sn	0·53	KCl	0·52
Pb	0·56	KBr	0·51
Sb	0·59	KI	0·50
Bi	0·59	RbBr	0·51
Mo	0·20	RbI	0·48
W	0·16		

probable, because of the similarity of these figures, that the mechanism determining compressibility—that is, the nature of the atomic forces—cannot be very different in the metals and the ionic crystals, and in fact recent wave mechanics developments, in particular the work of Slater,[12] contain specific suggestions as to what this similarity may consist in.

After the magnitude of the compressibility itself, the next significant feature is the variation of compressibility with pressure and temperature. It is natural to expect that the compressibility will decrease with increasing pressure, and this is indeed usually the case, but there is no inner necessity in such a decrease of compressibility, contrary to opinions that have sometimes been expressed, and there are a few substances whose compressibility increases with pressure. A number of varieties of glass belong in this category; the magnitude of the effect is closely connected with the SiO_2 content, the effect being largest for pure SiO_2 glass, the compressibility of which is 8 per cent. larger at 12,000 kg./cm.^{2} than at atmospheric pressure. From this the increase of compressibility may vary all the way down to nothing, depending on the composition. It is not difficult to understand how there may be such an effect if the molecules are thought of as approximately like spheres at low pressures, perhaps because of rapid rotational motion, but at high pressures the rotational motion is inhibited by the increasing constraints, and the molecules assume a greater semblance of order, perhaps lying side by side like ellipsoids with the long axes all pointing in the same direction. It is significant that the effect is shown so prominently in a glass which is amorphous, and which is capable of existence in a crystalline form of smaller volume. It is to be emphasised, however, that although the amorphous phase may possibly at high pressures assume in some degree the order of the crystalline arrangement, nothing like crystallisation in the proper sense takes place; the relation between pressure and volume has no trace of hysteresis, and apparently is perfectly reversible.

Beside the glasses, one crystalline phase is known in which the compressibility increases with increasing pressure, the low-pressure modification of Ce. It may well be that there is something special in the mechanism here, as would be suggested by the fact that the low-pressure modification presently gives way at high pressure to another more stable phase.

The "instantaneous" compressibility, $-\frac{1}{v}\left(\frac{\partial v}{\partial p}\right)_\tau$, evidently has a greater tendency to increase at high pressure than the compressibility proper, $-\frac{1}{v_0}\left(\frac{\partial v}{\partial p}\right)_\tau$, which has been the subject of discussion above, for the factor $\frac{1}{v}$ increases with increasing pressure, whereas $\frac{1}{v_0}$ is of course constant. The instantaneous compressibility of potassium increases with pressure at the upper end of the pressure range; this will be discussed in greater detail later.

Except for these few exceptional cases, the compressibility does decrease with increasing pressure as would be expected. This can be seen by glancing through Table V, for the change of compressibility can be at once obtained in terms of the coefficients listed. The compressibility, which we abbreviate as a, we defined as

$$-\frac{1}{v_0}\left(\frac{\partial v}{\partial p}\right)_\tau;$$

the coefficients a and b listed in the table are the coefficients in the relation

$$\Delta v/v_0 = -ap + bp^2.$$

Since

$$\left(\frac{\partial v}{\partial p}\right)_\tau = \left(\frac{\partial \Delta v}{\partial p}\right)_\tau,$$

we have at once

$$a = a - 2bp, \quad \text{and} \quad \left(\frac{\partial a}{\partial p}\right)_\tau = -2b,$$

or the change of compressibility with pressure is the negative of twice the second coefficient. Since b is positive in nearly all cases, we see at once that compressibility decreases with increasing pressure in nearly all cases.

Furthermore, it is obvious at once on superficial examination that those substances for which a is large—that is, those which have a high compressibility—also have a high b, or a high rate of change of compressibility with pressure. This fact has been fairly well known; Richards [13] observed that most of those liquids which he measured over a pressure range of 500 kg./cm.² show a similar effect to such a pronounced degree that if $2b$ is plotted against a, the points for different liquids lie very nearly on a single smooth curve. Adams [14] has remarked the same relation for those solids which were compressible enough to allow the change of compressibility with pressure to be determined by his method, and he has suggested the use of the statistical relation between a and b disclosed by his measurements as a means of finding a probable value for the change of compressibility with pressure in those cases where it cannot be determined by direct experiment. He has used such values in discussing the probable change in velocity of elastic disturbances deep in the earth's crust. It must be remarked that this relation has a strong tendency to become less valuable as the compressibility becomes smaller.

If we go further and consider the ratio of the change of compressibility with pressure to the initial value of the compressibility—that is, $\dfrac{1}{a_0}\left(\dfrac{\partial a}{\partial p}\right)_\tau$, which by the above is merely $\dfrac{2b}{a}$, we shall find that, excepting the very compressible alkali metals, the order of magnitude of this ratio is 10^{-5} for the metallic elements. It is interesting that the dimensions of the reciprocal of $\dfrac{1}{a}\left(\dfrac{\partial a}{\partial p}\right)_\tau$, which numerically is of the order of 10^5, has the dimensions of p, and 10^5 kg./cm.² is the order of the internal pressure, which according to various theories, in particular that of Richards,[15] is supposed to characterise the interior of solids. It is perhaps of significance that 10^5 is also the order of the pressure which would be exerted by various metals if the atoms were completely disintegrated into a gas of electrons and protons, acting according to the classical gas laws, and occupying the actual volume of the metal.

If we go a step further and form the ratio $\dfrac{1}{a_0{}^2}\left(\dfrac{\partial a}{\partial p}\right)_\tau$, a number will be found which is now smallest for the most compressible metals, and in fact varies from 2·18 for Sr to 36 for Ir and 39 for Rh. It is to be noticed that $\dfrac{1}{a_0{}^2}\left(\dfrac{\partial a}{\partial p}\right)_\tau$ is dimensionless, and it is therefore to be expected that it will be of the order of magnitude of a not large number, and that any simple theory of the equation of state and the compressibility of solids will give a result of the correct order. Such was in fact the prediction of Grüneisen's theory of solids, based on more or less classical conceptions of the forces between atoms, and also the prediction of Born's theory of crystal lattices, with its inverse second power of attraction and its inverse ninth power of repulsion.

The theoretical explanation of the decrease of compressibility with pressure has given considerable difficulty. Thus the theory of Born, although capable of giving fairly good values for the compressibility, did not reproduce at all its change with pressure. Pauling [16] in a comparatively recent paper emphasises the inadequacy of the wave theory to reproduce this feature, but apparently it is now on the point of yielding to treatment in the case of simple ionic lattices, and fairly good values have already been obtained by Birch in a paper yet unpublished. The theory has not yet, however, given a satisfactory treatment of the change of compressibility with pressure or even of the compressibility itself for metals.

The values given in the table provide the material for finding the change of thermal expansion with pressure; it must be recognised, however, that the accuracy of the temperature variations given by the table is probably not so high as the accuracy of the pressure variations. It follows from the various equations of definition that

$$\left(\frac{\partial \beta}{\partial p}\right)_{p=0} = -\frac{a(75°) - a(30°)}{45},$$

where β is the thermal expansion defined by $\beta \equiv \dfrac{1}{v_0}\left(\dfrac{\partial v}{\partial \tau}\right)_p$. It is at once evident on inspection that for nearly all metals β decreases with increasing pressure. In other classes of materials, particularly in special directions in a number of single crystals, there are exceptions, and the thermal expansion increases with increasing pressure. The relative change of thermal expansion with pressure—that is, $\dfrac{1}{\beta_0}\left(\dfrac{\partial\beta}{\partial p}\right)_{p=0}$, may be computed, and proves to be for the metallic elements of the order of 10^{-5}, like $\dfrac{1}{a_0}\dfrac{\partial a}{\partial p}$. For most metals the relative change of thermal expansion with pressure is somewhat smaller than the change of compressibility, but the rule is by no means universal, as shown, for example, by Cu, Al, and Ir. From the dimensional point of view, it is not surprising that $\dfrac{1}{a}\dfrac{\partial a}{\partial p}$ and $\dfrac{1}{\beta}\dfrac{\partial\beta}{\partial p}$ should be nearly the same in numerical magnitude, because the dimensions of both are the dimensions of pressure. In the case of $\dfrac{1}{a}\dfrac{\partial a}{\partial p}$, we have seen this to correspond to the internal pressure.

From one point of view it might be expected that the effect of pressure on β would be greater than on a, because at 0° abs. at atmospheric pressure, where in some of these cases the volume may be considerably larger than at room temperature at 12,000 kg., a remains finite, but β vanishes. We saw, however, that in the case of liquids the behaviour is the exact opposite of this expectation, and the thermal expansion remains large. The metals approach more closely to this expectation, but the departures are still large.

The internal energy of a substance is a significant feature, which may be calculated from a knowledge of a and β. The change of internal energy with pressure is given by the equation

$$\left(\frac{\partial E}{\partial p}\right)_\tau = -\tau\left(\frac{\partial v}{\partial \tau}\right)_p - p\left(\frac{\partial v}{\partial p}\right)_\tau.$$

Ordinarily $\left(\dfrac{\partial E}{\partial p}\right)_\tau$ starts with a negative value at low pressures —that is, as a substance is compressed energy flows out in the form of heat faster than it is put in in the form of mechanical work. The interpretation is that the interatomic forces are on the average attractive under these conditions, and the potential energy of these forces decreases as the atoms are brought closer together by an amount greater than the rise of potential energy of the repulsive forces. But all our pictures would suggest that ultimately, when the atoms are brought close enough together, the repulsive forces must preponderate, so that $\left(\dfrac{\partial E}{\partial p}\right)_\tau$ must eventually become positive. The equation above shows that the pressure at which this reversal of sign takes place is given by

$$p = -\tau\left(\frac{\partial v}{\partial \tau}\right)_p \bigg/ \left(\frac{\partial v}{\partial p}\right)_\tau.$$

In Table VII are shown the values for this pressure for a number of metals. In calculating these figures, due account has been taken of the variation of both $\left(\dfrac{\partial v}{\partial p}\right)_\tau$ and $\left(\dfrac{\partial v}{\partial \tau}\right)_p$ with pressure. The error made by not taking account of this variation is of the order of 5 or 10 per cent., the uncorrected pressure being in general too low. The significance of the pressure found in this way may be taken to be that at the volume corresponding to this pressure the repulsive and attractive forces are in balance. To a first approximation one would expect this balance of the two kinds of forces to depend only on the volume. Now at 0° abs. at atmospheric pressure the two forces are in balance, as shown at once by the equation, for $\left(\dfrac{\partial E}{\partial p}\right)_\tau = 0$ when $p = 0$ if $\tau = 0$. It is instructive to compare the volume at which $\left(\dfrac{\partial E}{\partial p}\right)_\tau = 0$ at room temperature with the volume at 0° abs. and atmospheric pressure. In Table VII are shown the volume decrements : (1) on increasing pressure at 30° C. to the value at which $\left(\dfrac{\partial E}{\partial p}\right)_\tau$ vanishes, and (2) on decreasing temperature at atmospheric pressure from 30° C. to 0° abs. We see that the first of the volume decrements is always materially larger than the second—that is, the volume at 30° C. and high pressure

TABLE VII

Metal	P, =pressure at which $\left(\dfrac{\partial E}{\partial p}\right)_\tau =0$ at 30° C. kg./cm.^{2}	$-\triangle V/V_0$	
		Between atmospheric pressure and P at 30° C.	On cooling at atmospheric pressure from 30° C. to 0° abs.
Ag	18,000	0·0163	0·012
Al	15,500	0·0161	0·012
Cu	20,700	0·0138	0·0090
Pb	10,900	0·0239	0·0188
Fe	20,000	0·0109	0·0065
Na	6,000 ±	0·082 ±	0·045 ±
Ir	21,700	0·0052	0·0033

at which the attractive and repulsive forces balance is less than the volume at 0° abs. where they balance. The obvious explanation of this is that the atoms have themselves been deformed by the high pressure, so that the effective boundary of the atom—that is, the region in which the repulsive forces increase rapidly in intensity—is closer to the centre of the atom at high than at low pressures. These considerations contain an approximation in that we have assumed that the potential energy of the mutual forces between rigid atoms is a function of the volume only. This is not strictly true, but it does not seem probable that this assumption could be accountable for the large discrepancies of volume found experimentally.

One would like to be able to get the specific heat of the metal as a function of pressure, as in the case of liquids, by use of the formula $\left(\dfrac{\partial C_p}{\partial p}\right)_\tau = -\tau\left(\dfrac{\partial^2 v}{\partial \tau^2}\right)_p$. This, however, is quite beyond the accuracy of these measurements, and in any event would have demanded measurements at three temperatures at least, whereas only two temperatures were attempted. However, the initial value of $\left(\dfrac{\partial^2 v}{\partial \tau^2}\right)_p$ at atmospheric pressure is known for a number of metals, so that the order of magnitude of the change of C_p with pressure may be found. $\left(\dfrac{\partial^2 v}{\partial \tau^2}\right)_p$ is almost always positive, so that initially C_p decreases with increasing pressure. The initial rate of decrease is usually so small, however, that a pressure of 10^8 or 10^9 kg./cm.^{2} would be required to reduce C_p to zero, assuming the decrease to remain independent of pressure at the initial rate. Even in the case of so deformable a metal as potassium, a pressure of 1.4×10^7 kg./cm.^{2} would be required. These pressures are so very far beyond those attainable in the laboratory that the conclusion seems justified that in the experimental range the change of C_p is negligibly small.

The difference between C_p and C_v, given by $\tau \left(\dfrac{\partial v}{\partial \tau}\right)_p^2 \Big/ \left(\dfrac{\partial v}{\partial p}\right)_\tau$, involves only first derivatives, and may be computed as a function of pressure in terms of quantities listed above. It will be found on actually making the computations that there is no universal rule, but $C_v - C_p$ increases with pressure for some metals and decreases for others.

The entropy, s, of an ordinary substance decreases with pressure according to the relation $\left(\dfrac{\partial s}{\partial p}\right)_\tau = -\left(\dfrac{\partial v}{\partial \tau}\right)_p$. If the total entropy content of a body is limited and is equal to zero at $0°$ abs., as demanded by the third law, then the total decrease of entropy brought about by the application of any pressure, no matter how high, cannot be greater than the total entropy content of the body as given by $\int_0^\tau \dfrac{C_p}{\tau} d\tau$. That is, $-\int_0^\infty \left(\dfrac{\partial s}{\partial p}\right)_\tau dp \left(\equiv \int_0^\infty \left(\dfrac{\partial v}{\partial \tau}\right)_p dp \right)$ must not exceed a certain well-defined limit. If $\left(\dfrac{\partial v}{\partial \tau}\right)_p$ remained constant with pressure, this limit could obviously be exceeded, so that the decrease of $\left(\dfrac{\partial v}{\partial \tau}\right)_p$ with increasing pressure appears as a necessity, imposed by thermodynamics. The restriction imposed by the finite value of $\left(\dfrac{\partial v}{\partial \tau}\right)_p$ is most exacting for those substances for which

$\left(\dfrac{\partial v}{\partial \tau}\right)_p$ is large and C_p comparatively small.   These substances
are the low melting, mechanically soft, substances, and in
particular the alkali metals.   For K the data are known
with sufficient accuracy to permit an evaluation of the

integral, taking account of the variation of $\left(\dfrac{\partial v}{\partial \tau}\right)_p$ with pressure, and it will be found that the entropy decrease brought
about by an application of 12,000 kg./cm.² at 30° C. is only
one-third of the entropy decrease on cooling from 30° C. to
0° abs. at atmospheric pressure.   That is, assuming the
average rate over the first 12,000 kg./cm.², about 40,000
kg./cm.² would be necessary to reduce the entropy to 0.

But $\left(\dfrac{\partial v}{\partial \tau}\right)_p$ actually becomes smaller at the high pressures, so
that probably pressures of the order of 100,000 kg./cm.² must
be reached before the restriction imposed by the entropy
condition becomes of dominating importance.   If, however,
the atom itself should begin to disintegrate at pressures of
this order, then changes of entropy much greater than
allowed by the third law would be possible.

There are now a number of special topics connected with
compressibility to be considered.   A special discussion of the
compressibility of the alkali metals is desirable in view of the
facts that their compressibility is so large, and that the constitution of the alkali metals is now becoming understood from
the wave mechanics point of view.   The numerical values of
compressibility have already been given in Table V.   The
compressibility increases in the order of the atomic weight,
the compressibility of Cs being greatest, and very nearly the
same as that of the most compressible liquids, such as ether.
There is this difference between the compressibility of Cs
and ether : the compressibility of Cs tends to decrease less
with increase of pressure than does that of ether.   In fig. 43
is shown the instantaneous compressibility, defined as $\dfrac{1}{v}\left(\dfrac{\partial v}{\partial p}\right)_\tau$

as already stated, in distinction from the ordinary compressibility $\frac{1}{v_0}\left(\frac{\partial v}{\partial p}\right)_\tau$, for the five alkali metals as a function of pressure. The most striking feature of the curves is the exceptional position of K. Its instantaneous compressibility drops with increasing pressure much less than would be

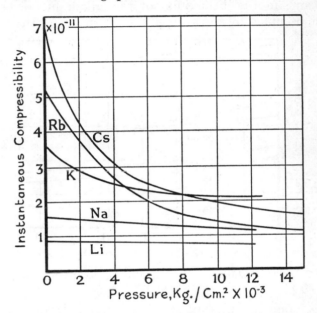

FIG. 43.—The instantaneous compressibility $\frac{1}{v}\left(\frac{\partial v}{\partial p}\right)_\tau$ in abs. C.G.S. units of the alkali metals as a function of pressure.

expected from the behaviour of the other metals, and in fact actually begins to rise again at 12,000 kg./cm.²; furthermore, it crosses the curves of Rb and Cs, so that at pressures above 8000 kg./cm.² K is the most compressible of the alkali metals. This constitutes one piece of evidence for an abnormally open structure in the atom of K, if we accept the idea that at high pressures the major part of the compressibility arises from the compression of the atoms. Another bit of evidence in the same direction is the atomic volume itself of K at atmospheric pressure, which lies entirely out of the sequence of the other alkali metals when atomic volume is plotted against atomic number.   Fig. 43 also suggests that at high pressures the curve for Na will similarly cross that of Rb, although this effect is not nearly so pronounced.

The volume occupied per discrete electrical charge, assuming the Bohr atom, seems to be as significant, if not more significant, in the behaviour of the alkali metals, as the volume occupied by the atomic units.   In fig. 44 the volume

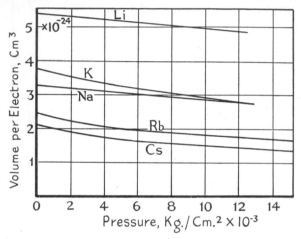

FIG. 44.—The volume per electron in c.c. of the alkali metals as a function of pressure.

occupied by one discrete electrical charge is plotted as a function of pressure.   This volume may be called for convenience the 'volume per electron' or the 'electronic volume,' to which it is approximately equal, although it is actually the volume per discrete charge, irrespective of whether the charge is negative or positive.   Specifically, it is obtained by dividing the volume per atom, calculated from ordinary density and atomic weights, by 4 for Li, 12 for Na, 20 for K, 38 for Rb, and 56 for Cs.   The 'electronic volume' varies through the series by a factor of 2·6 on passing from Cs to Li, the electronic volume of Li being the greatest, whereas the atomic volume varies by a factor of 5·3 through the series, the atomic volume of Cs being the greatest.   The exceptional position of K with respect to the electronic volume is shown by its curve crossing that of Na. At the high pressure end of the range the effect of pressure on electronic volume becomes strikingly the same for all these metals except K. In Table VIII is given for the five alkali metals the decrease in

TABLE VIII

Volume Compression per Electron per Kg./cm.^{2}

Metal	Electronic compression	
	Atms. pressure	12,000 kg./cm.^{2}
Li	$4 \cdot 7 \times 10^{-29}$	$3 \cdot 5 \times 10^{-29}$
Na	$5 \cdot 1$	$3 \cdot 2$
K	$13 \cdot 5$	$5 \cdot 9$
Rb	$12 \cdot 5$	$2 \cdot 2$
Cs	$14 \cdot 5$	$2 \cdot 5$

cm.^{3}, produced by a pressure increment of 1 kg./cm.^{2} in the volume occupied by one electron, both at atmospheric pressure and at $12,000$ kg./cm.^{2}. The initial compressibilities differ by a factor of threefold; an important part of the initial effect doubtless arises from the pressure taking the 'slack' out of the free space between the atoms, as has already been suggested in connection with liquids. The slack is greatest in Cs. At $12,000$, where most of the slack has been taken out, the compressibility per electron varies through the series only by 50 per cent. (except for K), and now Li is the most compressible.

It is significant that the volumes and the compressibility per electron are the same order of magnitude as they would be if electrons and nuclei together constituted a perfect gas, the kinetic unit of this gas being the electron as distinguished from the atom. The pressures which would be required to compress a perfect gas containing the equivalent number of electrons and nuclei to the volumes actually occupied by the alkali metals at atmospheric pressure and room temperature are about 7000 kg./cm.^{2} for Li, $10,000$ for K, $11,500$ for Na,

15,500 for Rb, and 18,000 for Cs, all pressures of an order experimentally attainable.   If now the external pressure on the alkali metal increases, decreasing its volume, the pressure required to compress the corresponding gas of electrons and nuclei to the same volume increases also, but by a small amount.   In the case of Li under 12,000 kg./cm.², the elec-

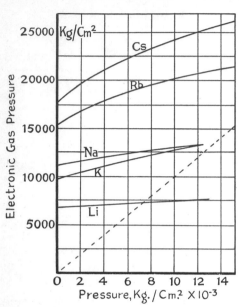

tronic gas pressure has increased only from 7000 to 7500, so that if it were not for the quantum forces maintaining the electrons more or less rigidly in position in the atoms, the structure of Li would collapse under a pressure of 12,000 kg./cm.².   At low pressures the volume of the metal is less than it would be if a perfect gas, so that at low pressures the quantum forces may be thought of as exerting a con-tracting tendency.   At high pressures, on the other hand, it is ob-

FIG. 45.—The 'electronic gas pressures' (see text) of the alkali metals at 0° C. as a function of external pressure.

vious that the rôle of the quantum forces changes and distends the structure against the collapsing tendency of the external pressure.   The pressure at which the change of sign takes place is about 14,000 for Na and K, and possibly 20,000 or 30,000 for Rb and Cs.   The intensity of the distending action of the quantum forces increases as the external pressure increases.

In fig. 45 is plotted the pressure that would be exerted by the material in a perfect gas condition, or the 'electronic gas pressure' against the external pressure.   If the material were actually a perfect gas, these two pressures would be equal to each other, and the plot would be a 45° line. The closeness with which the actual line approaches this 45° line may be taken as a measure of the closeness of the metal to the perfect gas condition. It is evident that the closeness is greatest for K; not only is the slope more nearly equal to 45° for this, but beyond 8000 kg./cm.² there is a slight reversal of curvature, barely perceptible on the diagram, so that above 8000 the curve for K is concave upward, as it must be if it is eventually going to assume a slope of 45°.

Consider next the compressibility that the alkali metals would have if they were in the condition of a perfect electron gas. The compressibility of a perfect gas is $\frac{1}{v}\left(\frac{\partial v}{\partial p}\right)_\tau = -\frac{1}{p}$. Call v_e the volume occupied by one electron, which by hypothesis is the kinetic unit. Then for the perfect gas $pv_e = 1\cdot35 \times 10^{-16}\tau$. Solve this equation for $\frac{1}{p}$ and substitute, obtaining

$$\frac{1}{v_e}\left[\frac{1}{v}\left(\frac{\partial v}{\partial p}\right)_\tau\right] = \frac{-1}{1\cdot35 \times 10^{-16}\tau} = -2\cdot7 \times 10^{13} \text{ at } 300° \text{ abs.}$$

That is, for a perfect gas of electrons, the function $\frac{1}{v_e}\left[\frac{1}{v}\left(\frac{\partial v}{\partial p}\right)_\tau\right]$ has the value $-2\cdot4 \times 10^{13}$ at 300° abs., *independent of pressure.* But the function $\frac{1}{v_e}\left[\frac{1}{v}\left(\frac{\partial v}{\partial p}\right)_\tau\right]$ may be computed for the actual alkali metal at different pressures. If we call this function the "gas function," then the closeness with which the gas function for the alkali metals approaches at 300° abs. the constant value $-2\cdot4 \times 10^{13}$, independent of pressure, is a measure of the closeness with which the actual compressibility of the alkali metals approaches the compressibility of the equivalent perfect gas of electrons. In fig. 46 are plotted the gas functions for the five alkalies as a function of external pressure; the gas function is far from constant and certainly has not the value $2\cdot4 \times 10^{13}$. But the gas function is of the right order of magnitude, and in the case of K there is a reversal of direction above 6000 kg./cm.², in that above 6000 kg./cm.² the gas function for K is rising toward the theoretical value.   Now at excessively high pressures it is not unnatural to think that there may be some sort of atomic collapse, the quantum forces which maintain the structure being incapable of rising indefinitely to meet the demands put upon them, so that eventually the atoms may break into a gas of electrons and nuclei.   This sort of atomic disintegration is to be found at high pressures and low temperatures, and is not the sort

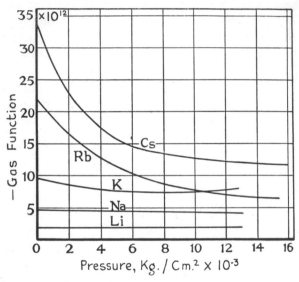

Fig. 46.—The 'gas function' (see text) of the alkali metals as a function of external pressure.

of disintegration usually considered in astronomical specu-lations like those of Eddington, for example.   His sort of collapse is produced by the temperature in spite of the pressure; this is a collapse at high pressure in spite of the low temperature.   If there is eventually a collapse of this new sort, it will begin to show itself by the volume and compressibility curves approaching the curves for an elec-tronic gas.   The fact that both these curves for K have shown the beginning of a tendency in this direction may perhaps be taken as indicative of the first beginnings of an atomic collapse, which cannot be expected to become complete or even important except under pressures far beyond the reach of present experiment. Of all the elements examined, K is the one which shows this very small tendency to the most marked extent. This is what might be expected, in view of the abnormally open structure of the K atom as shown by its abnormally high atomic volume, and to this extent increases the probability of this point of view.

If a substance can be forced to approach a perfect electron gas by exceedingly high pressures, then its thermal expansion, defined now as $\dfrac{1}{v}\left(\dfrac{\partial v}{\partial \tau}\right)_p$, must approach the value for a perfect gas, that is $\dfrac{1}{\tau}$. We have already commented on the fact that the thermal expansion of solids and liquids does not drop with increasing pressure as much as might have been expected, but remains large.

Leaving now the alkali metals, the compressibility of alloys has received no systematic investigation. There are results by Lussana [17] for a number of brasses; perhaps the greatest number of alloys has been investigated by Mehl and Mair,[18] who measured about a dozen primary alloys by the method of Richards to 500 kg. In general, the compressibility is less than that computed by the rule of mixtures from the pure components. This is what would be expected in view of the fact that the observed density is greater than that given by the rule of mixtures. These facts are qualitatively in accord with what would be expected according to the views of Richards on internal cohesive pressures; the matter has been elaborated by Mehl.[19]

The compressibility of single crystals next engages us. Only non-cubic crystals concern us, since the linear compressibility of a cubic crystal is the same in all directions, so that as far as hydrostatic pressure is concerned there is no difference between a single cubic crystal and a microscopic aggregate of crystal grains. The linear compressibility of six non-cubic elements in the single crystal form has been determined: Zn, Bi, Sb, Te, Sn, and Mg. The most striking characteristic of the compressibility of most of these substances is the very great differences in different directions. In Zn the compressibility parallel to the hexagonal axis is nearly seven times greater than perpendicular to the axis, and in Cd it is nearly nine times greater. There is also strong evidence that the compressibility of single crystal As differs greatly in different directions. It is a general rule that the compressibility is greatest perpendicular to the best-developed cleavage plane, or to the plane of easiest slip, in those cases where the cleavage plane is not well developed— that is, the compressibility is greatest in those directions in which the atomic separation is greatest, as would be expected. This difference between different directions is most marked in the case of Te; here the cleavage plane is not perpendicular to the crystal axis, but is parallel to it, so that the linear compressibility would be expected to be greater perpendicular to the axis.   This is the case, and, furthermore, the difference between the two directions is so great that the compressibility parallel to the axis is actually negative. That is, when a Te crystal is subjected to a uniform hydrostatic pressure, it increases in length parallel to the axis. Of course the shortening in the perpendicular direction must be great enough to more than compensate, for the total volume must decrease under pressure.   The fact that the cleavage plane in Te runs parallel to the axis instead of perpendicular to it is connected with the lattice structure, the atoms being arranged in tightly wound helices in the direction of the crystal axis. A connection has been suggested by Mehl [20] between the negative compressibility along the axis and this peculiarity in the structure.

Magnesium in the single crystal form is of considerable interest. This crystallises in the hexagonal close-packed arrangement, and the axial ratio is such as to correspond to a close-packed arrangement of spheres. In this respect it is different from Zn and Cd, which also are hexagonal close packed, but their axial ratio corresponds to considerably elongated ellipsoids rather than spheres.   Now the hexagonal close-packed arrangement of spheres does not differ greatly from the cubic close-packed arrangement; in fact, if one builds up the lattice by adding planes perpendicular to the crystal axis, then the relative position of any two planes is the same in the hexagonal and cubic pilings, the only difference being a small relative change of the third planes. It is to be expected, therefore, that a hexagonal close-packed arrangement of spheres will not differ greatly in physical properties from a cubic close-packed arrangement, and, in particular, that the linear compressibility of such a metal as Mg will be nearly the same in all directions, because the linear compressibility of a cubic crystal is the same in all directions. This actually turns out to be the case, as shown by the figures in Table V.

The temperature coefficient of compressibility, or the pressure coefficient of thermal expansion, has a rather strong tendency to abnormality in these crystals. The normal behaviour is for the compressibility to increase with rising temperature, and this is the case for Zn and Sn. But the linear compressibility in both directions of Bi and Sb decreases with rising temperature, and that of Te increases perpendicular to the axis, but decreases parallel to the axis by an amount sufficient to more than compensate, so that the total volume compressibility of Te decreases with increasing temperature. The behaviour is still more complicated for Mg; the temperature coefficients of linear compressibility have different signs in different directions, and, furthermore, the second degree terms in the pressure behave differently, so that the temperature coefficient of volume compressibility is negative at low pressures, but positive at high pressures.

The compressibility of non-cubic crystals of compounds behaves qualitatively in the same way as that of elements. There is much room for future work here in determining the connection between X-ray structure and the elastic constants, in particular the linear compressibility in different directions. In general, the difference of compressibility in different directions is not so marked in the case of compounds as for metals; among all the non-metallic substances examined there are none in which the ratio of compressibility in different directions is as high as for Zn and Cd, and, in fact, ratios as high as 2 are not common. There is no obvious connection between the compressibilities in different directions and the axial ratios of the crystallographer, there being examples both of larger and smaller compressibility in the direction of greater axial ratio. It does not follow necessarily, however, that the compressibility is not greatest in the direction in which the atoms are separated by the greatest amount, because there is no immediate connection between the axial ratios of the crystallographer and the atomic distances of X-ray analysis. It is not unnatural to think, however, that the intensity of atomic forces cannot be related to atomic distance in as simple a way in a complicated structure composed of many different kinds of atoms as in the simple metallic lattices, so that departures from the rule are to be looked for. In general, compressibility decreases in all directions with rising pressure and falling temperature for the more complicated crystals as well as for the metals, but there are a number of exceptions to the temperature effect, and there is a much greater tendency to irregularity, there being a good number of cases in which the temperature coefficient of compressibility has different signs in different directions.

One result that came out of the measurements is that crystals as they are found in nature are not well-defined objects, but have individual variations, which may be important. Even a substance apparently as well defined and reproducible as quartz has fairly large variations. Thus a specimen of quartz measured by me (B. 56) had an initial compressibility 6 per cent. greater than a sample measured by Adams and Williamson,[21] and showed marked peculiarities in different directions, the compressibility perpendicular to the trigonal axis increasing with pressure above 6000 kg./cm.^{2}. To check the method, I later (B. 63) measured a small rod cut from the identical specimen of Adams and Williamson, and obtained agreement of initial compressibility to 0·2 per cent. The average compressibility between atmospheric pressure and 12,000 of the two determinations differed, however, by considerably more than this, 1·7 per cent. Too much significance should not be attached to this discrepancy in view of the possible lack of perfect homogeneity of the large crystal of Adams and Williamson, and in any event it is evident that different specimens may differ by much more than the possible error of measurement.

Among the single crystals of compounds measured are several organic substances. The differences of compressibility in different directions shown by those members of this class of substance investigated up to now is perhaps not as great as might be expected from the strongly developed crystalline properties. Thus diphenylamine, which has a very strong tendency to crystallise from the melt in thin plates, differs in linear compressibility in different directions by less than 12 per cent. The most extreme variation yet found is shown by tartaric acid, the initial compressibility of which in the 'a' direction is only one-sixth of that in the 'b' direction. This substance is also unusual in the very high value of its temperature coefficient of compressibility, which is approximately 0·0018 between 30° and 75° C.

[1] A. MALLOCK, *Proc. Roy. Soc.*, **74**, 50 (1904).

[2] E. GRÜNEISEN, *Ann. Phys.*, **33**, 1239 (1910) ; **39**, 284 (1912).   *Verh. D. Phys. Ges.*, **13**, 491 (1911).

[3] J. Y. BUCHANAN, *Trans. Roy. Soc. Edin.*, **29**, 589 (1880) ; *Proc. Roy. Soc. Edin.*, **73**, 296 (1904).

[4] E. H. AMAGAT, *C.R.*, **108**, 727 (1889).

[5] S. LUSSANA, *Nouv. Cim.*, **7**, 1 (1904).

[6] T. W. RICHARDS, *Carnegie Inst. Washington*, No. 76, p. 44 (1907).

[7] W. C. ROENTGEN und J. SCHNEIDER, *Wied. Ann.*, **31**, 1000 (1881).

[8] T. W. RICHARDS, *Jour. Amer. Chem. Soc.*, **37**, 1643 (1915).

[9] L. H. ADAMS, E. D. WILLIAMSON, and J. JOHNSTON, *ibid.*, **41**, 12 (1919).

   L. H. ADAMS, *Jour. Wash. Acad. Sci.*, **11**, 45 (1921).

   L. H. ADAMS and E. D. WILLIAMSON, *Jour. Frank. Inst.*, **195**, 475, (1923).

   L. H. ADAMS and R. E. GIBSON, *Proc. Nat. Acad. Sci.*, **12**, 275 (1926).

   L. H. ADAMS, *Jour. Wash. Acad. Sci.*, **17**, 529 (1927).

   L. H. ADAMS and R. E. GIBSON, *Proc. Nat. Acad. Sci.*, **15**, 713 (1929).

[10] Second reference under 9.

[11] T. W. RICHARDS, *Jour. Amer. Chem. Soc.*, **37**, 1643 (1915) ; *Proc. Nat. Acad. Sci.*, **4**, 388 (1918) ; *Jour. Amer. Chem. Soc.*, **50**, 3290 (1928).

[12] J. C. SLATER, *Phys. Rev.*, **35**, 509 (1930).

[13] T. W. RICHARDS, W. N. STULL, J. H. MATTHEWS, and C. L. SPEYERS, *Jour. Amer. Chem. Soc.*, **34**, 971 (1912).

[14] Fifth reference under 9.

[15] T. W. RICHARDS, *Jour. Amer. Chem. Soc.*, **46**, 1419 (1924).

[16] L. Pauling, *ZS. f. Krist.*, **67**, 377 (1928).

[17] S. Lussana, *Nouv. Cim.*, **19**, 182 (1910).

[18] R. F. Mehl and B. J. Mair, *Jour. Amer. Chem. Soc.*, **50**, 55 (1928).

[19] R. F. Mehl, *ibid.*, **50**, 73 (1928).

[20] R. F. Mehl, *ibid.*, **49**, 1892 (1927).

[21] L. H. Adams and E. D. Williamson, *Jour. Frank. Inst.*, **195**, 475 (1923).

CHAPTER VII

MELTING PHENOMENA UNDER PRESSURE

THE main points in the historical development have already been sufficiently indicated in the first chapter. It is enough to recall here that the outstanding experimental question was as to the character of the melting curve—whether it ends in a critical point or rises to a maximum or behaves otherwise, and that the chief experimental work in this field had been by Tammann, presented in his two books, which he regarded as confirming his theory that the melting curve rises to a maximum temperature.

My own first experiments in this field were published in 1912 (B. 5, 6) and included measurements of the freezing curves of mercury up to 12,000 and water up to 21,000 kg./cm.². This was followed in 1914 (B. 14) by a paper on the melting of eleven substances under pressure, and in 1915 (B. 19) by a second paper on the melting of eleven more substances. This completed the work specially devoted to this subject, but since then a number of other melting curves have been observed incidentally; a number were observed in the course of measurements of the effect of pressure on polymorphic transitions (B. 20, 24, 25), to be described presently, and other melting curves have been more or less partially determined in connection with measurements of the effect of pressure on electrical resistance or thermal conductivity or viscosity (B. 35, 46, 58). In a couple of cases, mercury and gallium, the melting pressure has been determined as a function of temperature by observing the point of discontinuity in electrical resistance, but in practically all other cases the method of volume discontinuity used by Mack and extensively exploited by Tammann was employed. A second method used considerably by Tammann, that is, of slowly varying temperature at constant volume, plotting pressure as a function of temperature, analogous to the ordinary method of the arrest point in determining melting or transition points at atmospheric pressure, proved not to be sufficiently trustworthy, because of the difficulty of properly correcting for the fact that the system is always to a certain extent out of equilibrium.

It was now possible for the first time, because of the complete freedom from leak obtainable with the new packing, to make the method of volume discontinuity give accurate results, not only for the temperature and pressure of melting, but also for the change of volume during melting. In fig. 47 is reproduced a set of experimental points giving the position of the piston at constant temperature as a function of pressure during the freezing of the high-pressure modification of ice. By making a series of readings like this at different temperatures,

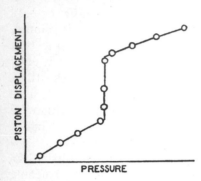

FIG. 47.—Experimentally determined points showing the sharp change of volume during freezing of a pure liquid.

it is possible to completely determine all the ordinary parameters which specify the thermodynamics of melting. This can be done through Clapeyron's equation, $\dfrac{d\tau}{dp}=\dfrac{\tau\Delta v}{L}$. p is known as a function of τ along the melting curve, so that $\dfrac{d\tau}{dp}$ is also known at every point, and this may be combined with the experimentally determined Δv to give L, the latent heat. Furthermore, the difference of slope of the curve of piston displacement above and below the discontinuity obviously gives the difference of compressibility of solid and liquid phases. By combining the difference of compressibility with other knowledge implied in a knowledge of the melting curve, by equations which will be written down explicitly later, it is also possible to calculate the difference of thermal expansion and specific heat between liquid and solid.

In many cases, unfortunately, the ideal state of affairs shown in fig. 47 does not hold, because when there are impurities present the melting temperature is depressed by an amount depending on the concentration of the impurity. If the solid phase separates pure, as it usually does, the concentration of the impurity left in the liquid increases as freezing progresses, so that instead of having a curve with sharp discontinuities, as shown in fig. 47, a rounded curve as in fig. 48 is obtained. From a curve like this it is obviously impossible to get the difference of compressibility between solid and liquid, and the difference of volume can only be obtained by an extrapolation, as indicated by the dotted line. If the impurity is only

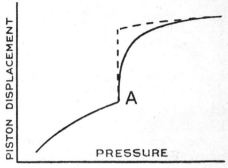

FIG. 48.—Shows the effect of impurity on the sharpness of the discontinuity at freezing.

slight, the error in Δv caused by such rounding is not important. In more than half my cases the amount of impurity was sufficient to cause perceptible rounding, and in these cases the difference of compressibility could only be obtained by an approximate method to be described later.

In connection with the question of impurity, an important point is whether the initial depression of the melting-point, determined by the point at which the pure solid just starts to separate from the liquid, is a strong function of pressure or not. If it does depend importantly on pressure, an error may be introduced into the general shape of the melting curve by impurities. The ordinary thermodynamic formula for the depression of the freezing-point with impurity, $\dfrac{d\tau}{dn} = -\dfrac{R\tau^2}{L}$, remains valid at high pressure. In general, L in this formula increases with temperature along the melting curve, but by an amount less than τ, so that in general the depression of the melting-point increases at high pressure. This means that the melting curve of an impure substance will rise less rapidly than the curve of a pure substance, but numerical consideration shows that the effect is not important, and is without doubt negligibly small for those liquids actually investigated.

In those cases in which the amount of impurity is so considerable that the initial part of the discontinuity shown as A in fig. 48 departs by a measurable amount from the vertical, it is theoretically possible to apply a correction for the depression of the melting-point, as was shown by Tammann. In practice, however, this does not work well, since the time required to reach complete equilibrium at every intermediate stage of the freezing is very long, because of the slowness of diffusion, which is accentuated by the large effect of pressure on viscosity.   It is therefore better to spend the time that might have been spent in getting adequate data in increasing the purity of the liquid.

The general set-up of the apparatus is as follows : The substance whose melting is to be examined is placed in the lower cylinder, suitably separated from the transmitting liquid, and this lower cylinder is kept at the desired temperature by a regulated bath.   Pressure is transmitted in the regular way to the lower cylinder through a pipe leading into the upper cylinder, which is at the temperature of the room.   The displacement of the piston into the upper cylinder is measured in a suitable way; in most of the experiments four micrometer readings were made of the position of the piston, so spaced as to avoid error from any warping of the frame of the press.   The manipulation of pressure was not always a straightforward matter because of the sub-cooling which most liquids pass through before freezing, and which in some cases may be very high.   It is usually necessary to freeze the liquid completely by running several thousand kilograms beyond the freezing pressure, and then to make the measurements of piston discontinuity with decreasing pressure, which is always possible, because the solid phase will practically never superheat with respect to the liquid.   If the melting-point desired was near the upper end of the pressure range, it was in many cases not possible to induce freezing by raising the pressure; in these cases freezing had to be induced by the more inconvenient method of lowering the temperature by a large amount below the expected freezing temperature, and then bringing it back after freezing was completed.

In converting the measured discontinuity into volume change a correction has to be applied for the distortion of the pressure cylinder, which was less than 1 per cent. for the pressure range of 12,000 kg./cm.², and also a second correction for the change of volume of the transmitting liquid on passing from the lower to the upper cylinder, arising from the temperature difference between the two cylinders. This second correction may rise to the order of 10 per cent. when the temperature difference is of the order of 200°, and has to be determined by independent experiment in which the p-v-t relations of the liquid are determined by the methods of Chapter V.

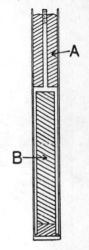

FIG. 49. — Container for substances whose melting curve under pressure is to be determined.

The method by which the substance under investigation is separated from the transmitting liquid is a matter of some importance. Since the transmitting liquid —kerosene or petroleum ether—is in nearly all cases miscible with the liquid under investigation, it is necessary to prevent all contact between the two substances. I believe that some of Tammann's results are affected by an error arising from the partial mixing of the two liquids. Mercury was used in my experiments as the separating medium; for those substances which are solid at room temperature the arrangement was as shown in fig. 49. The substance B is placed in a cylindrical shell inverted under mercury A in a deeper outer shell. The vertical stem, combined with a washer at the top not shown, prevents the inner shell from floating to the surface of the mercury and facilitates the removal of the shell. The use of steel containers proved to be almost essential, because the stress developed in a substance when crystallising under pressure is almost always so great as to shatter a glass container. The necessity of using steel containers and mercury results in some restriction in the kind of substance that can be investigated; for example, some of the stronger organic acids could not be investigated because of chemical action, and there are other substances which are too active chemically. If the substance to be

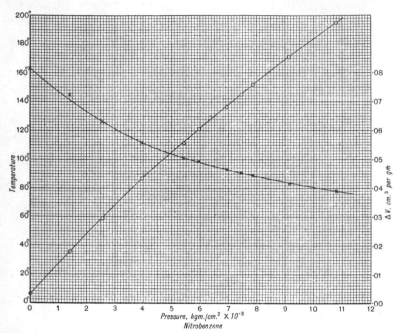

FIG. 50.—The freezing curve and $\triangle V$ curve of nitrobenzene. The observed freezing temperatures are shown by the circles, and the observed $\triangle V$'s by crosses.

investigated is liquid at room temperature, then it is convenient to modify the inner cup of fig. 49 by adding at the lower end a capillary stem to facilitate filling.   In measuring the melting curve of CO_2, a bulb with an inward opening valve was used, much like the containers described in Chapter V for the compressibility of gases.

The general experimental accuracy of the measurements is indicated in fig. 50, which gives the experimentally determined points for nitrobenzene, the p-t points as circles with the scale at the left, and the Δv points as crosses with the scale at the right.   The Δv given in this diagram is in cm.[3] per grm.   In fig. 51 the melting curves are drawn for a number of substances, and in fig. 52 the values of Δv.   Here, in order to afford a better comparison between different substances, the fractional Δv is shown, that is, Δv in cm.[3]

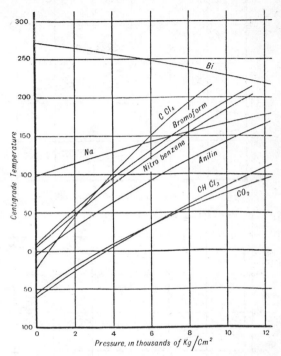

FIG. 51.—Melting curves of a number of typical substances.

per grm. multiplied by the density at 0° C. at atmospheric pressure.   In fig. 53 are shown the latent heats along the melting curve of a number of substances, calculated by Clapeyron's equation from the slope of the melting curve and Δv, in kg./m. units per grm. molecule of substance. The numerical values of melting temperature and Δv as a function of pressure are given in Table IX for all those substances for which I have obtained good data over a wide pressure range.

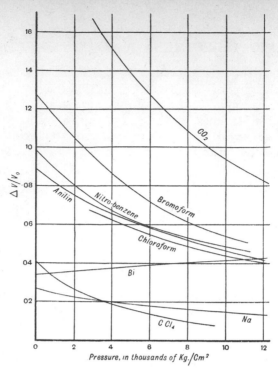

FIG. 52.—Curves of relative $\triangle V$ for a number of typical substances.

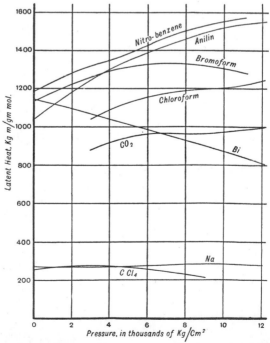

FIG. 53.—Latent heat along the melting curve of typical substances.

TABLE IX

SUMMARY OF MELTING DATA

Substance		Pressure Kg./cm.^{2}			
		1	4000	8000	12,000
42·71 Li	t	178·4°	191·8°	201·6°	°
22·32 Na	t	97·62	128·8	155·1	177·5
	Δv	·02787	·02072	·01711	·01398
16·186 P	t	44·2	99·3 (2000)	148·2 (4000)	191·9 (6000)
	Δv	·01927	·01667	·01436	·01218
16·155 K	t	62·5	115·8	152·5	179·6
	Δv	·02680	·01676	·01073	·00642
42·107 Ga	t	29·85	21·4	12·6	3·55
69·396 Rb	t	38·7	74·5 (2000)	95·9 (3500)	
	Δv	·0185	·0119	·0101	
69·403 Cs	t	29·7	70·2 (2000)	98·5 (4000)	
	Δv	·0136	·0088	·0072	
7·432 Hg	t	−38·8	−18·4	+1·8	+21·9
	Δv	·002534	·002512	·002445	·002290
22·28 Bi	t	271·0	256·0	238·6	218·3
	Δv	·00345	·00378	·00407	·00429
16·161 CO$_2$	t	−56·6	8·5	55·2	93·5
	Δv	··	·0979	·0697	·0531
22·7 SiCl$_4$	t	−10·0(2000)	42·6	139·4	205·4 (11,000)
	Δv	·0522	·0428	·0330	·0297
16·175 CCl$_4$	t	−22·6	75·8 (3000)	149·5 (6000)	211·9 (9000)
	Δv	·02580	·01401	·00862	·00538
22·5 CHBr$_3$	t	7·78	94·7	163·2	209·1 (11,000)
	Δv	·0391	·0266	·0188	·0157
16·163 CHCl$_3$	t	−61·0	3·4	58·6	107·9
	Δv	··	·0498	·0389	·0321
29·95 C$_2$H$_4$O$_2$ acetic acid I	t	16·68	54·3 (2000)		
	Δv	·1560	·0887		
II	t	54·1 (2000)	83·4	129·6	156·6 (11,000)
	Δv	·1003	·0828	·0598	·0520
29·101 C$_2$H$_5$NO acetamide I	t	81·5	119·0		
	Δv	·1098	·0429		
II	t	··	111·5	151·1	173·5 (11,000)
	Δv	··	·0746	·0492	·0383

TABLE IX—*continued*

Substance			Pressure Kg./cm.²			
			1	4000	8000	12,000
23·120 C₃H₇NO₂ urethane	I	t	47·90°	64·2 (2000)°	°	°
		Δv	·05990	·02774		
	II	t	67·3 (2500)	75·8		
		Δv	·03130	·01880		
	III	t	..	80·3 (4500)	119·0	156·7
		Δv	..	·06294	·05000	·03780
22·22 C₄H₆O₄ methyloxalate		t	54·24	132·6	196·8	
		Δv	·1453	·0798	·0606	
22·10 C₆H₅Br		t	−31·1	35·9	85·7	127·9
		Δv	..	·0345	·0248	·0205
22·9 C₆H₅Cl		t	−45·5	16·7	64·0	103·6
		Δv	..	·0469	·0349	·0285
16·167 C₆H₅NO₂ nitrobenzene		t	5·6	87·6	153·8	198·6 (11,000)
		Δv	·0814	·0555	·0442	·0386
22·17 C₆H₅NO₃ p-nitrophenol		t	42·4	159·8 (2000)	198·8 (4000)	
		Δv	·0891	·0614	·0445	
16·171 C₆H₆ benzene		t	5·4	96·6	162·2	204·2 (11,000)
		Δv	·1317	·0675	·0485	·0394
23·114 C₆H₆O phenol	I	t	40·87	63·3 (2000)		
		Δv	·0567	·0280		
	II	t	62·1 (2000)	99·8	158·8	209·2
		Δv	·0831	·0714	·0564	·0468
16·165 C₆H₇N amlin		t	−6·4	64·5	119·1	165·3
		Δv	·0854	·0631	·0502	·0405
16·181 C₇H₈O o-cresol	I	t	30·8	84·8	118·1	
		Δv	·0838	·0406	·0264	
	II	t	..	102·7 (6000)	129·3	175·9
		Δv	..	·0559	·0499	·0422
22·19 C₇H₉N p-toluidine		t	43·6	131·3	204·6	
		Δv	·1413	·0852	·0647	
16·169 C₁₂H₁₁N diphenylamine		t	54·0	144·9	212·9	
		Δv	·0958	·0586	·0448	
22·13 C₁₂H₁₀O		t	47·77	142·0	213·7	
		Δv	·0904	·0571	·0442	

In addition to these measurements over a wide pressure range, there are a large number of observations of Tammann up to 3000 kg., and a few by other observers to lower pressures, covering in all about forty substances. These will be found collected in *International Critical Tables*, vol. iv, p. 9, in which are given the two constants of the parabolic formula $t-t_0=ap+bp^2$, in terms of which Tammann was able to reproduce his results.

In almost all cases the melting curve rises with increasing pressure, corresponding, of course, to the fact that the liquid usually has a larger volume than the solid. Only three cases of falling melting curves have been investigated under pressure: bismuth, gallium, and water. The curvature of all these is in the same direction, concave downward, which means that as pressure increases the abnormal drop of melting temperature becomes accentuated. This suggests some sort of instability in the structure, and in fact in the case of water there is such an instability, and beyond 2200 kg./cm.² ordinary ice is replaced by another more stable form, which will be discussed in greater detail later. If, however, the downward curvature of the melting curves of Bi and Ga is similarly evidence of approaching instability, the appearance of the expected new modification is deferred to much higher pressure, since gallium shows nothing new up to 12,000 kg./cm.², and special experiments on bismuth have disclosed nothing new up to nearly 20,000 kg./cm.². The increase of the slope of the melting curve of these metals is not very large, however, the fractional increase in $\dfrac{d\tau}{dp}$ produced by 12,000 kg./cm.² being only 12 per cent. for gallium and 60 per cent. for bismuth. The increase for water at 2200 kg., the pressure at which the new modification appears, is nearly 100 per cent., so that it is evident that the whole pressure scale of the phenomena for water is very much smaller than for the metals, and it is not surprising if considerably higher pressures than those yet reached are necessary to produce analogous effects for the metals.

Except for bismuth, gallium, and water, the melting curve, then, rises with pressure. The amount by which it rises varies greatly with the material, being least, as would be expected, for metals and greatest for some of the complex organic compounds.   Among the substances examined by me, the smallest effect is shown by lithium, the melting temperature of which rises 3·7° for 1000 kg./cm.^{2};   from here up it increases to such typical values as 27·3° per 1000 kg./cm.^{2} for SiCl$_4$ and 62° for CBr$_4$.   The greatest effect ever recorded is that of Hulett [1] on camphor, the melting curve of which rises initially at the rate of 130° per 1000 kg./cm.^{2}.   In spite of this comparatively wide variation, it is a fair generalisation that the effect of pressure on melting is much less than on vaporisation, and that, roughly, thousands or tens of thousands of kilograms per sq. cm. are required to produce changes of melting temperature equal to changes of vaporisation temperature produced by hundreds of kilograms per sq. cm.

The curvature of the rising melting curves is, without exception, the same as that of the falling curves—that is, they are concave downward, or $\dfrac{d^2\tau}{dp^2} < 0$.   Furthermore, for all rising curves $\dfrac{d^2\tau}{dp^2}$ becomes numerically less at the higher pressures, or $\dfrac{d\tau}{dp}$ plotted against p is convex downward.   The falling curves for water and bismuth, on the other hand, are such that $\dfrac{d\tau}{dp}$ is concave downward, or $\dfrac{d^2\tau}{dp^2}$ is positive, thus suggesting the possibility of a vertical tangent at some finite pressure.   The experimental accuracy is not great enough to give the curvature of $\dfrac{d^2\tau}{dp^2}$ against pressure.   It is obvious that if $\dfrac{d^2\tau}{dp^2}$ is not constant, the relation between temperature and pressure cannot be represented with sufficient accuracy by a second-degree formula in the pressure.

The behaviour of Δv on all rising curves has important features in common.   Examination of the data of the figures shows that in nearly all cases Δv decreases numerically with increasing pressure, the rate of decrease itself always becoming less at the higher pressure, so that Δv is convex toward the pressure axis.   There are two exceptional cases.

The initial stages of the Δv curve for the high-pressure modification of ice has abnormal curvature, but this reverses and becomes normal as pressure rises. This is, doubtless, to be connected with the low-pressure abnormalities of liquid water, which are known to disappear at high pressure. The other exception is liquid mercury. The curvature of Δv against pressure is also abnormal for this, but the total change of Δv over the entire pressure range is only 10 per cent., and little weight is to be attached to this behaviour in view of the known abnormalities of mercury in other respects.

The curve plotting latent heat against pressure has no such uniformities as the curves for melting temperature or Δv, as is evident from fig. 53. In the large majority of cases, however, the latent heat rises with pressure by a moderate amount, and in those cases where it falls, as potassium, CCl_4, p-nitrophenol, and methyl oxalate, the fall is by only a small percentage amount. The falling latent heat curve for bismuth is perhaps anticipatory of a new modification. In spite of the irregularities, this generalisation may be made: except for two or three isolated points which may be explained by experimental error, the rise of latent heat with pressure along the melting curve is less rapid than the rise of temperature with pressure, so that $\dfrac{d}{d\tau}\left(\dfrac{\mathrm{L}}{\tau}\right)<0$.

The question of the ultimate behaviour of the melting curves as the pressure rises indefinitely is one on which there has been a large amount of discussion, and is one which I believe can now be definitely answered in view of the wide experimental range now open. At first it was the general expectation that the melting curve would end in a critical point like the vaporisation curve, which would mean that by a proper choice of path in the p-t plane it would be possible to pass around the end of the curve, and thus achieve a continuous transition from a liquid to a solid (or crystalline) phase. One of Planck's [2] earliest papers dealt with the possible co-ordinates of such a critical point between liquid and solid water, and there were also early papers by Poynting [3] and Lodge.[4] Van Laar [5] actually set up a possible equation of state containing two regions like the region of the ordinary van der Waal's equation, in which three real roots gradually coalesce, leaving a single real root and two imaginary ones in the region above the critical point. Van Weimarn [6] and Eucken [7] have advanced arguments for this possibility. The second important theory is that of Tammann; this theory has passed through various stages of development, in which different opinions have been held about the probable behaviour in the low-temperature region; the part which essentially concerns us has to do with the behaviour at higher temperatures, and has been held by Tammann consistently without change from the beginning until now. This theory is that the melting curve rises to a maximum temperature and then falls, so that if an ordinary material is heated above a definite temperature it will not be possible to make it crystallise by any pressure, no matter how high, whereas at a temperature lower than this it may be made to freeze by the application of sufficient pressure, and then by the application of a still higher pressure it may be made to melt again. There is also a theory of Schames,[8] which has played a less important rôle, namely, that the melting curve rises to an asymptotic temperature at infinite pressure.

As far as direct experiment goes, neither of these presumed eventualities has ever been observed. In some of my published phase diagrams, melting curves are indicated as coming to an end; this never means a critical point, but means only that it was impossible to follow the melting curve further for one reason or another: either the temperature or pressure get beyond the experimental range, or the substance decomposes at the high temperature, or the viscosity becomes so great that it is too difficult to follow the process further. It is necessary, therefore, in all cases to try to extrapolate the curves into the region as yet untouched, and estimate whether either a critical point or a maximum is indicated in this region by the phenomena in the obscrvable range. There are certain criteria which can be applied to answer this question.

Consider first the possibility that there is a critical point. It is to be said at once that the curvature of the melting curve is the same in all cases as that of the vapour-liquid curve, so that as far as this feature is concerned the situation is not unfavourable.   There are, however, other important characteristics at the critical point.   Here Δv and latent heat vanish together, while $\dfrac{d\tau}{dp}$ remains finite.   The necessity for this is evident at once on inspection of Clapeyron's equation. The question which we have to ask is therefore this: Is the shape of the Δv and the latent heat curves in the experimental range such as to indicate that they will cross the axis at the same pressure (or temperature) beyond the experimental range ?   I believe that mere inspection of figs. 52 and 53 is sufficient to answer this question in the negative; unless there is some reversal of the trend in the present experimental range at pressures beyond those now attainable, there is no reason to think that the melting curve can end in a critical point.   Such a critical point between liquid and solid would violate our present ideas of the nature of the crystal phase; it is not easy, although not impossible, to imagine that the regular arrangement of the atoms in the lattice of the crystal can change continuously to the haphazard arrangement of the liquid.   Eucken has recently seriously revived a conception of van Weimarn as to an ultimate critical point, which is that the crystals may pass continuously into the liquid by breaking up into a colloidal suspension of crystal grains as the critical point is approached, the size of the grains becoming smaller and ultimately vanishing at the point itself. Eucken has supported this theory by considering my data for CCl_4 and potassium, the two substances among all those which I have examined most favourable to his theory, and showing that the measurements in the actual pressure range are not inconsistent with a possible extrapolation such that both Δv and latent heat vanish together.   It seems to me, however, that this sort of graphical extrapolation may be very dangerous with curves that are capable of as much irregular variation as the latent heat curves, as shown in fig. 53 by the examples of nitro-benzene and CO_2.   Apart from extrapolation of latent heat, extrapolation of Δv is itself hazardous enough, in spite of the fact that Δv for both

CCl_4 and K has dropped at the end of the range to 0·24 of its initial value. For example, if a parabolic curve is passed through the Δv values of the initial, mid, and final points for both CCl_4 and K, it will be found that the constants are such that the curve never crosses the axis. CCl_4 and K are, furthermore, not typical substances; CCl_4 has two high-pressure modifications, one of which is likely to rise to the melting curve and supplant the ordinary modification at pressures sufficiently high, and K is abnormal with respect to compressibility and other features, as already explained in Chapter VI. There is another argument. Any ordinary substance is more stable in large than in small aggregates if both are present together; the large grows at the expense of the small, as shown, for example, by the disappearance of small drops of condensed liquids and their collection into large drops. One would therefore not expect the colloid phase to be more stable than the ordinary solid phase, but the reverse. All things considered, it seems to me that the probability that there is a critical point between liquid and solid is so remote that it can be dismissed without further discussion.

Consider next the evidence with respect to Tammann's theory that there is a maximum melting temperature. If there is such a maximum point, $\dfrac{d\tau}{dp}$ must vanish, Δv must vanish at the same point, and L must remain finite. What now is the evidence that $\dfrac{d\tau}{dp}$ will ever vanish? To answer this it is more advantageous to consider $\dfrac{d\tau}{dp}$ as a function of temperature along the melting curve than as a function of pressure. In fig. 54, $\dfrac{d\tau}{dp}$ is plotted as a function of temperature for a number of substances. These curves are all convex toward the temperature axis, and evidently need not cross the axis at any finite temperature. Furthermore, if there is a maximum temperature, $\dfrac{d\tau}{dp}$ must approach this temperature with a vertical tangent; this demands concavity toward the temperature axis instead of convexity. The evidence, therefore, from the shape of the melting curve itself, is against the possibility of a maximum; this evidence holds without exception for all curves determined, and seems to correspond to some fundamental property of melting curves. Next

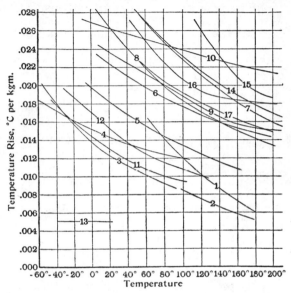

FIG. 54.—The slope of the melting curve of a number of substances as a function of temperature. The figures on the curves refer to the substances as follows : (1) potassium, (2) sodium, (3) carbon dioxide, (4) chloroform, (5) anilin, (6) nitrobenzol, (7) diphenylamine, (8) benzol, (9) bromoform, (10) silicon tetrachloride, (11) monochlorbenzol, (12) monobrombenzol, (13) mercury, (14) benzophenone, (15) paranitrophenol, (16) paratoluidin, (17) methyl oxalate.

consider the behaviour of Δv. Plotted against pressure, Δv is convex toward the pressure axis for all substances investigated, with the two exceptions previously noted, and an inspection of fig. 52 shows that the shape of these curves is not such as to lead to the expectation that they will cross the axis at any finite pressure. But a better method of extrapolating is to plot Δv against temperature as was done for $\dfrac{d\tau}{dp}$. Here again, if Δv vanishes at a finite temperature it must cross the axis with a vertical tangent, and therefore be concave towards the axis, whereas actually the curvature is the reverse of this.   This holds true even in the case of K; the curve for its Δv is nearly linear, but still is curved in the universal direction.   The behaviour of K is by far the most difficult to extrapolate satisfactorily of all substances examined, but even here the most casual inspection shows that, granted all the will in the world to make both $\dfrac{d\tau}{dp}$ and Δv vanish at some finite temperature, they cannot possibly be made to vanish at the *same* temperature, which would be necessary for a maximum.

The evidence, therefore, seems to be unequivocally against the possibility of a maximum temperature.   It is more difficult to rule out the possibility of an asymptotic tempera- ture, as supposed by Schames.   This would demand that the curves for $\dfrac{d\tau}{dp}$ and Δv against temperature cross the axis at the same finite temperature.   It is not impossible that this should happen, but there is no indication at present that there is much probability of it, and I believe that unless there should be some good theoretical reasons for expecting such a course, we have to take the present experimental evidence as making it exceedingly probable that the melting curves of all substances rise indefinitely, at a continuously decreasing rate, until pressures are reached so high as to introduce entirely new kinds of phenomena, such as fore- shadowed by the behaviour of potassium with its suggestion of atomic breakdown.

The improbability of a maximum in any pressure range in which the atoms are not affected is much heightened by recent work of Simon and co-workers [9] on the melting curve of helium.   Using the method of Keesom [10] of the plugging of a pipe connecting two pressure gauges he has followed the melting curve of He to nearly 6000 kg./cm.[2].   An em- pirical equation has been set up which fits the facts well over this range, and which is of such a character as to indi- cate an indefinite rise of temperature with pressure.   As Simon emphasises, the pressure effects in He take place on such a small scale that 6000 kg./cm.^{2} on it is equivalent to a very much higher pressure on ordinary substances, assuming that they all have the same type of behaviour.

Special Melting Phenomena.   There are a number of miscellaneous phenomena connected with melting that demand special discussion.   We have seen that there is similarity in only a few respects between the melting and vaporisation curves, so that, physically, melting and vaporisation appear as entirely different processes.   This is emphasised by the fact that apparently there is no connection between the critical point liquid-vapour and the melting curves.   In the case of several substances the melting curve has been followed to a temperature higher than the critical temperature between liquid and vapour.   This was first done by Tammann, who selected phosphonium chloride, a substance whose critical temperature and ordinary melting temperature lie unusually close together.   By the application of 2050 kg./cm.^{2}, Tammann raised the melting temperature to 102° C., whereas the critical point is at 75 kg. and 50° C. The next example is that of CO_2, the melting curve of which I have followed to 93° C. and 12,000 kg./cm.^{2}, whereas the critical point is at 30° C. and 75 kg./cm.^{2}.   A number of other such substances have recently been added by Simon,[11] who has followed the melting curves of several of the so-called permanent gases to pressures of several thousand kilograms, which is sufficient to raise the melting temperature to well above the critical temperature.   Of course it is just here that we would expect to find the effect most easily. It is highly probable, therefore, that any gas may be condensed directly into a crystalline phase by the application of sufficient pressure at any temperature above the critical temperature.

There is another sort of effect that emphasises the difference between melting and vaporisation.   The vaporisation of all substances is governed by so nearly the same type of equation that there is a law of corresponding states.   It is easy to show that there can be no such law of corresponding states for melting.   If there were such a law, then the ratio of the temperature on the melting curve at which Δv has dropped to, say, one-half its initial value to the temperature of melting at atmospheric pressure, would be the same for all substances.

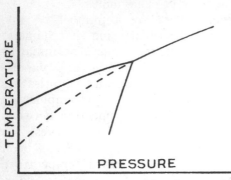

FIG. 55.—Melting and transition diagram of a substance with a phase unstable at atmospheric pressure, which becomes stable at high pressure.

That it is not true is shown, for example, by the fact that the ratio is 1·17 for o-kersol, and 1·64 for nitrobenzene. When it is considered that the melting phenomena are partially determined by the properties of the solid phase, and that these in turn are determined by the lattice in which the solid crystallises, such a law of corresponding states is not to be expected.

There are a number of cases in which the melting curve runs into a triple point, above which the ordinary low-temperature modification is replaced by a second high-pressure modification. This suggests the question as to whether the unstable forms which a number of substances are known to possess at atmospheric pressure may not become stable at higher pressures, so that the known lower melting-point of the unstable forms under ordinary conditions

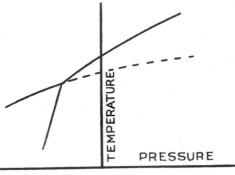

FIG. 56.—Melting and transition diagram of a substance with a phase unstable at atmospheric pressure, which becomes still more unstable at high pressures.

can be described as arising from a prolongation backward into the unstable region of a melting curve stable at higher pressures, as indicated in fig. 55.   The effect of pressure on the melting of some of the more common unstable forms has been investigated by Hulett,[12] Körber,[13] and Wahl,[14] and they have found in all cases examined the exact opposite, namely, that the melting curves draw apart with rising pressure, instead of approaching, so that it would appear that for most substances the triple point between liquid and two solid forms is to be found rather at negative pressures, as indicated in fig. 56.

This naturally raises the question of the possibility of prolonging a melting curve beyond a triple point into the region in which the other solid phase is the more stable form.   It seems that there is no general rule here; there are a number of cases in which such an extension is possible, but there are also cases in which it is not possible, at least under the actual experimental conditions.   Thus in the case of water, it is possible to prolong the melting curve of ice I into the region of stability of ice III, and also at the same triple point to prolong the melting curve of ice III into the region of stability of ice I; similarly at the triple point between liquid, ice III, and ice V the melting curve of either ice III or ice V may be carried into the region of stability of the other.   But at the triple point between ice V, ice VI, and the liquid, although it is possible to prolong the melting curve of ice VI past the triple point to lower pressures and temperatures into the region of stability of ice V, I was not able in many attempts to prolong the melting curve of ice V to higher temperatures and pressures into the region of stability of ice VI.   o-cresol affords another sort of example.   At the triple point liquid-I-II, it is easy to carry the melting curve of I to higher temperatures and pressures into the region of stability of II, but not possible to carry the melting curve of II backward to lower temperatures and pressures into the region of stability of I. It would appear that the example of o-cresol is like that of the majority of substances, except that for most substances the triple point between stable and unstable modifications is displaced to negative pressures.

There is an obvious criterion as to whether the unstable form may become stable at higher pressures; if its volume is less than that of the stable form, then we may expect it to become stable at high pressures, but otherwise the application of pressure will only increase the relative instability of the unstable form.

It may be worth while to mention an unpublished experiment by which I attempted to get some further light on the mechanism of superheating and subcooling. It is well known that although the ordinary liquid may be subcooled by large amounts, the ordinary solid cannot be superheated at all. There is one possible exception to this in the behaviour of some of the very viscous silicates examined at the Geophysical Laboratory, but these are recognised to be highly unusual. The same sort of phenomenon is also always found under pressure; the solid can never be superheated, whereas sometimes the liquid may withstand so much subcooling as to make the determination of the melting phenomena difficult. The interesting question is: what is the mechanism which makes it impossible to superheat the solid? In attempting to answer this question experimentally under ordinary conditions, the difficulty is encountered that apparently the solid always melts first at the exterior boundary, which makes the interpretation of the results difficult because of surface complications and the thermal effects of surface melting. Under ordinary conditions there seems to be no way of avoiding this situation, because of necessity when a solid is warmed the exterior surface will get warmed before the interior. By the use of pressure, however, this effect may be avoided; if pressure is suddenly released on a normal solid on the high-pressure side of the melting curve by an amount sufficient to carry it into the region of stability of the liquid, it is evident that the solid is carried at once as a whole into the region of stability of the liquid with the velocity of propagation of an elastic disturbance, and the surface effect is avoided. Conversely, if the solid is ice, by raising the pressure it may be carried as a whole and at once into the region of stability of the liquid. The experiment consisted in watching through glass windows the effect of increasing the pressure on a block of ice. No trace of interior melting was observed under these conditions; it is therefore highly probable that the impossibility of superheating an ordinary solid is some sort of surface phenomenon.

The melting phenomena of the alkali metals are important enough to merit special mention.   In fig. 57 the melting curves of the five alkali metals are shown as far as they have been followed experimentally. The most striking feature is the crossing of the curves of Na and K, with the result that at pressures above 9000 kg./cm.^{2} the melting temperature of K is higher than that of Na.   Further, it appears as if there will be similar crossing by the other curves so that it is not improbable that at pressures above perhaps 30,000 kg./cm.^{2} the order of melting may be completely reversed, Cs having the highest melting point and Li the lowest.

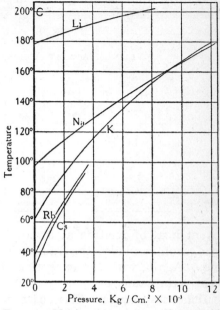

FIG. 57.—Melting temperatures of the alkali metals as a function of pressure.

With the exception of Li, the melting of the alkali metals in the lower range of pressure and temperature is governed roughly by a law of corresponding states.   This is shown in Table X,

TABLE X

Metal	Ratio of temperatures (see text)
Na	1·123
K	1·138
Rb	1·103
Cs	1·124

which gives the ratio of the absolute temperature at which Δv has dropped to two-thirds its initial value to the melting temperature at atmospheric pressure.   This ratio would be constant if there were an exact law.   This similarity is further shown by other melting data in Table XI, which gives the

TABLE XI

MELTING DATA AT ATMOSPHERIC PRESSURE

Metal	Latent heat kg.m./gm. atom	Fractional change of volume
Li	49	·0060
Na	271	·0271
K	215	·0231
Rb	236	·0284
Cs	219	·0256

latent heat of melting at atmospheric pressure in kg.m. per grm. atom, and also the fractional change of volume on melting.   Again, with the exception of Li, the quantities are all nearly the same.

Difference between Thermodynamic Properties of Solid and Liquid.   The data discussed above for the melting curves may be combined with the difference of compressibility between solid and liquid, determined from the difference of slopes of the curve of piston displacement against pressure before and after melting, as already explained, to give the difference of thermal expansion between solid and liquid, and also the difference of specific heats at all points of the melting curve. This may be done by the following equations.   Abbreviate the difference of compressibility by Δa, difference of thermal expansion by $\Delta \beta$ and difference of specific heat by ΔC_p, where specifically:

$$\Delta a = -\left\{ \left(\frac{\partial v_l}{\partial p} \right)_\tau - \left(\frac{\partial v_s}{\partial p} \right)_\tau \right\}$$

$$\Delta \beta = \left(\frac{\partial v_l}{\partial \tau} \right)_p - \left(\frac{\partial v_s}{\partial \tau} \right)_p$$

$$\Delta C_p = C_{pl} - C_{ps} \,,$$

the subscript l denoting the liquid and s the solid.   Then we have the relations:

$$\frac{d\Delta v}{dp}=\frac{d\tau}{dp}\Delta\beta-\Delta\alpha \quad . \qquad . \qquad . \qquad . \qquad (1)$$

and

$$\frac{d\mathrm{L}}{dp}=\frac{d\tau}{dp}\Delta C_p+\Delta v-\tau\Delta\beta \qquad . \qquad . \qquad (2)$$

These equations may be simply deduced, and need not be derived here.   The first of these gives a means of calculating $\Delta\beta$ in terms of known quantities, and then the second gives ΔC_p.

Unfortunately, however, in many cases it is not possible to obtain $\Delta\alpha$ by direct experiment, because of the rounding of the corners of the discontinuity by the effect of impurity, as already discussed.   In such cases it is possible to get approximate information by the use of certain inequalities which experiment shows to hold universally.   In fig. 58, $\Delta\alpha$, $\frac{d\tau}{dp}\Delta\beta$, and $\frac{1}{\tau}\left(\frac{d\tau}{dp}\right)^2\Delta C_p$ are indicated as standing in certain relative positions to each other. It is easy to prove that the relative positions shown in the diagram always hold—that is, $\Delta\alpha$ always stands above $\frac{d\tau}{dp}\Delta\beta$, which in turn always stands above $\frac{1}{\tau}\left(\frac{d\tau}{dp}\right)^2\Delta C_p$.   Furthermore, the distance of separation between $\Delta\alpha$

$\Delta\alpha$

$-\dfrac{d\Delta v}{dp}$

$\dfrac{dT}{dp}\Delta\beta$

$\dfrac{1}{T}\dfrac{dT}{dp}\left[\Delta v-\dfrac{dL}{dp}\right]$

$\dfrac{1}{T}\left(\dfrac{dT}{dp}\right)^2\Delta C_p$

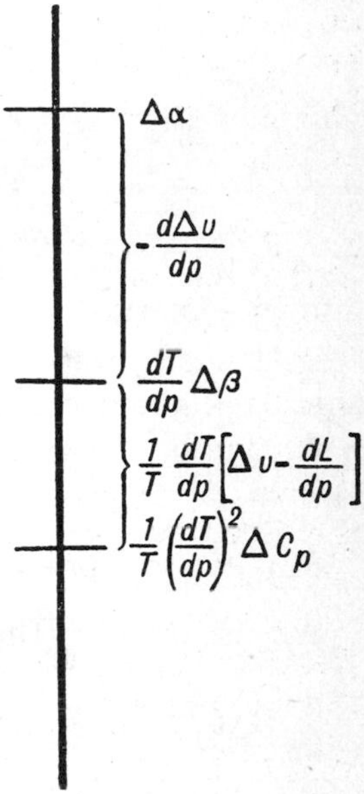

FIG. 58.—Illustrates the relations between the difference of compressibility, thermal expansion, and specific heat of solid and liquid on the melting curve.

and $\frac{d\tau}{dp}\Delta\beta$ is greater than the distance of separation between the latter and $\frac{1}{\tau}\left(\frac{d\tau}{dp}\right)^2\Delta C_p$.   The proof of this is as follows: By equation I the distance between Δa and $\frac{d\tau}{dp}\Delta\beta$

is $\Delta a - \frac{d\tau}{dp}\Delta\beta = -\frac{d\Delta v}{dp}$. But by experiment $\frac{d\Delta v}{dp}$ is always negative, so that this difference is always positive, as drawn. Furthermore, equation 2 above can be written, multiplying by $\frac{\text{I}}{\tau}\frac{d\tau}{dp}$:

$$\frac{d\tau}{dp}\Delta\beta - \frac{\text{I}}{\tau}\left(\frac{d\tau}{dp}\right)^2\Delta C_p = -\frac{\text{I}}{\tau}\frac{d\tau}{dp}\left[\frac{dL}{dp} - \Delta v\right]$$
$$= -\left(\frac{d\tau}{dp}\right)^2\left[\frac{\text{I}}{\tau}\frac{dp}{d\tau}\left\{\frac{dL}{dp} - \Delta v\right\}\right].$$

But now there is the well-known thermodynamic relation

$$\frac{d}{d\tau}\left(\frac{L}{\tau}\right) = \frac{\text{I}}{\tau}\frac{dp}{d\tau}\left[\frac{dL}{dp} - \Delta v\right],$$

and by experiment $\frac{d}{d\tau}\left(\frac{L}{\tau}\right)$ is always negative. Substituting this back at once establishes the relative magnitude of $\frac{d\tau}{dp}\Delta\beta$ and $\frac{\text{I}}{\tau}\left(\frac{d\tau}{dp}\right)^2\Delta C_p$.

Finally the proof that the difference between Δa and $\frac{d\tau}{dp}\Delta\beta$ is greater than the difference between $\frac{d\tau}{dp}\Delta\beta$ and $\frac{\text{I}}{\tau}\left(\frac{d\tau}{dp}\right)^2\Delta C_p$ is given by the equation

$$\left(\Delta a - \frac{d\tau}{dp}\Delta\beta\right) - \left(\frac{d\tau}{dp}\Delta\beta - \frac{\text{I}}{\tau}\left(\frac{d\tau}{dp}\right)^2\Delta C_p\right) = -\frac{\text{I}}{\Delta v}\frac{dp}{d\tau}\frac{d^2\tau}{dp^2} \quad (3)$$

This equation may be obtained by differentiating Clapeyron's equation for $\frac{d\tau}{dp}$ with respect to pressure, and using the value of $\frac{dL}{dp}$ in equation 2. But now the right-hand side of equation 3 is positive, Δv and $\frac{dp}{d\tau}$ being positive and $\frac{d^2\tau}{dp^2}$ always negative, so that the left-hand side is always positive, which was to be proved.

It now follows, since for the class of substances which we are considering $\dfrac{d\tau}{dp}$ is positive, and also $\dfrac{1}{\tau}\left(\dfrac{d\tau}{dp}\right)^2$ is intrinsically positive, that whenever ΔC_p is positive, $\Delta\beta$ and $\Delta\alpha$ must also be positive, and that if $\Delta\beta$ is positive, $\Delta\alpha$ must also be positive, but possibly ΔC_p may be negative. In other words, if we have a new substance of which we know nothing except that we are to assume that the universal relations found above on all melting curves hold for it also, then the chances are greatest that $\Delta\alpha$ is positive, intermediate that $\Delta\beta$ is positive, and least that ΔC_p is positive. It is therefore a more fundamental and universal fact that the compressibility of the liquid is greater than that of the solid than that the specific heat of the liquid is greater than that of the solid, although we might have been plausibly predisposed to take the other view. It is highly probable that in most cases ΔC_p is positive; the argument above shows that $\Delta\alpha$ must be greater than $\dfrac{1}{\tau}\left(\dfrac{d\tau}{dp}\right)^2\Delta C_p$.

We can use these inequalities to get approximate numerical information. Consider a substance for which $\Delta\alpha$ cannot be found experimentally because of impurities, but for which the co-ordinates of the melting curve can be satisfactorily found. Then in equations (1) and (2) we know only two of three unknown quantities, but the differences between $\Delta\alpha$, $\dfrac{d\tau}{dp}\Delta\beta$, and $\dfrac{1}{\tau}\left(\dfrac{d\tau}{dp}\right)^2\Delta C_p$ are known in terms of melting-curve data. If now we assume that ΔC_p is zero and calculate $\Delta\beta$ and $\Delta\alpha$ in terms of the known differences, we shall obtain lower limits. Furthermore, it is probable that the percentage error in the lower limit for $\Delta\alpha$ is less than 30 per cent., as may be shown by a rough calculation, using average values for $\dfrac{d\tau}{dp}$ and probable values for ΔC_p, as suggested by known specific heats of solid and liquid. The percentage error in the lower limit found for $\Delta\beta$ in this way is evidently greater than the percentage error in $\Delta\alpha$.

The validity of the approximate values obtained by neglecting ΔC_p may be tested in those cases in which Δa can be measured experimentally. Richards [15] has determined by direct experiment that Δa for solid and liquid benzene is 4.7×10^{-5}, whereas the value computed, neglecting ΔC_p, is 5.0×10^{-5}. For o-cresol, Richards [15] found $\Delta a = 1.9 \times 10^{-5}$, which agrees exactly with the calculated value.

In some cases there is other evidence giving the third relation necessary to complete equations 1 and 2. Thus, for a number of substances, direct measurements have been made of $\Delta \beta$ at atmospheric pressure, and for others there are direct measurements of ΔC_p. It is to be emphasised, however, that a direct measurement of Δa or $\Delta \beta$ or ΔC_p at atmospheric pressure is subject to the same error from premature melting from impurities as has been discussed for high pressures. ΔC_p is particularly sensitive; a very small amount of impurity is sufficient to change the apparent sign of ΔC_p. I (B. 19) have discussed the matter in detail in connection with measurements of Griffiths [16] of ΔC_p for Na; so far as I know there is no substance for which it is probable that ΔC_p is negative, when due account is taken of the effect of small impurities.

In Table XII are collected some values for Δa and $\Delta \beta$ at various pressures on the melting curves of a number of substances. These values have been obtained in various ways; by assuming ΔC_p is zero and using the melting data when there is no other information, and in other cases by combining with known values of Δa or $\Delta \beta$ or ΔC_p, the latter always at atmospheric pressure. A detailed discussion of the way in which many of these values were obtained will be found in *Physical Review*, vi, 100 (1915). The values for Δa in the table are doubtless much better than those for $\Delta \beta$, which are to be taken merely as suggestive. The very large decrease of Δa along the melting curve is striking, this decrease being much greater than the decrease of Δv. That is, solid and liquid approach equality of compressibility more rapidly than equality of volume. This is in line with ideas already suggested by other phenomena, namely, that at high pres-

TABLE XII

Substance	Difference of compressibility (= Δa)					Difference of expansion (= $\Delta\beta$)	
	1	3000	6000	9000	12000	1	12000
Potassium	$0\cdot0_547$	$0\cdot0_534$	$0\cdot0_527$	$0\cdot0_521$	$0\cdot0_516$	$0\cdot0_475$	?
Sodium	$0\cdot0_533$	$0\cdot0_518$	$0\cdot0_511$	$0\cdot0_68$	$0\cdot0_68$	$0\cdot0_489$	$0\cdot0_54$
Carbon dioxide	..	$0\cdot0_410$	$0\cdot0_59$	$0\cdot0_56$	$0\cdot0_535$	..	$0\cdot0_46$
Chloroform	..	$0\cdot0_537$	$0\cdot0_539$	$0\cdot0_530$	$0\cdot0_516$	..	$0\cdot0_44$
Anilin	$0\cdot0_582$	$0\cdot0_548$	$0\cdot0_536$	$0\cdot0_529$	$0\cdot0_530$	$0\cdot0_43$	$0\cdot0_48$
Nitrobenzol	$0\cdot0_411$	$0\cdot0_57$	$0\cdot0_532$	$0\cdot0_527$	$0\cdot0_524$[1]	$0\cdot0_311$	$0\cdot0_46$[1]
Diphenylamine	$0\cdot0_424$	$0\cdot0_57$	$0\cdot0_54$	$0\cdot0_53$[2]	..	$0\cdot0_328$	$0\cdot0_439$[2]
Benzol	$0\cdot0_450$	$0\cdot0_413$	$0\cdot0_56$	$0\cdot0_544$	$0\cdot0_538$[1]	$0\cdot0_348$	$0\cdot0_48$[1]
Carbon tetrachloride {L–I	$0\cdot0_412$	$0\cdot0_533$	$0\cdot0_533$	$0\cdot0_514$	..	$0\cdot0_312$	..
{L–II	$0\cdot0_419$	$0\cdot0_58$	$0\cdot0_536$	$0\cdot0_521$	..	$0\cdot0_441$	..
o-cresol	..	..	$0\cdot0_530$	$0\cdot0_527$	$0\cdot0_514$	..	$0\cdot0_54$
Phosphorus	$0\cdot0_443$	$0\cdot0_528$	$0\cdot0_514$	..	..	$0\cdot0_310$	..
Bromoform	$0\cdot0_549$	$0\cdot0_538$	$0\cdot0_528$	$0\cdot0_519$	$0\cdot0_516$[1]	$0\cdot0_46$	$0\cdot0_454$[1]
Silicon tetrachloride	..	$0\cdot0_587$	$0\cdot0_537$	$0\cdot0_517$	$0\cdot0_512$[1]	..	$0\cdot0_421$[1]
Chlorobenzol	$0\cdot0_411$	$0\cdot0_558$	$0\cdot0_536$	$0\cdot0_526$	$0\cdot0_515$	$0\cdot0_413$	$0\cdot0_44$
Bromobenzol	$0\cdot0_413$	$0\cdot0_549$	$0\cdot0_534$	$0\cdot0_517$	$0\cdot0_510$	$0\cdot0_321$	$0\cdot0_44$
Benzophenone	$0\cdot0_429$	$0\cdot0_413$	$0\cdot0_57$	$0\cdot0_53$[2]	..	$0\cdot0_347$	..
p-nitrophenol	$0\cdot0_420$	$0\cdot0_416$[4]	$0\cdot0_412$[5]	..	..	$0\cdot0_311$	..
p-toluidine	$0\cdot0_431$	$0\cdot0_413$	$0\cdot0_59$	$0\cdot0_57$[2]	..	$0\cdot0_344$	$0\cdot0_318$[2]
Methyl oxalate	$0\cdot0_467$	$0\cdot0_599$	$0\cdot0_56$	$0\cdot0_57$	..	$0\cdot0_341$	$0\cdot0_319$[3]
Bismuth	$0\cdot0_757$	..	..	..	..	$0\cdot0_48$	..

[1] 11000 kg./cm.².   [2] 8000 kg./cm.².   [3] 9000 kg./cm.².   [4] 2000 kg./cm.².   [5] 4000 kg./cm.².

sures the most important part of compressibility arises from the compression of the atoms or molecules, which involves, of course, that liquid and solid approach the same compressibility. The difference of volume, however, is partly due to structural differences, which persist, and therefore are more nearly constant as long as the solid retains the same lattice arrangement. The values of $\Delta\beta$ in the table roughly bear out the results already found for liquids, namely, that thermal expansion tends to decrease less at high pressure than does compressibility.

Melting in Two-component Systems. There is no reason, except of convenience, for giving a special treatment to melting phenomena in systems of two components apart from a general treatment of p-v-t relations in two-component systems—that is, in binary mixtures. However, as a matter of experiment, practically no complete investigation has been made of the effect of pressure on binary mixtures, but limited aspects only have been treated. If, for example, an electrolyte is added to water, temperature lowered until ice separates, and then the effect studied which pressure exerts on this temperature, the experiment may be conveniently described as one on the effect of pressure on melting temperature in a two-component system. But if the solution is made so concentrated that the solid salt separates, and the effect of pressure measured on the conditions under which this occurs, then the experiments may be conveniently described as on the effect of pressure on solubility. Merely for convenience, the latter phenomena and certain related phenomena in which two liquid phases separate will be described in the last chapter on miscellanies, and the former here. This scheme of classification evidently fails when both components separate together—that is, at the eutectic point. Somewhat arbitrarily, the eutectic phenomena will be described here. As a matter of fact, the work done in the entire subject of binary mixture is not at all extensive, either in range of materials or of pressure, and the method of classification is not of great importance.

The effect of pressure on the freezing-point of dilute solutions of sugar and NaCl has been measured by Lampa [17] up to 200 kg./cm.^{2}. In this range the relation between temperature and freezing pressure was linear. The temperature-pressure slope of the freezing curve was found to be greater than that of pure water, and also greater than would be calculated from the ordinary thermodynamic formula for the depression of the freezing-point. Such a failure might be accounted for by an effect of pressure on the association of the dissolved substance.

There seems to be no other work on the effect of pressure on the effect of simple substances in solution. There are, however, other more complicated systems which have been studied under pressure, and in which the phenomena may be described as melting. For example, if solid $SrCl_26H_2O$ is heated at atmospheric pressure to 62·6°, it decomposes into solid $SrCl_22H_2O$ and its saturated solution. This may be described as a melting of the hexahydrate. The temperature of decomposition or 'melting' is a function of pressure, Clapeyron's equation obviously applies to this phenomenon as well as to melting in a one-component system. The effect of pressure on this 'melting' has been studied by Tammann in three different systems. The decomposition temperature of $SrCl_26H_2O$ was found by him to be a linear function of pressure up to 2800 kg./cm.^{2}, temperature rising in this range from 62·1° to 75°. $Na_2CrO_410H_2O$ decomposes at atmospheric pressure at 19·6°, with decrease of volume, to Na_2CrO_4 and its saturated solution. Because of the sign of the volume change, we would expect that the temperature of decomposition would fall with increasing pressure. Tammann has in fact followed it to 0° and 3050 kg./cm.^{2}. Finally, $Na_2SO_410H_2O$, which is isomorphous with $Na_2CrO_410H_2O$, decomposes in the same way to Na_2SO_4, and its saturated solution at 32·6° at atmospheric pressure, but with a very small increase of volume, so that we would expect, since the latent heat is not especially small, a very small increase of the temperature of decomposition with increasing pressure. Tammann, in fact, found a very small initial rate of increase, but with a maximum temperature only 0·1° higher at 460 kg./cm.^{2}. Tammann has laid great emphasis on this, which he regards as confirming his views regarding a maximum

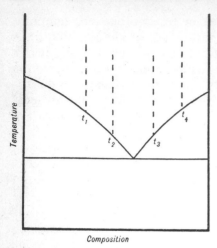

FIG. 59.—Suggests the method of finding the eutectic point under pressure from several cooling curves.

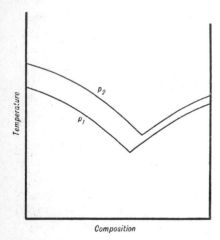

FIG. 60.—Suggests the way in which the eutectic point is displaced by pressure when the effect of pressure on the melting-points of the two pure components is dissimilar.

melting temperature for ordinary pure substances. It should be evident, however, that there is no parallelism between these two phenomena, for here the liquid phase is one of variable composition because of the effect of pressure on solubility. In fact, Pushin [18] has taken the trouble in a recent paper to go into the matter in complete detail, and show that the maximum is to be entirely accounted for in terms of the effect of pressure on solubility, and that no inferences can be made to pure one-component systems.

The effect of pressure on the eutectic point was first investigated by Roloff,[19] who found that the eutectic temperature of naphthalene and diphenylamine is raised 0·67° by a pressure of 23 kg./cm.². The subject has lately been investigated by Pushin and Grebenshikov [20] up to pressures of 4000 kg./cm.² on the systems: urethane - diphenylamine, urethane - nitroanisole, urethane - benzene, and Na—Hg. The method consisted in finding with thermocouples the point of temperature arrest when liquid solutions of different compositions were cooled at various constant pressures.   Thus suppose the ordinary binary mixture diagram at some definite constant pressure is like that shown in fig. 59.   Then by finding the temperature arrest points on the cooling curves of mixtures of various compositions, the points t_1, t_2, etc., may be located, and the eutectic temperature located by the intersection of the curves joining these points.   In this way the effect of pressure can be found both on the eutectic temperature and the composition at the eutectic point.   The facts were found to be exactly as might be expected from the geometrical considerations suggested in fig. 60.   In general, the effect of pressure will be different on the melting temperatures of the two pure components.   The diagram shows that we would expect in general the composition at the eutectic point to be changed in such a direction that the eutectic mixture becomes richer in that component whose melting temperature is increased by pressure by the smaller amount.   Large changes of eutectic composition may be brought about in this way; in the system urethane-benzene the eutectic composition at atmospheric pressure is 4·4 per cent. urethane, but at 4050 kg. it has risen to 80 per cent. urethane.   In the system Na—Hg the effect of pressure on eutectic composition was not great enough to detect.

[1] G. A. HULETT, ZS. f. Phys. Chem., **28**, 629 (1899).

[2] M. PLANCK, Wied. Ann., **15**, 446 (1882).

[3] J. H. POYNTING, Phil. Mag., **12**, 32 and 232 (1881).

[4] O. LODGE, Nat., **20**, 264 (1881).

[5] J. VAN LAAR, Proc. Amst., **11**, 765 (1909) ; **12**, 130, 133 (1909); **13**, 454, 636 (1910).

[6] P. VAN WEIMARN, Jour. Russ. Phys. Chem. Ges., **42**, 56 (1910).

[7] A. EUCKEN, Müller-Pouillets Lehrbuch der Physik, **3**, 1, 492, Braunschweig, Vieweg (1926).

[8] L. SCHAMES, Ann. Phys., **38**, 830 (1912) ; **39**, 887 (1912).

[9] F. SIMON, M. RUHEMANN, und W. A. M. EDWARDS, ZS. f. Phys. Chem., **2**, 340 (1929) ; **6**, 62 (1929).

[10] W. H. KEESOM, Leiden Communications, No. 184 (1926).

[11] F. SIMON, M. RUHEMANN, und W. A. M. EDWARDS, ZS. f. Phys. Chem., **6**, 331 (1930).

[12] S. A. HULETT, ibid., **28**, 629 (1899).

[13] F. KÖRBER, ibid., **82**, 45 (1913).

[14] W. WAHL, Phil. Trans. Roy. Soc., **212**, 142 (1912).

[15] T. W. Richards, *Jour. Amer. Chem. Soc.*, **37**, 1648 (1915).

[16] E. Griffiths, *Proc. Roy. Soc.*, **89**, 561 (1914).

[17] A. Lampa, *Wien. Ber.*, **111**, 2A, 316 (1902).

[18] N. A. Pushin, *Proc. Serbian Roy. Acad. of Nat. Sci.*, Feb. 1927, pp. 161–169.   [Text in Serbian, with *résumé* in French.]

[19] Roloff, *ZS. f. Phys. Chem.*, **17**, 325 (1895).

[20] N. A. Pushin und E. V. Grebenshikov, *ibid.*, **118**, 276 (1925).

[21] N. A. Pushin, *ibid.*, **118**, 447 (1925).

CHAPTER VIII

POLYMORPHIC TRANSITION UNDER PRESSURE

EXPERIMENTAL investigation of the effect of pressure on polymorphic transitions does not go back quite as far as investigations of melting. Apparently the first experiment in this field was by Mallard and le Chatelier [1] in 1883, who studied, by the method of volume discontinuity, the effect of pressure on the transition of AgI. It was known that there is a transition, at atmospheric pressure and 140° C., of the ice type, the high-temperature phase having the smaller volume, so that it was to be expected that the transition would take place at a lower temperature at high pressure. Mallard and le Chatelier did find a transition at 20° C. at 2475 kg./cm.[2].

However, if the initial slope, $\dfrac{d\tau}{dp}$, of the transition line is calculated by Clapeyron's relation from the known volume change and latent heat, it will be found that a depression of the transition temperature by this pressure of only 30° instead of 120° is to be expected. Another difficulty was that the volume discontinuity found by Mallard and le Chatelier was very much larger than was to be expected. Tammann later cleared up the matter by showing that Mallard and le Chatelier had actually found a transition to a third, hitherto unknown, modification. In the same year, 1883, Reicher [2] published as a thesis an examination of the effect of pressures up to 12 kg./cm.[2] on the transition between rhombic and monoclinic sulphur, and showed that Clapeyron's equation applies. Hulett [3] in 1889 studied the effect of pressure to 300 kg./cm.[2] on the transition point between the anisotropic liquid phase and the ordinary liquid phase of so-called liquid crystals. Lussana [4] studied in 1895 by the method of arrested temperature during cooling the transitions of NH_4NO_3 up to 250 kg./cm.[2]. This method is not capable of giving very accurate results. Tammann began in 1897 an extensive series of investigations over his maximum range of 3000 kg./cm.[2]. This work continued, with the aid of his pupils, over a number of years, and is fully described in his two books. It is not necessary to go into these results in detail here because I have repeated nearly all of the measurements over my wider range of pressure, at the same time determining the volume changes, and these will be discussed in full detail presently. One of the phase diagram investigations not repeated by me is that of Tammann and Hollmann on ethylene iodide. They found a very complicated diagram; I believe that this work should be checked before it is finally accepted.

Other work in this field is an investigation by Wahl [5] in 1912 by an optical method up to 1500 kg./cm.[2] of the transitions of CBr_4 and $\alpha\beta$—bibrompropionic acid. Jaenecke [6] also published investigations in 1915 of the transitions of $AgNO_3$, NH_4NO_3, and KNO_3 by the method of pressure discontinuity on cooling through a transition point at constant volume. As used by Jaenecke, this method gives results more qualitative than quantitative, because the pressure was not truly hydrostatic.

In principle there is no difference thermodynamically between heterogeneous equilibrium between a liquid and a solid phase and between two solid phases. The transition from one solid to another takes place with change of volume and with a latent heat, just as the passage from liquid to solid, and both sorts of transition are controlled by Clapeyron's equation, $\dfrac{d\tau}{dp} = \dfrac{\tau \Delta v}{L}$. There are, however, certain differences in the details of the transitions which may involve important differences in the physical manipulations. One important physical difference is that in general any solid phase may be carried into the region in which it is unstable with respect to another phase, without the reaction starting to run, whereas a solid cannot be carried into the region of stability of the liquid without melting taking place. This means that inconvenient manipulation may sometimes be necessary in the case of two solids to obtain points on the equilibrium lines. If the method of volume discontinuity is used, which is the method of all my work, in many cases one phase has to be carried so far into the region of stability of another before the new phase will appear that the volume change during the transition is not sufficient to carry pressure back automatically to the equilibrium value, but the transition would run to completion.   In such cases the pressure has to be artificially changed after the transition has been initiated, and the secondary change of pressure observed after every such change of pressure, until the pressure is found at which the secondary change reverses in sign.   The pressure at which this reversal takes place is the equilibrium pressure. The use of this method is further complicated by the fact that there are a number of transitions between solids of such a nature that the transition will not run within a certain range, centring about the point of thermodynamic equilibrium, even when the two phases are in contact.   The phenomena here are entirely different from the phenomena of transition between a liquid and vapour or between liquid and solid, and will be discussed in considerably greater detail later.   The point of immediate significance is that there are transitions with such a 'region of indifference,' and that for these all that can be done is to map out the boundaries of the region by observing the p-t loci on which the automatic pressure reaction has one or the other direction.   The region of indifference seldom has a width of more than a few hundred kilograms per sq. cm.; in such cases the only possible assumption is that the line of thermodynamic equilibrium is situated in the middle of this region.   The percentage error in such an assumption cannot be great when dealing with pressures of the order of several thousand kg. per sq. cm.

Another inconvenient phenomenon met when measuring the equilibrium co-ordinates between solid phases is the possibility of false equilibrium.   Imagine, for example, that the high-pressure phase is in the apparatus, and the pressure is lowered below its equilibrium value into the region of stability of the low-pressure modification with the larger volume.   If the transition to the new modification begins at some point in the interior of the high-pressure phase, then the low-pressure modification is immediately subjected to increase of pressure because of the volume differences, and this pressure on the new phase need not be the same as the pressure in the transmitting liquid, because the high-pressure modification which surrounds the nucleus of the low-pressure modification is itself capable sometimes of supporting rather high mechanical stresses. The result is that the transition in the interior of the solid mass will run under pressure conditions which are not the same as those in the transmitting liquid, so that incorrect values for the equilibrium pressures may be obtained. Such an effect is shown by the transition points not lying on a smooth curve. Fortunately the transitions usually start at the exterior surface, just as we have seen already that melting starts, so that the effect is not a common one. If an irregular point is found, a redetermination is usually sufficient to correct the matter. Not only may incorrect equilibrium co-ordinates be obtained when there is this sort of protecting action, but it is very easy to obtain incorrect values for Δv, as of course must happen if the transition does not run to completion, but an untransformed mass of one modification remains buried in a protecting covering of the other. The effect is sometimes found when the reacting substance is tightly held in a steel container, due to the constraints exerted by the walls of the container. The pressure which can be exerted during such transitions is quite considerable. Sometimes thick-walled containers of Cr—Va steel have been ruptured in this way; the pressure estimated required to produce such rupture may be several thousand kilograms per sq. cm. I believe that some of the effects found by Tammann—in particular, transition curves of supposedly unstable modifications of ice paralleling the transition lines of the stable forms—may be explained by this sort of false equilibrium.

To counterbalance these effects just discussed, which often make it more difficult to locate a transition line than a melting line, there are other effects favouring the accuracy of the transition determinations. It is very seldom, in fact I have never come across more than two cases, that impurities present in the solid phase are in a state of solid solution; this means that there is no alteration of the transition point by impurities, so that there is no premature rounding of the discontinuity. This means that not only can good values of Δv be obtained, but also good values for the difference of compressibility between different solid phases. The difference of compressibility can be combined with other information obtained from the transition curve in the manner already described in connection with melting, to give values for the difference of thermal expansion and specific heat. In general, then, we can get more accurate information for these differences for solid-solid transitions than for melting.

In a large number of cases the phase diagrams of solids contain triple points; in such cases the conditions of self-consistency of the data in the neighbourhood of the triple point afford valuable checks on the correctness of the results. At a triple point there are three or more conditions that must be satisfied by any internally self-consistent set of data. In the first place, the three transition lines, which are determined by three independent sets of experiments, must have such relative locations as to pass through a common point; secondly, the three independently determined volume changes at the triple point must satisfy the additive relation $\Delta v_{12} + \Delta v_{23} = \Delta v_{13}$; and thirdly, the three latent heats calculated by Clapeyron's equation for each of the three independent transition lines must also satisfy the additive condition $L_{12} + L_{23} = L_{13}$. It is almost always necessary to make slight readjustments in the directly determined quantities in order to satisfy these conditions. The readjustments have to be made by a process of trial and error, but it is evident that any readjustment in the slope of the transition line demanded by the condition that the three transition lines pass through a common point entails changes in the corresponding latent heats which may or may not satisfy the additive conditions. In practice there was never any question as to the way in which the adjustments should be made, and in general the mere fact that it is possible to make the necessary adjustment without exceeding possible experimental error is of itself sufficient evidence that the experimental points are essentially correct. In particular, one may be sure that data which satisfy the conditions at a triple point cannot be in error because of any false equilibrium phenomenon, either with regard to the p-t co-ordinates, or with regard to Δv.

In some cases a transition line is terminated at either end by a triple point; if the adjustments are possible at both triple points within experimental error, one may be particularly confident of the correctness of the results.

The occurrence of a triple point is, furthermore, most welcome, because of the information it can give about Δa, $\Delta \beta$, and ΔC_p for the three phases. We have already seen that at an ordinary point on a transition line the transition data give us two conditions on these three quantities, so that independent experimental evidence, usually a direct measurement of Δa, is necessary to completely determine the three quantities. At a triple point, on the other hand, the transition data alone (that is, a knowledge of t and Δv as a function of pressure along the transition line) are sufficient to completely determine the Δa, $\Delta \beta$, and ΔC_p. The reason is easy to see. There are in all three Δa's, three $\Delta \beta$'s, and three ΔC_p's at the triple point, one for each of the three transition lines meeting at the point; that is, in all nine quantities to be determined. Each transition line by itself gives two equations of condition on its corresponding set of three quantities. But there are in addition at the triple point the three additive relations $\Delta a_{12} + \Delta a_{23} = \Delta a_{13}$, etc., making in all nine relations from which the nine quantities may be completely determined. If in addition, any one of the Δa's or $\Delta \beta$'s should be determined by independent experiment in the neighbourhood of a triple point, we have a further check.

The experimental method for measuring the transition points is essentially the same as that for determining melting points, and differs only in unessential details. A large number of the solids investigated, mostly inorganic salts, are insoluble in the transmitting liquid, kerosene, and these could, accordingly, be directly exposed to the kerosene without the complication of an intermediary of mercury. The usual method was to compress with a ramrod the powdered material into a thin steel shell, which was then directly exposed to the kerosene in the lower cylinder.   The shell was perforated with many fine holes to allow pressure access to all parts of the interior, and to minimise the possibility of false equilibrium.   Most of the organic substances investigated are soluble in kerosene, and these had to be isolated from contact with it by a mercury seal, very much as in the case of liquids, except that the capillary entrance of the liquid container was not necessary, but the substance could be rammed dry into a cylindrical container open on one end.   The walls of these containers had to be made fairly heavy, perhaps $\frac{1}{16}$ in. wall thickness with an inside diameter of 0·5 in., and the container itself was made of heat-treated chrome vanadium steel, in order to resist the rupture which is often a consequence of the volume changes accompanying the transition.

The experimental procedure is different according as a transition already known to exist is being investigated, or the transition is a new one.   If the transition is a known one, and if the difference of volume between the two phases is known, it is known whether the transition temperature will rise or fall with increasing pressure.   The first process is to locate the approximate equilibrium point by the method of volume discontinuity at a temperature above or below the atmospheric transition temperature, as the case may be, then to determine the transition data accurately at this point.   The approximate slope of the transition line is then determined in terms of these two points, so that the new temperature at which to locate a new point is suggested. The number of points measured on the transition line depends, of course, on the curvature of the line and on the variation of the thermal parameters, such as Δv, along the line.   As a rough average, points were measured perhaps every 1000 kg./cm.², or a dozen points to completely fix a curve.

If the substance was one for which no polymorphic transitions were known, then it had to be subjected to a preliminary exploration to locate such possible points.   The total number of substances examined for new forms has been about 150; the range of exploration was up to 12,000 or 13,000 kg./cm.², and the temperature range from room temperature to 200° C. The method is merely to plot on a large scale piston displacement as a function of pressure at room temperature and at 200° C. The pressure points were taken close enough together, so that a volume discontinuity of 0·01 per cent. of the initial volume of the material should not have escaped attention. If no discontinuity was found on the two isotherms at 20° and 200° it was assumed that there are no transitions in the range, and the substance was not examined

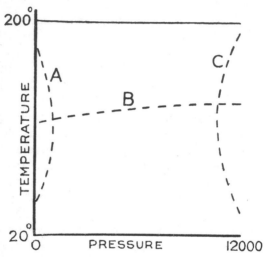

Fig. 61.—Illustrates the method of exploration for new modifications.

further. In order to be rigorously certain that there are no transitions in the range, it would have been necessary to go completely around the boundary of the region—that is, to add to the exploration an examination of the two isobars at atmospheric pressure and at 12,000 kg., reaching from 20° to 200°. This was dispensed with, largely to save time, for any manipulation in which temperature is changed as the arbitrary variable is much more tedious than one in which pressure is the independent variable, because of the long time required to reach thermal equilibrium in cylinders of the size used. Furthermore, an exploration which is restricted to the two isotherms can miss only very unusual types of transition line, such as cut into and out of the region on the unexplored part, as indicated by the dotted lines in fig. 61.   The two transition lines A and B are ruled out with a certain probability by the absence of previous observations of transitions at atmospheric pressure, so that the only remaining possibility of overlooking a transition is when it is of the type C. This sort of transition is certainly very rare, but there are examples, of which the high-pressure modification of benzene is one.   If this transition had been somewhat displaced to lower temperatures, I might have missed it, and the possibility must be recognised that some transitions of this type may have escaped, but I believe the probabilities are very low.   It must further be kept in mind that it is always possible that some favourably situated transitions were suppressed because of viscosity or similar effects.

The method is not restricted by sensitiveness to those transitions with a volume change of 0·01 per cent., but if the transition were known to exist, it could be located by taking the experimental points sufficiently close together, even if the volume change were considerably smaller than 0·01 per cent., and some such transitions have actually been studied.   However, to have made the exploration for all substances to the limit of sensitiveness would have demanded a very much greater amount of labour, and the very small probability of finding new transitions did not seem to justify the large decrease in the number of substances that could have been examined.

Of the 150-odd substances examined, some 40 show the phenomenon of polymorphism; the phase diagrams of 36 of these are reproduced on a small scale in figs. 62 and 63. It is not possible to reproduce here all the numerical results, for which the original papers must be consulted.   On the diagrams are given, however, symbols indicating in a qualitative way certain aspects of the behaviour.   An arrow on a transition line indicates the direction in which Δv decreases numerically, and an α, β, or C_p on one or the other side of a transition line means that the phase stable on that side of the line has a greater compressibility, or thermal expansion, or specific heat than the other phase.   In addition, there are

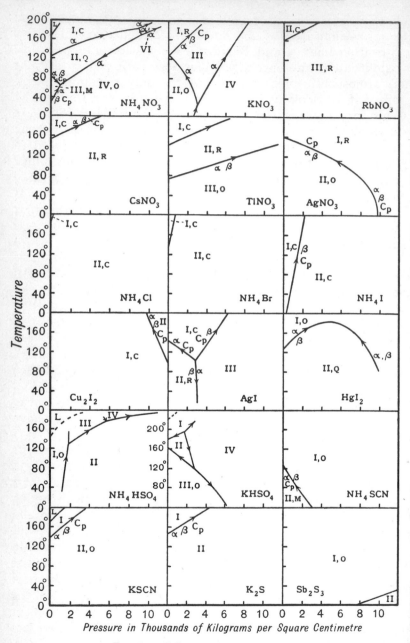

FIG. 62.—Collected phase diagrams.

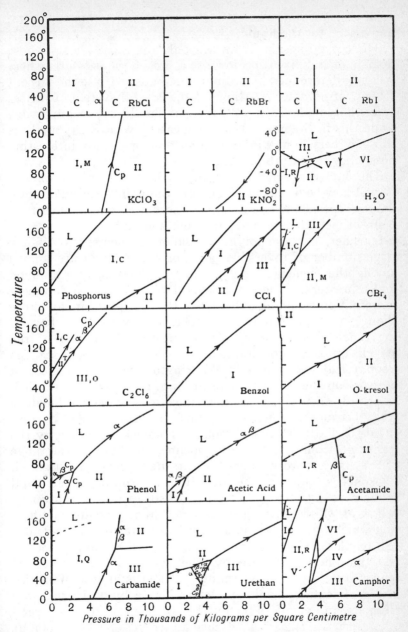

FIG. 63.—Collected phase diagrams.

symbols indicating the crystal system, when this is known, and also L for the liquid.

The crystal system is known only for those phases whose domain of stability runs down to atmospheric pressure, where it can be determined by ordinary methods.   Up to the present I know of no determinations of the crystal system of phases stable only under high pressures; it should not be impossible to do this in the most favourable cases by an X-ray analysis through beryllium windows, and this opens an interesting possibility for future investigation.

The data contained in the phase diagrams in figs. 62 and 63 will now be discussed, first from the point of view of noteworthy special substances, and second from a statistical point of view to determine any common features.

Consider, first, the polymorphism of the elements.   A fairly large number of reversible polymorphic changes are known among the elements; only a few of these were examined under pressure.   There are several reasons why a complete examination would be difficult; in the first place, many of the changes take place only at temperatures very much beyond the range of this work.   An interesting beginning of the attack on this problem, however, has been made at the Geophysical Laboratory in Washington, where the effect of pressure on the magnetic transition of iron has been studied, as already mentioned.   Other transitions may be difficult to study because of slowness of transition or because of complicated solubility or dissociation phenomena, as shown by the transition from rhombic to monoclinic sulphur.   In the range of the present work four elements have been investigated. Phosphorus, shown as the third from the top in the left-hand column of fig. 63, is a well-marked example.   Ordinary yellow or white phosphorus has a new high-pressure modification which was first found at room temperature in the course of a regular exploration, and was later picked up at $-80°$ C. at atmospheric pressure by the conventional method of temperature arrest on warming.   The crystal system of this new modification should be determinable by ordinary X-ray methods, and ought to be found.   In addition to this reversible transition, P may experience an irreversible transition at high pressures and temperatures; this irreversible transition will be discussed in Chapter XIII.

Cd. (B. 53) has two new polymorphic forms at high pressures, shown separately in fig. 64. These transitions are obviously of a somewhat unusual kind; they were first dis-

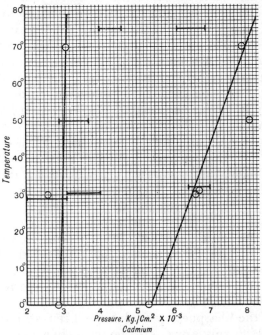

Fig. 64.—The phase diagram of the two high-pressure modifications of Cadmium. The circles indicate readings obtained with the two phases simultaneously present, while the horizontal lines show the range within which the transition was shut by other readings obtained with only one phase present, and therefore not under equilibrium conditions.

covered in the course of measurements of the linear compressibility of single crystals. The change of linear dimension is of such a character that there is approximate compensation in different directions, so that the volume of the two phases is practically the same. The discontinuity was checked by measurements of the electrical resistance as a function of pressure for a single crystal with hexagonal axis parallel to the length, the most favourable orientation. It is not possible to detect any discontinuity in the resistance or the compressibility of an ordinary poly-grain piece of Cd. There seems to be no regular connection between the orientation of the original crystal and that into which it is transformed; the discontinuity in length may be either an increase or a decrease. If the transition is reversed the single crystal condition is not recovered, but apparently there are likely to be several grains of the original modification produced in random orientations. The result is that if the transition is made to go back and forth a number of times, all semblance of the original single crystal condition is lost, the discontinuity becoming less at each successive transition, until it is finally completely lost, and the original single crystal has assumed the ordinary multi-crystal condition.

Cerium (B. 62) has a new high-pressure form. The existence of this was discovered during measurements of the compressibility, and was later verified by the discontinuity in electrical resistance. The transition pressure is approximately 7600 kg./cm.^{2} at 30° C. and 9400 at 75° C. The transition is apparently sensitive to impurities; my first measurements on Ce were made on a sample known to be somewhat impure; this showed no trace of the transition. It was not until a much purer sample was available that the transition was found. Even then it was not perfectly sharp. It is not unlikely that if the purity could be still further increased the sharpness would have been still further improved. If so, this is one of the few cases in which an impurity is soluble in the solid phases in such a way as to affect the transition. The low-pressure modification of Ce is anomalous in two respects; its compressibility increases with rising pressure, and its temperature co-efficient of compressibility is negative.

The last element found to be polymorphic under pressure is As (B. 72). Here again the phenomena can be detected only in the single crystal form; I had previously examined ordinary polycrystal As for new forms without result, both by the method of volume discontinuity and resistance discontinuity. A later study of a sample which had been slowly lowered from an electric furnace while melted under its own vapour pressure in a heavy quartz capillary, and in which crystallisation took place throughout in approximately a single grain, disclosed transition phenomena which the multicrystalline piece did not show.   The existence of a new form was verified both by compressibility and by resistance measurements.   As in the case of Cd, perfectly sharp results were not obtained.   The transition at 30° is in the neighbourhood of 5500 kg./cm.^{2}; it is almost certain that the low-pressure modification varies greatly in compressibility with direction, and it is also probable that the pressure coefficient may have different signs in different directions.

After the elements, the simplest substances which have been examined are the alkali halides.   Slater [7] found in the course of his compressibility measurements of RbBr and RbI new polymorphic forms at high pressures; his compressibility apparatus was not adapted to measuring the volume change.   I later (B. 64) took up the matter with a more suitable apparatus, the piston piezometer already described, and was able to locate the transition points more accurately, and also to determine the change of volume; I found in addition that RbCl also has a high-pressure transition.   The phase diagrams are collected in fig. 63.   The transition lines run nearly vertical, and there is a progressive change in the transition pressure through the series.   The transitions were not perfectly sharp, and, furthermore, showed an unusual tendency to hysteresis; the way in which these effects changed when the purity of the material was changed makes it probable that the perfectly pure material would not show hysteresis. The most interesting question in connection with these simple salts is as to the nature of the high-pressure lattice.   CsCl— Br—I are known to crystallise in the body-centred cubic arrangement, whereas all the other alkali halides crystallise simple cubic.   It was natural to expect, therefore, that the high-pressure modifications of the Rb salts would have the crystal structure of the Cs salts, particularly since the volume change is in the right direction.   At first, however, it seemed to me that the quantitative volume relations made this improbable, as the volume of the high-pressure modification does not have at all the value that would correspond to this lattice, assuming that the atoms behave like rigid spheres during the transition. It was soon pointed out, however, by Goldschmidt in private correspondence and by Pauling,[8] that in other cases the atomic radii must be different in body-centred and in simple cubic lattices, and that when this factor is taken into consideration the volume relations come out almost exactly right. There can be little doubt, therefore, that these new high-pressure forms are body centred, like the body-centred Cs salts. It is interesting in this connection that KI and KBr showed no similar transition to nearly 20,000 kg./cm.[2], and that CsF, which crystallises simple cubic, has no transition up to 12,000 kg./cm.[2]. The stability relations of the simple cubic and body-centred cubic lattices are evidently very sensitive to slight variations in the crystal parameters. The matter has been discussed by Pauling,[9] who concludes that we are not yet in a position to calculate from other data where to expect a transition, even in such very simple cases as these. It is somewhat surprising, in view of the extreme sensitiveness to slight changes in the parameters, that the transition pressures of the three Rb salts should be so nearly the same.

Perhaps the next most important substances examined were the ammonium halides. NH_4Cl and NH_4Br were already known to have two forms at atmospheric pressure. The two forms belong to the same sub-group of the cubic system, and, furthermore, have a very large volume difference of 15 per cent. or more. The transition lines of these two substances run nearly vertical, so that the phenomena in the temperature range up to 200° are confined to a very low pressure range. The ordinary form of NH_4I was known to have the same structure as the high-temperature form of NH_4Cl and NH_4Br, and search had been made at atmospheric pressure [9] for another form of NH_4I corresponding to the low-temperature modification of the chloride and bromide, but without success. In this search temperature was lowered to −16° C., which seemed ample, in view of the comparative closeness of the transition temperatures of the chloride and bromide, 184·3° and 137·8° C. The expected modification was found almost at once on making the pressure exploration, and the complete transition line runs very nearly parallel to that of the two others, being merely displaced to somewhat higher pressures. The transition point at atmospheric pressure turns out to be $-18°$; but it is not a simple matter to get the new form on cooling at atmospheric pressure, because of various lag effects.

Since my measurements of the pressure transitions of the NH_4 salts were made, the X-ray structure has been worked out, and it appears that there is complete parallelism between these salts and the Rb salts. The high-temperature, or low-pressure, modifications of the three NH_4 salts are like NaCl in structure, and the low-temperature, or high-pressure, phases are like CsCl. Furthermore, the character of the transition lines in the NH_4 and the Rb series is very nearly the same, in each case running very nearly vertical, with very small latent heat. The order of the transition pressures is different in the two series, however ; in the Rb series the transition pressure at constant temperature decreases from Cl to I, whereas in the NH_4 series it increases.

This completes the list of simple substances, either elementary or with simple lattices, and also completes the list of substances in which the phase diagrams of similar substances are similar. There are no other examples of such thoroughgoing similarity as shown by the Rb and NH_4 halides. In fact, in all other cases chemical similarity seems to offer no clue as to similarity of phase diagram, but the phase diagrams of closely related substances may be very different.

A striking example of dissimilar phase diagrams of chemically closely related substances is afforded by CCl_4 and CBr_4. Each has three solid phases, but the phase diagram of CBr_4 is apparently the exact inverse of that of CCl_4, the phase of CBr_4 stable at high temperature corresponding to the phase of CCl_4 stable at low temperature. However, until the crystal structure of these modifications is worked out, we cannot be sure that we have set up the correspondence between these modifications in the right way, so that it is possible that the lack of correspondence may not be as extreme as just suggested. In any event, the phase diagrams will be greatly dissimilar.

Other examples of chemically similar substances with completely different phase diagrams are the two pairs NH_4SCN, $KSCN$, and NH_4HSO_4, $KHSO_4$.

The univalent nitrates are an interesting group, in which there are many tempting suggestions of resemblances which cannot at present be worked out.   The three nitrates with the simplest diagrams, $RbNO_3$, $CsNO_3$, and $TlNO_3$ contain one transition line which is so similar in all three diagrams that we may suspect a deep-seated structural similarity. This line is the II–III line of $RbNO_3$, the I–II line of $CsNO_3$, and the I–II line of $TlNO_3$.   In all cases this line is situated in approximately the same part of the phase diagram, it rises to higher temperatures with increasing pressure, and it marks a transition between a cubic form stable at the higher temperature and a rhombic form stable at the lower temperature.   But the II–III line of $TlNO_3$ has no known analogue in the other diagrams; perhaps it may be found at lower temperatures than have yet been explored.   There is, however, a complication in identifying the forms in this way, in that it is stated [10] that $RbNO_3$ has a transition at atmospheric pressure at 219° C. (above the temperature range of the diagram) which does not appear in the other materials. I was unable, however, to find this high temperature form of $RbNO_3$ by my usual method of exploration.

There is a further resemblance between the diagrams of these three nitrates and the high - pressure part of the NH_3NO_3 diagram between the phases I, VI, and IV.   Here the I–VI transition of NH_4NO_3 would be the analogue of the I–II transition of $TlNO_3$, and the VI–IV transition of NH_4NO_3 the analogue of the II–III transition of $TlNO_3$. An argument in favour of this identification is that the crystal system of the known forms is the same—that is, I of NH_4NO_3 and I of $TlNO_3$ are both cubic, and IV of NH_4NO_3 and III of $TlNO_3$ are both orthorhombic; this would involve that VI of NH_4NO_3 is rhombic, like II of $TlNO_3$.

There are other superficial resemblances in other diagrams of this group which may or may not correspond to anything real.   Thus the I–II line of $AgNO_3$ suggests the II–III line of KNO_3; the low temperature phase is in each case orthorhombic, and, furthermore, the high-temperature phase of $AgNO_3$ is known to be rhombic. This is also with high probability the crystal system of III of KNO_3, because this is known to be the crystal system of a form unstable at atmospheric pressure,[11] which occurs only in the neighbourhood of the I–II transition point, and which must almost certainly be the high-pressure modification carried a couple of hundred kilograms into its region of instability. But, on the other hand, there is the difficulty that $AgNO_3$ will not form mixed crystals with any of the other nitrates; this difficulty may perhaps be explained as meaning a rather large difference in the size of the structural units rather than as a difference of lattice type.

A further possibility seems tempting at first, suggested by the obvious resemblance between the entire low-pressure part of the diagrams of NH_4NO_3 and KNO_3. The II–III–IV triple point of KNO_3 is the analogue of the II–III–IV triple point of NH_4NO_3, and the I–II–III point of KNO_3 would be the analogue of an unrealised triple point I–II–III of NH_4NO_3 existing at negative pressures. The difficulty with this is that the known crystal systems do not correspond; the corresponding phases for NH_4NO_3 and KNO_3, I and I, are cubic and rhombic respectively, the corresponding II and III are respectively tetragonal and rhombic, and III and II are respectively monoclinic and orthorhombic.

The phase diagram of NH_4NO_3 contains an interesting feature beyond the temperature range of the diagram. There is still another transition point to a modification V at about $-20°$ at atmospheric pressure. V has a larger volume than IV, so that the transition is of the ice type; this has been studied by Behn.[12] V is known to belong to the same crystal system as II, and the interesting question arises as to whether II and V are not really the same phase, a thing which is thermodynamically possible, but which has never been established actually. The study of Behn does not rule out this possibility, but makes it less likely than had appeared at first.

Another interesting feature about this diagram is, that a point on the unstable prolongation of the II–IV line to atmospheric pressure has been realised by Bowen [14] at the Geophysical Laboratory.

One of the most interesting phase diagrams is that of water. Tammann first found the two new forms II and III, followed the melting curve of III over a short pressure range, and followed the transition lines between I and III and between I and II to low temperatures.   He found the crossing of the transition lines I–II and I–III, but could not find the transition line II–III which thermodynamics demands must start from the crossing point.   He also found a number of other transition points in the neighbourhood of these lines, which he explained by assuming other less stable forms, with properties very closely similar to those of the stable forms, and which he made the basis of an elaborate theory.   I could not find these forms, although I did find phenomena which were similar to those found by Tammann, and I am convinced (B. 13, 17) that the explanation of all these irregularities is merely false equilibrium arising from internal stresses, as already discussed.   Tammann made the very interesting experiment of cooling the modification II to liquid air temperature, and then releasing pressure and taking II out of the apparatus at atmospheric pressure.   As long as it remains cold, the transition to the stable form I does not run because of the great internal viscosity, but as II slowly warms into the region of appreciable reaction velocity, it puffs up with 20 per cent. increase of volume, falling apart into a powder of the ordinary form of ice.

The modifications V and VI of ice were beyond Tammann's range.   I first found VI, as already explained, in making compressibility measurements of liquid water near room temperature.   The melting line between liquid and Ice VI has been followed beyond the range of the diagram to nearly 21,000 kg., with no indication of another form. Apparently both ice and liquid water approach complete normality at high temperature.   On the low-pressure end of the liquid-VI line there are certain abnormalities, in particular the curvature of Δv against pressure, which is concave toward the pressure axis, but this reverses in the neighbourhood of 9000 kg./cm.², and from here on the behaviour is normal.

The conditions under which the different modifications of ice appear are somewhat capricious, and often inconvenient manipulation is necessary to arrive in the part of the phase diagram desired. The behaviour is particularly striking in the neighbourhood of the V–VI–liquid triple point, say, between 0° and −10° and between 5000 and 6000 kg./cm.². With the ordinary form of apparatus, water in a steel piezometer with pressure transmitted to it by mercury, it is very difficult to produce the modification V. For example, if liquid water at −10° is compressed across the melting curve, the liquid will persist in the sub-cooled condition for an indefinite time, without freezing to V, the stable form. But if the compression is carried further, to the unstable prolongation of the liquid-VI line, freezing to the form VI will take place almost at once on crossing the line, in spite of the fact that VI is unstable with respect to V. Now in order to make V appear, VI may be cooled 30° or 40°, when it will spontaneously change to V, the stable form. If pressure is now released back to 5000, say, and temperature raised to the melting line of V, melting takes place at once, and so the co-ordinates of the melting line may be found. Suppose now after the melting is completed the liquid water be kept in the neighbourhood of the melting line for several days, and then the pressure increased again across the melting line at −10°; it will be found that the instant the melting line of V is crossed the liquid freezes to V. This suggests that there has persisted in the liquid phase some sort of structure, not detected by ordinary large-scale experiments, favourable to the formation of V. It is now known, of course, from X-ray analysis that structures are possible in the liquid; this experiment suggests a specificity in these structures that might well be the subject of further study.

The formation of the nucleus of V may be favoured by the proper surface conditions. Thus if there is any glass in contact with the liquid, either a fragment of glass wool purposely introduced into the liquid, or by enclosing the water in a glass instead of a steel bulb, freezing to V takes place at once without the slightest hesitation immediately on carrying the virgin liquid across the freezing line.

Turn now from this consideration of special cases to general considerations. Enough cases have been examined to give some significance to a statistical study of all the transitions. There is apparently only one generalisation justified by all these cases; no critical point between solid phases has ever been found. Whenever in the diagrams a transition line apparently ends, the explanation is to be found in some other effect. Thus, in many cases, the substance decomposes when carried to high temperature; examples of this are the I–III and the III–IV lines of NH_4SO_4, the I–III and III–II lines of CBr_4 or the I–II line of CCl_4. The VI–IV line of camphor illustrates another possibility. Here the transition could not be followed further, because its velocity had become impossibly low; it is very common to find this state of affairs in following the line to low temperature, but it is more uncommon on a rising line. However, on a rising line it must be recognised that there are two opposing tendencies; increasing temperature tends to increase the transition velocity, and increasing pressure tends to decrease it, in part because of the increase of viscosity produced by pressure. Sometimes, as in the case of camphor, the retardation produced by pressure is greater than the acceleration by temperature.

A critical point between two solids with the accompanying possibility of a continuous transition from one solid to another is not at all to be expected, because such a change from one type of lattice to another is contrary to all that very wide experience which is expressed in Neumann's law, namely, that the external crystal shape gives the most complete expression to the symmetry of the structure. For example, if a cubic crystal could be changed continuously into a tetragonal one, there would have to be a preferred direction in the original cubic structure, that is, the direction of the final tetragonal axis, which would be inconsistent with its complete cubic symmetry.

Apart from the impossibility of a critical point, there seems to be no sort of restriction to which the phase diagrams of solids are subject. The situation here is entirely different from that with regard to the melting curves, which show a number of regularities.   Nearly all melting curves rise to higher temperatures with increasing pressure; there are only three of the ice type—that is, those falling to lower temperatures with increasing pressure—which have been investigated, and of these that of ice itself is stable only over a range of 2000 kg./cm.^{2}.   Casual inspection of the collected phase diagrams of solids shows that there are a comparatively large number of falling transition lines, and in fact nearly one-quarter of those investigated are of this type.   Furthermore, a falling curve need not carry in itself any evident seeds of inherent instability, such as was evident in the falling melting curve of ice, which is terminated by the appearance of a solid form of more normal properties.

The melting curves all rise indefinitely, but there is one example, HgI_2, of a substance with a transition line that rises to a maximum and then falls.   There seems no intrinsic reason why there should not also be curves with a minimum; perhaps resorcin, studied by Dennecke,[13] may prove to be an example of this sort if some method can be devised for overcoming the very rapidly increasing sluggishness of the transition at low temperatures and high pressures.   At a maximum temperature on a transition line, at which the transition runs with no change of volume, there is an interesting relation between the compressibility of the two phases. A moment's consideration will show that to the left of the maximum point of HgI_2 the phase I, which here has a larger volume than II, is more compressible than II.   This larger compressibility continues to characterise I even to the right of the maximum, where I is now the phase of smaller volume. This means that the mere existence of a maximum temperature involves the necessity of there being conditions under which the phase of smaller volume has the higher compressibility.

Not only may the transition lines have horizontal tangents, but there are examples of vertical tangents.   Neither of these show well on the scale of the diagram, but the transition line between ice I and III contains a vertical tangent, at which pressure is a maximum, and the transition line between the two varieties of benzene has a vertical tangent with minimum pressure. The mere existence of such vertical tangents, at which the transition runs with no latent heat, means that there are conditions under which the phase stable at the lower temperature has the higher specific heat. Thus the modification I of benzene, below the temperature of the vertical tangent, has a higher specific heat than II. This may be proved by a simple cyclic process carried out at a pressure slightly above the minimum pressure, starting with I below the transition line, carrying I across the transition line and transforming to II, raising temperature on II at constant pressure until the transition line is reached again, carrying II across the line with transition to I, and then carrying I directly back to the starting-point at constant pressure through the region of stability of II, the transition to II this time being suppressed.

The melting lines, and vapour lines as well, always are curved in the same direction—that is, they are concave toward the pressure axis. There are a number of transition lines between solids with the opposite curvature; KNO_3 and NH_4SCN afford examples of rising and falling lines respectively in which the curvature is very great.

In Table XIII the results of a statistical examination of all the transition lines are collected. The classification is according to the various significant features of the curves; first with regard to the slope, whether rising or falling; second, with respect to the curvature, whether concave toward the pressure axis, called normal, or convex, called abnormal; third, with respect to the variation of Δv along the transition line, called normal if Δv decreases with rising temperature along the line and abnormal if Δv increases; fourth, with respect to the difference of compressibility between the phases, called normal if the phase of larger volume is the more compressible, otherwise abnormal; fifth, with respect to the difference of thermal expansion, called normal if the phase stable at the higher temperature has the greater expansion, otherwise abnormal; and finally with regard to the difference of specific heats, called normal if the phase stable at the higher temperature has the greater specific heat, otherwise abnormal. In those cases where the varia-

TABLE XIII

STATISTICAL STUDY OF TRANSITION PARAMETERS

Slope of transition line	Curvature		Δv		Δa		$\Delta \beta$		ΔC_p	
	Normal	Abnormal	Normal	Abnormal	Normal	Abnormal	Normal	Abnormal	Normal	Abnormal
Rising 53 .	18		34		8	7	4	4	4	5
				4		1				
		9	4			2	1	2	1	2
				3						
Falling 16 .	8		6			4		1	3	
				8	1	3	2	2	3	
		1	1		1			1		1

tion has been too small to detect, the normal classification has been made.

The significant thing about the table is the comparatively large number of cases of 'abnormal' behaviour. The number of ice-type curves and curves with abnormal curvature has already been commented on. The behaviour of Δa, difference of compressibility, is most significant; here there are ten normal cases and seventeen abnormal cases. That is, in more than half the known cases the phase of smaller volume is the more compressible. Such behaviour was not so difficult to understand in the case of liquid water and ice, the smaller compressibility of ice being evidently connected with the rigidity to be expected of a lattice structure, but such a relation between different lattices is more unexpected.

Similarly there are more abnormal than normal cases of thermal expansion, 10 against 7. This perhaps is not so surprising when it is considered that the thermal expansion involves a departure of the restoring forces from linearity, and the restoring forces must vary in very complicated ways, as shown by the multiplicity of the transition phenomena. The difference of specific heat is normal in 11 cases and abnormal in 8. One is inclined to think of the specific heat as a measure of the number of internal degrees of freedom, and to expect the greater number of degrees of freedom in the high-temperature form, and therefore the larger specific heat here also. The proportion of cases of the opposite behaviour seems high, although not more than half. Any substance for which the specific heat of the high-temperature form is less than that of the low-temperature form must have further abnormalities in the specific heats as a function of temperature at atmospheric pressure. If we assume that the third law of thermodynamics is applicable to any two crystal phases, then at a transition point the low-temperature phase will pass into the high-temperature phase with increase of entropy. But the high-temperature phase cannot have a greater entropy than the low-temperature phase if its specific heat is lower all the way down to the absolute zero. It follows the specific heat curves must cross, and, furthermore, must cross at a temperature high enough to permit the total area under the curve of $\dfrac{C_p}{\tau}$ against τ to be greater for the high- than the low-temperature phase.

In addition to the data covered in the table, a consideration of the crystal system of the various forms is of significance. Here, unfortunately, the crystal systems of only those phases are known which can be realised at atmospheric pressure. The close-packed arrangements occur in the cubic and hexagonal systems, and it is therefore perhaps natural to expect that there will be a tendency for high-pressure forms, which are produced from the low-pressure forms with decrease of volume, to crowd into the cubic and hexagonal systems. The exact opposite turns out to be the case; in general that one of the two forms which has the higher symmetry has also the greater volume.   Thus in the examples above, there are seventeen cases in which one of the two phases is cubic; in ten of these cases the cubic form is the form of greater volume, so that the effect of pressure is to compel a transition from a form of high symmetry to a form of lower symmetry.   Similarly, in four of the five cases in which one of the reactants is hexagonal, the phase of greater volume has the greater symmetry.   An explanation of this at once suggests itself.   The close-packed arrangement, either cubic or hexagonal, referred to above, is a close-packed arrangement of spheres.   It is difficult to think that pressure will not produce a more close-packed arrangement; the fact that it does not tend to produce the cubic or hexagonal arrangement means that the molecules are not in general spherical, so that the close-packed arrangement must follow a more complicated pattern.   At low pressures, on the other hand, where the molecules are not crowded so closely together, the parts of the fields of force which determine the relative positions in the lattice approach more closely to spherical symmetry, just as in electrostatics the field about any complicated charged system approaches the field of a point charge at great distances, and the close-packed spherical structures predominate.

Thus far we have considered only the mutual relations between two phases at a transition line.   At a triple point three phases may coexist.   The relative positions of the three transition lines at a triple point depend on the relative volume and energy relations of the three phases.   Roozeboom[14] divided all possible triple points into eight classes, depending on the relative positions of the transition lines in the four quadrants of the p-t plane consistent with the demands of thermodynamics.   At the time that he wrote, an example of only one of these eight types of triple point was known.   The transitions given in figs. 62 and 63 above now bring the number of types of triple point that have been realised up to six, so that there are only two cases missing.   There is no reason to think that there is any essential reason why these cases will not also sometime be found, if enough substances are examined.   The details of Roozeboom's classification of triple points is described in one of my papers (B. 25). It is not significant enough to demand further discussion here; the important point for us is that there is here additional evidence of the greatest possible variety in the relations between polymorphic forms.

Up to this point we have considered only the thermodynamic parameters of the reacting solids.   There is another group of phenomena, equally significant in giving a complete picture of the physical situation; these are the phenomena of the velocity of transition from one phase to another, or the various phenomena of retardation or suppression of transitions which are thermodynamically possible as judged by the criterion of equality of thermodynamic potentials.   In connection with many of the transition measurements described above, I made a number of observations of velocity phenomena, which have been collected into a separate paper (B. 24).   One of the most striking of these phenomena is the enormous temperature coefficient of velocity of some transitions.   A particularly striking example is that of ice I–III.   Near the upper end of this transition line, which is terminated by the triple point with the liquid, the transition runs with almost explosive rapidity; in fact pressure so rapidly follows the change of volume produced by moving the piston that the pressure gauge will show no change of pressure. At a temperature only 20° lower the transition is so slow that hours are required to run to completion, and below −70° the transition line cannot be followed at all. This reaction is quite different from the ordinary chemical reaction, which doubles in velocity for a temperature rise of about 10°.   The fact that the latent heat of the transition from ice I to III is practically zero makes it possible for the phenomena to be especially striking, since the apparent transition velocity need never be limited by the necessity for removing the heat of transition by conduction.   Except for this complication there is no reason to think that there are not other substances with as large a temperature coefficient of transition velocity.

The method of measuring transition velocity in these cases is very simple; it consists merely in observing the rate at which pressure spontaneously changes at constant volume. The known difference of volume between the reacting phases enables this at once to be turned into cm.³ or grm. of transition per minute. This may be further changed into speed of motion of the surface separating the two phases, if desired, because in nearly all cases one could be sure that the transition progressed by a surface of separation of the two phases sweeping lengthwise through the material, which was in the form of a cylindrical rod.

A great many observations were made of pressure as a function of time as the transition was progressing. The speed of the transition can be found from the slopes of these curves, and this speed then may be plotted against pressure. The general behaviour to be expected is that the velocity of the transition, in one or the other direction, becomes greater as the pressure displacement from the equilibrium pressure increases. This turns out to be true, unless the displacement is so great that a new phenomenon enters, analogous to one shown by liquids on freezing, namely, the velocity may become less again because a region is entered in which viscous effects predominate. This viscous effect was found on a few occasions, notably with $AgNO_3$, but in general the effect is difficult to observe, and the transition usually runs more and more rapidly at greater distances from the equilibrium curve.

As the pressure approaches the equilibrium value, many solid transitions, although not by any means all, show a phenomenon which has no analogue in melting phenomena, and which points to an essential difference of mechanism. If a solid and its melt are in contact, the reaction will run, either until one phase is exhausted or until the equilibrium pressure is attained, the direction from which the equilibrium pressure is attained, whether from above or below, being a matter of indifference. The same is true, of course, of equilibrium between a liquid and its vapour. In these two cases equilibrium is doubtless set up by a sort of dynamic mechanism, there being streams of molecules going in both directions, some leaving the vapour phase and condensing on the liquid, and others leaving the liquid and vaporising. At equilibrium the velocities of the two streams are equal, and if the equilibrium point is left, the velocity of one or the other stream becomes greater, and equilibrium is automatically restored. There are many solid transitions in which the behaviour is entirely different. This may be illustrated by an imaginary typical example. Suppose equal amounts of the two phases A and B in contact with each other at the pressure and temperature of thermodynamic equilibrium. Keeping temperature constant, pressure is raised 500 kg./cm.^{2} into the region in which A is the stable form.   The transition to A now runs with decrease of volume, and pressure drops back at a rate which can be measured, and which becomes so rapidly less that the transition ceases to run when pressure is still 100 kg./cm.^{2} above the initial pressure, and while there is still a considerable amount of B left. The same sort of thing happens if pressure is lowered 500 kg./cm.^{2} below the initial pressure; B, with the larger volume, is now the stable phase, and the transition runs, carrying the pressure back toward the initial value, but at a rate which rapidly becomes less until the transition ceases to run entirely at a pressure 100 kg./cm.^{2} below the initial pressure.   This means that there is a region 200 kg./cm.^{2} wide within which the transition will not run with perceptible velocity although the two phases are in contact.   This may be checked by setting the pressure artificially at any point within the region, when it will be found that no transition will occur.   It might be thought that this cannot be a legitimate effect, and is to be explained by supposing that the transition velocity merely becomes too low to measure.   The answer to this is given by the actual curves plotting transition velocity as a function of the pressure displacement.   A typical example for the transition between the two modifications of yellow phosphorus is given in fig. 65.   There can be little uncertainty, I think, in extrapolating the curves to cut the axis, which means that the transition velocity is zero in the region between the two intersections.   Of course to prove this by direct experiment would demand experiments lasting for an indefinite time; as it was, a range of 5000-fold in the transition velocity was sometimes reached, with no trace of any bending of the curves at the lower end, which must occur if the objection is valid.

The region within which the transition will not run, even when the two phases are in contact, I have called the 'region of indifference.' The width of the region varies from point to point along the transition line in a way depending very greatly on the particular substance. Many substances have no region of indifference at all ; others have a region which may either increase or decrease in width on passing along the transition line to points at high temperatures, or even there may exist a region at one end of the line which may fade out to zero width in a finite distance on the transition line. In most cases, the width of the region does not exceed a few hundred kilograms, although in one extreme case, $AgNO_3$, a width as great as 1500 kg./cm.^{2} was observed. In the detailed paper will be found curves for a large number of substances giving the width of the region as a function of temperature along the transition line; the fact that the observed points lie on smooth curves lends additional strength to the contention that we are here dealing with a real phenomenon.

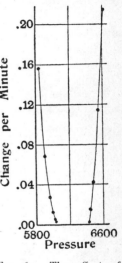

FIG. 65.—The effect of pressure on the transition velocity of the two modifications of yellow phosphorus at o°.

The curves of transition velocity against pressure, exemplified by fig. 65, have another interesting property, in that they are not symmetrical. It is obvious from the figure that the left-hand curve plunges into the axis less abruptly than the right-hand curve. The two limiting slopes where these curves strike the axis are characteristic of something in the transition mechanism; a number of these limiting slopes are given in the detailed paper as a function of position on the transition line. In practically all cases the behaviour is the same as that of the figure—that is, the slope is greatest on the high-pressure side. This is perhaps what might be expected; on this side of the transition line the transition runs with decrease of volume, which is the direction in which it would be urged by the pressure, so that as pressure is increased beyond the indifferent limit the transition can more readily acquire a high velocity than on the low-pressure side of the region, where it is running against the pressure. A similar asymmetry is to be expected on the melting line; it is not usually observed, however, because the speed of melting or crystallising is usually so great as to be entirely controlled by the necessity for conducting the latent heat away from the system, and this is a symmetrical phenomenon for not too great temperature differences. If, however, the liquid is one in which crystallisation takes place so slowly that it is not entirely controlled by thermal conduction, asymmetric effects are to be expected, and have in fact been observed in benzophenone, which crystallises very slowly.

The existence of a region of indifference means that the co-ordinates of the transition line on which the two phases are thermodynamically in equilibrium cannot be exactly determined by the method of volume discontinuity. The best that can be done under the circumstances is to locate the transition line in the middle of the region of indifference. The error so arising cannot be large, because as already stated the band is seldom more than 100 or 200 kg./cm.^{2} wide, which is small in comparison with total pressures of thousands of kilograms.

These velocity effects, so different for solid transitions from the corresponding melting or vaporisation phenomena, suggest that there must be a fundamental difference in the mechanism. It does not seem possible that the transition between two phases which have a region of indifference can be a matter of dynamic equilibrium between two streams of molecules going in opposite directions, but there must be something in the situation more static in character. The following gives a crude suggestion in a very much over-simplified case of what may possibly be involved. Imagine that the type of lattice structure can be specified by a single parameter, which is capable of continuous variation. Plot the potential energy of the lattice at constant pressure and temperature as a function of this parameter. Any lattice which is capable of physical existence will correspond to a minimum on the potential energy curve. If the pressure and temperature which we have chosen correspond to the co-ordinates of a point on the transition line, then on the potential energy curve there will be corresponding to the two modifications two minima with the same absolute value of the potential energy, because the two phases are in equilibrium with each other. Now consider that the pressure is displaced at constant temperature, first toward higher values and then toward lower values.

The one phase becomes relatively more stable for the one displacement, and the other for the other. This means a displacement of the curves for potential energy against lattice parameter something as indicated by the dotted lines in fig. 66. As the pressure displacement increases, the tendency shown by the two curves becomes accentuated, until at great enough displacements one or the other minimum entirely disappears.

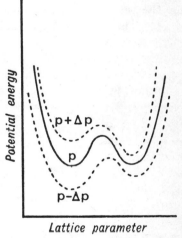

FIG. 66.—Schematisation of possible relative potential energies of two modifications in the neighbourhood of their transition pressure.

This picture suggests why it is that there may be a region of indifference. If one phase is able to pass into the other when the two are in contact at the point of thermodynamic equilibrium, it is necessary for some of the molecules in that phase to pass over the intermediate hill of potential which separates them from the other phase before they can settle down into their second position of equilibrium. The passage over the intermediate hill is facilitated by temperature agitation; if the hill is too high or temperature agitation too small, the hill cannot be surmounted, and the transition will not run in the neighbourhood of the equilibrium point. As pressure is displaced from the equilibrium point, however, the diagram suggests that the height of the hill of potential diminishes, until finally a point is reached where the haphazard energy of temperature agitation is sufficient to make the leap. If there were a truly Maxwell distribution of velocities, then some molecules would be capable of making the leap no matter how high the hill, and there would be no region of indifference. If we suppose that the definiteness with which the lattice parameter is distributed among the molecules varies with pressure and temperature along the transition line, as it must, then it is easy to see that the width of the region of indifference may change along the transition line. This picture of course is very crude. For example, I can see no necessary connection between the region of indifference and a Maxwell distribution, but I believe that qualitatively the actual situation must have features in common with that suggested above.

[1] E. Mallard and H. le Chatelier, *C.R.*, **97**, 102 (1883).

[2] L. Th. Reicher, *Rec. Trav. Chim. Pays Bas*, **2**, 246 (1883).

[3] G. A. Hulett, *ZS. f. Phys. Chem.*, **28**, 629 (1899).

[4] S. Lussana, *Nouv. Cim.*, **1**, 97 (1895).

[5] W. Wahl, *Trans. Roy. Soc.*, **212**, 117 (1912).

[6] E. Jänecke, *ZS. f. Phys. Chem.*, **90**, 257, 280, 313 (1915).

[7] J. C. Slater, *Phys. Rev.*, **23**, 488 (1924) ; *Proc. Amer. Acad.*, **61**, 135 (1926).

[8] L. Pauling, *ZS. f. Krist.*, **69**, 35 (1928).

[9] R. C. Wallace, *C. Bl. f. Min.*, p. 33 (1910).

[10] F. Wallerant, *Bull. Soc. Fr. Min.*, p. 311 (1905).

[11] F. C. Kracek, *Jour. Phys. Chem.*, **34**, 225 (1929).

[12] U. Behn, *Proc. Roy. Soc.*, **80**, 444 (1907–1908).

[13] W. Dennecke, *ZS. f. Anorg. Chem.*, **108**, 1 (1919).

[14] B. Roozeboom, *Die Heterogenen Gleichgewichte*, vol. i, p. 192. Vieweg und Sohn, Braunschweig, 1901.

CHAPTER IX

ELECTRICAL RESISTANCE OF METALS AND SOLIDS

REFERENCE has already been made in the historical introduction to measurements of the effect of pressure on electrical resistance. It will suffice here to very rapidly recapitulate this history, emphasising various points not previously made.

Wartmann,[1] in 1859, was apparently the first to search for the effect. He could find no change of resistance of a copper rod when subjected to a fluid pressure of 9 kg./cm.[2], but found an increase in resistance when the rod was compressed to an unknown pressure in sheets of gutta-percha between steel plates in a hydraulic press. His effect was doubtless mostly a temperature effect. Chwolson,[2] in 1882, found that the resistance of Cu, Pb, and brass decreases under a pressure of 60 kg./cm.[2], and he was able, furthermore, to establish that the resistance would also decrease when correction was made for the change of dimensions. Tomlinson,[3] in 1883, established the decrease of resistance of Cu and Fe to a pressure of a few hundred kilograms per sq. cm. Lussana,[4] in 1899, studied the effect of pressures to 1000 kg./cm.[2] on the resistance of a number of metals; he found two effects: a temporary effect immediately after the application of pressure, and a permanent effect, enduring during the application of pressure. No other observer, with one exception, has been able to repeat Lussana's results, and there can be little doubt that they are incorrect. A careful discussion of the various possible sources of error in Lussana's work will be found in the thesis of Lisell.[5] The pressure effects on liquid mercury were measured in 1882 by Lenz [6] to 60 kg./cm.[2], by Barus [7] in 1890 to 400 kg./cm.[2], and by de Forest Palmer [8] in 1897 to 2000 kg./cm.[2]. The latter measured the pressure with an Amagat free piston gauge, and his results were by far the most accurate in that field up to that time. Lisell,[5] in 1903, published results to 3000 kg./cm.² for six pure metals and a number of alloys, by far the most extensive work to that time. He reproduced the relation between pressure and resistance by a second-degree expression in pressure, and proposed the use of manganin as a secondary gauge. In 1907 Williams [9] investigated several metals to 700 kg./cm.², and first discovered the positive pressure coefficient of bismuth. In 1909 Lafay [10] measured manganin, platinum, and mercury to 3500 kg., and confirmed the linear relation found by Lisell between the resistance of manganin and pressure. In 1909 Montén [11] measured with Lisell's apparatus the resistance of Se and AgS. In 1911 Beckman [12] measured, also with Lisell's apparatus, the resistance of pyrites, iron glance (Fe_2O_3), and a number of alloys of Cd—Pb and Ag—Au. The pressure coefficient of pyrite was about $-2 \cdot 3 \times 10^{-5}$, and for Fe_2O_3 $-6 \cdot 8$ to $-8 \cdot 2 \times 10^{-6}$. Beckman has published other papers [13] on the same subject: in 1912 on the Au—Ni alloys, and in 1917 on Tl, Ta, Mo, and W to 2000 kg./cm.². In 1917 I published the first of a series of papers, and these will now be discussed in detail (B. 26, 27, 28, 32, 34, 35, 36, 37, 40, 45, 49, 53, 54, 62, 65, 66, 67, 72, 73).

The early measurements of the effect of pressure on resistance were very discordant, so that this came to be regarded as a difficult subject for physical investigation—much more difficult, for example, than the compressibility of a liquid. The chief cause of the early difficulty was the smallness of the effect, due to the narrowness of the pressure range; in fact, the effects were so small that they might be entirely masked by the effects arising from changes of temperature incidental to changing pressure. This can be realised when it is considered that there are several metals for which the effect on the resistance of a change of temperature of 1° is equal to that of a pressure change of 3000 or 4000 kg./cm.². But with increase of the pressure range to 12,000 kg./cm.² these difficulties disappear, and the measurement of the effect of pressure on resistance is now one of the easiest of pressure measurements.

There are several precautions to be observed in making such measurements. The temperature, of course, must be kept constant, or else must be measured and a suitable correction applied for its variation, to an accuracy corresponding to the pressure coefficient of the particular metal and the desired accuracy of the results. The metal, particularly if it is high melting and capable of holding internal strains, should be seasoned by preliminary temperature changes, and also by several preliminary applications of high pressure. Pressure should be transmitted by a liquid which does not freeze or become so viscous under pressure as to produce stresses in the metal. The terminals should be carefully attached, and the area over which attachment is made should be small enough so that slight relative motion of the various pieces of metal at the contact, arising from compressibility differences of the metals and solder, may not lead to measurable changes in the resistance.

Two different methods were employed, depending on the dimensions of the metals. A number of metals could be drawn into fine wire and insulated with silk. These were wound non-inductively into coils exactly like the manganin gauge coil, and the resistance measured in the same way on a Carey Foster bridge. Other metals, which could only be obtained in relatively coarse pieces of low resistance, were measured with a potentiometer. Four terminals were attached to the specimen, two current and two potential terminals, and the difference of potential at the potential terminals compared by a null method with the drop of potential produced by the same current flowing through a known combination of resistances. In using this latter method a special insulating plug has to be used in which three insulated terminals are carried into the pressure apparatus.

In addition to solid metals, a number of liquid metals have been measured. Most of these can be successfully measured by enclosing them in fine glass capillaries, provided with four sealed-in platinum terminals, and using the potentiometer method. There is a difficulty with the platinum contacts, which are almost certain to crack the glass because of unequal compressibility between glass and platinum. The difficulty may be minimised by using fine platinum wire, 0·001 in. in diameter; it is practically impossible to prevent the glass from cracking, but if leads as fine as this are used, the glass will usually hold together even after it has cracked.

In the thirteen years since my first measurements were published, the effect of pressure has been measured on the resistance of 52 pure metals, 5 of them in the form of single crystals, and 24 alloys. It is my purpose to keep the list as complete as possible, and I try to add to it whenever a new metal becomes available, or an old metal in a state of higher purity.

Resistance of Elements.—The first investigations of the effect of pressure did not suggest any notable variation in the behaviour of different metals. The early measurements were made on the common metals, which are most easily formed into wire, which have comparatively high melting-points, and which are comparatively incompressible, such as Fe, Ag, Cu, and Pt. The resistance of these was found to decrease with pressure, but by rather small amounts. In 1907 Williams discovered the abnormal positive pressure coefficient of Bi. This remained an exception for some time, other common metals, as they were added to the list, having also negative coefficients. But as the unusual places in the periodic table were explored, a wealth of behaviour at first unexpected has been disclosed, so that it is much more difficult to characterise the effect of pressure on resistance in a broad way than at first seemed possible. In the following discussion the method of classification will be from the simple to the complex; a more significant classification would be desirable. Classification according to the position in the periodic table might be thought to satisfy this condition, but this particular scheme would not be satisfactory, for the abnormalities of behaviour are scattered through the periodic table in an apparently haphazard manner. It will be necessary in this discussion to deal separately with those metals which crystallise in the cubic system and those which crystallise in some other system; the former have the same resistance in every direction, and therefore have only a single characteristic pressure variation, whereas the resistance of the latter varies with direction, and the pressure coefficient is also a function of direction.

As already intimated, the high melting, mechanically hard, strongly metallic elements all have pressure effects qualitatively alike, in that resistance decreases under pressure by something of the order of 1 or 2 per cent. for a pressure of 12,000 kg./cm.^{2}. The change of resistance is not linear with pressure, but as the pressure increases the change becomes less—that is, resistance plotted against pressure gives a curve convex toward the pressure axis as in fig. 67. In fig. 68 the actual resistance of sodium is shown as a function of pressure over a sufficiently wide temperature range to include in one diagram both solid and liquid metal. The curvature is very pronounced for sodium. The experimental accuracy is great enough, so that it can be definitely established in most cases that a second-degree expression in the pressure is not adequate to reproduce the results. No unique statement can

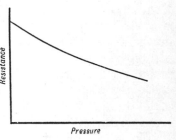

Fig. 67.—Shows the direction of curvature of practically all falling resistance curves.

be made about the third-degree term, there being examples of both positive and negative terms, but it is perhaps more common that it should be negative, so that in the majority of cases the change of resistance may be written as $\Delta R/R_0 = -ap + bp^2 - cp^3$, where a, b, and c are all positive. In the original papers the results may be found given in graphical form, which is more accurate than a power-series representation. It may be said that the reproduction of my results in *International Critical Tables* in two constant forms does not usually retain all the accuracy of the original results. The principal results for the elements and two compounds, TiN and TiC, are summarised in Table XIV, in which the relative resistances are given at 0, 4000, 8000, and 12,000 kg./cm.^{2} at usually two temperatures. In this table those substances are noted which satisfy a linear or quadratic relation.

The effect of pressure on a great many metals was determined at a number of temperatures between 0° and 100° C. It was found that there is comparatively little change of pressure coefficient with temperature. A consequence of this is that the temperature coefficient of resistance is nearly

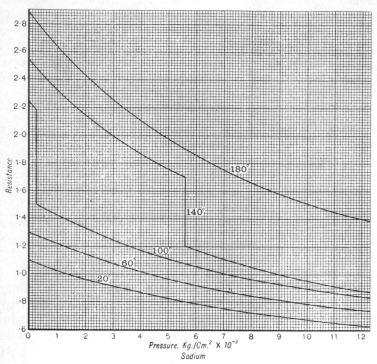

FIG. 68.—Relative resistances of sodium at constant temperature as a function of pressure.

independent of pressure. This might be expected in view of the fact that the temperature coefficient of so many metals is nearly the same and equal to $\frac{1}{\tau}$. Since a metal free from pressure, and the same metal exposed to pressure may be regarded as a special case of two different metals, it is not surprising that the temperature coefficient should be little affected by pressure.

TABLE XIV

RELATIVE RESISTANCES UNDER PRESSURE

Element				Pressure, kg./cm.^{2}			
			0	4000	8000	12,000
Li	Solid	0°	1·0000	1·0285	1·0594	1·0927
		50°	1·2122	1·2468	1·2842	1·3246
		100°	1·4580	1·4996	1·5446	1·5932
	Liquid, linear	202°	1·0000	1·0371	1·0742	1·1124
		237°	1·0507	1·0897	1·1287	1·1688
Be	Second degree	30°	1·0000	·99575	·99189	·98841
		75°	1·1332	1·12538	1·11850	1·11256
C	Graphite Amorphous, results irregular	50°	1·0000	1·0125	1·0179	1·0183 (?)
		30°	1·0000	·9332 (6000)		·8793
Na	Solid 12,000 kg. range	0°	1·0000	·7924	·6658	·5854
		80°	1·4227	1·0971	·9100	·7908
	Liquid	140°	2·5399	1·866	1·053	·885
		200°	3·0725	2·235	1·758	1·464
	18,000 kg. range	30°	1·0000	·7156 (6000)	·5757 (12,000)	·4974 (18,000)
Mg	Second degree	0°	1·00000	·98168	·96513	·95104
		100°	1·39000	1·36510	1·34286	1·33298
Al	Second degree	0°	1·00000	·98448	·96797	·95046
		100°	1·46680	1·44227	1·41886	1·39657
Si		0–100°	1·00	Rough average, material impure		0·86

TABLE XIV—continued

Element			0	Pressure, kg./cm.² 4000	8000	12,000
P (black)		0°	1·000	·372	·1079	·0297
		50°	·662	·239	·0766	·0238
		100°	·421	·1517	·0542	·0209
K	Measured resistance on bare wire	0°	1·000	·564	·372	·271
		30°	1·163	·648	·421	·305 { ·242 (16,000) ·211 (20,000)
	Measured in glass capillary. Results corrected to give relative specific resistance	60°	1·307	·696	·464	·330
		95°	2·387	·777	·508	·350
		130°	2·724	1·346	·551	·369
Liquid below dotted line		165°	3·040	1·586	·928	·389
Ca		0°	1·0000	1·0447	1·0963	1·1550
		50°	1·1552	1·2039	1·2599	1·3329
		100°	1·3327	1·3865	1·4470	1·5146
Ti, impure		20°	Mean coefficient $+ 1 \times 10^{-7}$ (?)			
Cr	Linear	30°	1·0000	·9977	·9954	·9930
Mn, not pure	Second degree	30°	1·00000	·97285	·94750	·92396
Fe	Second degree	0°	1·00000	·99057	·98152	·97287
		100°	1·62060	1·60535	1·59010	1·57484
Co	Linear	30°	1·0000	·9963	·9925	·9888
Ni	Second degree	0°	1·00000	·99255	·98523	·97804
		100°	1·63450	1·62204	1·60988	1·59804

TABLE XIV—*continued*

Element			Pressure, kg./cm.²			
			0	4000	8000	12,000
Cu		0°	1·00000	·99222	·98504	·97802
		100°	1·42930	1·41918	1·40906	1·39894
Zn, single crystal	Crystal axis 90° to length	0°	1·00000	·95515	·91406	·87672
		95°	1·39100	1·32938	1·27340	1·22304
	Crystal axis 7° to length	0°	1·03900	1·01814	·99927	·98287
		95°	1·41700	1·38547	1·35113	1·31399
Ga, linear	Solid	0°	1·0000	·9901	·9802	·9704
	Liquid	30°	0·6456	·6308	·6171	·6044
		100°	0·6824	·6667	·6522	·6386
As		30°	There is a transition at 5500 kg. The pressure coefficient of the low-pressure modification varies greatly with direction, and may even have different signs in different directions. An average coefficient between 0 and 5500 kg. is $-1·5 \times 10^{-6}$.			
Rb	Solid	0°	1·000	·583	·428	·360
		30°	1·176	·653	·474	·397 { ·367, 16,000; ·365, 18,000; ·369, 20,000 }
Values for liquid are relative specific resistances	Liquid	65°	2·195	1·579 (2000)	·565	
		95°	2·417	1·595 (2000) 820		
Sr		0°	1·0000	1·2220	1·4916	1·8160
		100°	1·3828	1·6214	1·9381	2·3144
Zr	Second degree	30°	1·00000	·99838	·99697	·99565
		75°	1·16340	1·16071	1·15824	1·15598

TABLE XIV—*continued*

Element			Pressure, kg./cm.²			
			0	4000	8000	12,000
Cb (unpublished) Linear		30°	1·0000	·9953	·9906	·9859
		75°	1·0925	·0870	1·0815	1·0761
Mo Second degree		0°	1·00000	·99475	·98960	·98457
		100°	1·43360	1·42622	1·41897	1·41184
Rh Second degree		35°	1·00000	·99320	·98672	·98054
		65°	1·12600	1·11818	1·11073	1·10364
Pd		0°	1·00000	·99221	·98469	·97726
		100°	1·31780	1·30784	1·29802	1·28834
Ag		0°	1·00000	·98588	·97259	·96002
		100°	1·40740	1·38767	1·36872	1·35063
Cd Single crystal at 0° C. Hex. axis 5° with length „ „ 87° „ „			Average pressure coefficient to 2000 kg. −1·450 × 10⁻⁵ „ „ „ „ „ −0·693 × 10⁻⁵			
In		30°	1·0000	·9451	·8974	·8550
		75°	1·1990	1·1300	1·0699	1·0186
Sn Resistance in the parallel orientation taken as unity	Tetragonal axis 11° to length	0°	·9897	·9499	·9130	·8791
		95°	1·4093	1·3497	1·2968	1·2474
	Tetragonal axis 74° to length	0°	·8546	·8228	·7936	·7671
		95°	1·2315	1·1861	1·1451	1·1066
Sb Figures are very nearly the actual specific resistances	Trigonal axis 7° to length	30°	34·60 × 10⁻⁶	37·78	41·70	46·20
		75°	42·47	45·94	50·02	54·60
	Trigonal axis 83° to length	30°	44·41	45·11	45·80	46·50
		75°	53·37	53·51	53·78	53·78

TABLE XIV—*continued*

Element			Pressure, kg/cm.²			
			0	4000	8000	12,000
Te,	Multi-crystalline	0°	1·000	·600	·370	·308
I			Resistance remains very high up to 8000 kg.			
Cs	Solid, bare wire	0°	1·000	·709	·794	·994
	Liquid in glass {	63°·4	2·099	1·721 (1500)		
		95°·8	2·280	1·851 (1500)	1·661 (3000)	
Ba		0°	1·0000	·9764	·9705	·9764
		75°	1·4380	1·3805	1·3586	1·3620
La	Linear	30°	1·00000	·99520	·99041	·98561
	Second degree	75°	1·08640	1·07871	1·07134	1·06433
Ce	Low - pressure modification, { linear	30°	1·0000	·9823	Coefficient of high-pressure modification is independent of temperature, and equals —1·42 × 10⁻⁵ in terms of resistance at 9000.	
Two modifications		75°	1·0485	1·0369		
Pr	Linear	30°	1·0000	·9988	·9975	·9963
		75°	1·0668	1·0633	1·0597	1·0562
Nd	Second degree	0°	1·00000	·99079	·98227	·97444
		50°	1·04000	1·02983	1·02043	1·01180
Ta	Second degree	0°	1·00000	·99413	·98841	·98284
		100°	1·29730	1·29044	1·28173	1·27417
W	Second degree	0°	1·00000	·99448	·98911	·98385
		100°	1·42090	1·41300	1·40522	1·39757

TABLE XIV—*continued*

Element			0	4000	8000	12,000
Hf	Linear	30°	1·0000	·9960	·9920	·9880
Ir	Second degree	30°	1·00000	·99465	·98943	·98434
		95°	1·18870	1·18240	1·17625	1·17025
Pt	Second degree	0°	1·00000	·99224	·98476	·97756
		100°	1·38680	1·37638	1·36618	1·35622
Au		0°	1·00000	·98791	·97650	·96554
		100°	1·39680	1·37995	1·36369	1·34789
Hg		0°	1·0000	·8894	·2409	·2192 Solid
	Liquid	50°	1·0480	·9268	·8377	·7684
		100°	1·0959	·9611	·8650	·7895
Tl		0°	1·0000	·9487	·9033	·8619
		100°	1·5170	1·4329	1·3592	1·2938
Pb		0°	1·0000	·9454	·8976	·8546
		100°	1·4207	1·3428	1·2722	1·2071
Bi Second degree, may extrapolate to 20,000	Trigonal axis 86° to length	30°	1·0000	1·0466	1·1027	1·1683
		75°	1·1788	1·2336	1·2985	1·3734
	Trigonal axis 8° to length	30°	1·2630	1·3710	1·5316	1·7150
		75°	1·4869	1·6189	1·7802	1·9707
Resis. at 271° at atmo. pressure unity	Liquid	275°	1·0019	·9584	·9167	
		260°		·9520	·9088	·8783
		240°	··	··		

TABLE XIV—*continued*

Element			0	4000	8000	12,000
				Pressure, kg./cm.²		
Th	Second degree	30°	1·00000	·98915	·97891	·96928
		75°	1·09690	1·08428	1·07241	1·06131
Ur	Linear	30°	1·0000	·9826	·9651	·9477
				Compounds		
TiN	Linear	30°	1·00000	·99626	·99252	·98878
		75°	1·07950	1·07519	1·07089	1·06659
TiC	Linear	30°	1·00000	·99450	·98800	·98350
		75°	1·00953	1·00408	·99863	·99318

Examples of metals to which the foregoing remarks apply are Fe, Co, Ni, Rh, Pd, Ir, Pt, Cu, Ag, Au, Zr, Hf, Th, Cb, Pr, La, Ta, Cu, Mo, W, and U. In this group the smallest pressure coefficient numerically is that of Zr, which is $-\cdot32 \times 10^{-6}$, and the largest is that of Ag, which is $-3\cdot83 \times 10^{-6}$. As might be expected from the approximate independence of the temperature coefficient of pressure, the pressure coefficient has a tendency to vary with temperature by a larger amount when the coefficient is absolutely small than when it is large. Thus the pressure coefficient of Zr is $-4\cdot3 \times 10^{-7}$ at $30°$ and $-6\cdot0 \times 10^{-7}$ at $75°$, while that of Ag is $-3\cdot62 \times 10^{-6}$ at $30°$ and $-3\cdot59 \times 10^{-6}$ at $75°$. In spite of the comparatively large percentage change in the pressure coefficient of Zr, the mean temperature coefficient between $30°$ and $75°$ decreases only from $0\cdot00351$ at atmospheric pressure to $0\cdot00346$ at $10,000$ kg./cm.^{2}.

An important fact to notice here is that the effect of pressure on these metals is a specific effect. The coefficient given is the coefficient measured in the ordinary way with terminals attached to the specimen. But the dimensions change under pressure, so that a correction must be applied to the pressure coefficient of measured resistance to obtain the pressure coefficient of specific resistance. The correction to be applied is obviously the linear compressibility, the measured resistance being too small by the linear compressibility because of the decrease of length, and too large by twice the linear compressibility because of the decrease of cross-section. In the case of Zr, where the correction is a maximum, the pressure coefficient of specific resistance at $30°$ is $-8\cdot30 \times 10^{-7}$ against $-4\cdot31 \times 10^{-7}$ for the pressure coefficient of measured resistance. For most of the other metals the effect is much less; thus the coefficient of specific resistance of Fe at $30°$ is $-2\cdot62 \times 10^{-6}$, and the coefficient of measured resistance is $-2\cdot42 \times 10^{-6}$.

After the group of the hardish metals with high melting-points, there is a group of softer metals with lower melting-points. The metals included in this list are Al, Mg, Zn, Pb, Cd, Tl, Sn, In, and Hg. Not all of these are cubic. Those which have been examined in the single crystal form will be discussed in detail later; in the meantime the average pressure coefficient for the multi-crystalline, haphazardly arranged metal will be understood. In this series the pressure coefficient runs from $-4 \cdot 1 \times 10^{-6}$ for Mg to $-15 \cdot 0 \times 10^{-6}$ and $-23 \cdot 6 \times 10^{-6}$ for In and solid Hg. The general tendency is for the curvature of the plot of resistance against pressure to become greater, the greater the absolute value of the coefficient. Thus for In the pressure coefficient $\dfrac{1}{R_0} \dfrac{dR}{dp}$ has the value $-15 \cdot 0 \times 10^{-6}$ at o kg. and $-9 \cdot 9 \times 10^{-6}$ at 12,000 kg./cm.², a decrease of 34 per cent., while for Al the corresponding figures are $-4 \cdot 16 \times 10^{-6}$ and $-3 \cdot 47 \times 10^{-6}$, a decrease of 16·6 per cent. This relation is not by any means universal; thus the change of the coefficient of Mg with pressure is nearly as large as that of In. It is true of these metals also that the temperature coefficient is nearly independent of pressure; there are slight variations and no uniformity about the sign; the pressure coefficient may either increase or decrease with temperature.

This completes the list of metals that one is tempted to describe as normal. There are besides these some fifteen or sixteen others that may be classified for one reason or another as abnormal. It is evident that when so large a proportional number must be described as abnormal no great significance can be attached to the designation 'normal.'

As already mentioned, Bi was the first metal abnormal with respect to its pressure behaviour to be measured; Williams found its pressure coefficient to be positive. I have extended the measurements to 12,000 kg./cm.², and in one case to 18,000 kg./cm.² for single crystal rods of different orientations. The

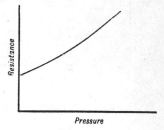

FIG. 69.—Shows the direction of curvature of practically all rising resistance curves.

first result is that the pressure coefficient becomes increasingly positive over this range; this means that the curve of resistance against pressure is convex toward the pressure axis as shown in fig. 69, as it was also in the cases of decreasing resistance.   The actual curves of resistance against pressure at several temperatures for strontium are shown in fig. 70; the upward curvature is very pronounced.   But whereas this was

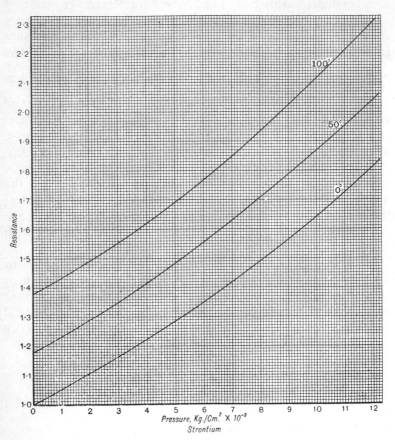

Fig. 70.—Relative resistances of strontium at several constant temperatures as a function of pressure.

entirely to be expected when resistance diminishes with pressure, it is quite the opposite of what might be expected when resistance increases.   One expects a sort of law of diminishing returns as pressure increases, the effect of equal increments of pressure diminishing as pressure increases, because of a kind of exhaustion of the possibilities.   This is what happens for those metals whose resistance diminishes with increasing pressure, but in the case of Bi the effect of equal increments of pressure increases as pressure increases, so that we have instead a law of increasing returns.

The effect of crystal orientation is very marked, the pressure coefficient when current flows perpendicular to the basal plane, which is also the plane of principal cleavage, being nearly twice as great as for flow in the basal plane. The following results can be found by extrapolation of experiments made at orientations of 8° and 80° :—

TABLE XV

Angle between basal plane and length	Temperature	
	30° C. $\Delta R/R$ (0 kg., 30°)	75° C. $\Delta R/R$ (0 kg., 75°)
0°	$1\cdot039 \times 10^{-5}p + 2\cdot94 \times 10^{-10}p^2$	$1\cdot047 \times 10^{-5}p + 2\cdot65 \times 10^{-10}p^2$
90°	$2\cdot026 \times 10^{-5}p + 8\cdot16 \times 10^{-10}p^2$	$1\cdot990 \times 10^{-5}p + 6\cdot19 \times 10^{-10}p^2$

It is thus seen that the resistance across the basal plane increases more rapidly with pressure than the resistance in that plane. But the resistance across the basal plane is at atmospheric pressure greater in the ratio of 139 to 109 than the resistance in the plane, so that this means that the difference between these two directions becomes accentuated at high pressures. In fact, the ratio of the resistance in the two directions increases from 1·275 at atmospheric pressure to 1·48 at 12,000 kg./cm.².

From one point of view the accentuation by pressure of the excess of resistance for flow perpendicular to the basal plane is not to be expected, because one explanation for this excess has been that the cleavage plane constitutes a sort of flaw in the crystal which makes the passage of current more difficult. This point of view would lead to the expectation that under pressure the separation of the cleavage planes would be diminished, and therefore the difference of resistance decreased. The fact that pressure works in the other direction suggests that the abnormal pressure effect is in some way connected with the lattice structure.   This is confirmed by the behaviour of liquid Bi under pressure.   Liquid Bi is quite 'normal,' in that its resistance decreases under pressure; furthermore, the curve of resistance against pressure is convex toward the pressure axis, and the temperature coefficient is nearly independent of pressure.   At $275°$ C. the initial pressure coefficient is $-12\cdot3 \times 10^{-6}$, which is in the same range of magnitude as for other low-melting metals, such as Pb or Sn.   This question of the resistance of the liquid will be referred to again later.

It was natural to expect that Sb would be abnormal as well as Bi, and this, indeed, turned out to be the case.   In 1917 I found that the resistance of ordinary polycrystalline Sb increases with pressure, the average coefficient to 12,000 kg./cm.^{2} being $12\cdot2 \times 10^{-6}$.   The effects were not regular, there being hysteresis and permanent change of zero after application of pressure.   Later, measurements were made on the effect of pressure on single crystal rods of various orientations.   Under these conditions the irregularities disappeared, and the resistance became a single valued function of pressure. Doubtless the irregular results for polycrystalline material are due to internal strains arising from the unequal compressibility of the single crystal grains in different directions.   Such effects are always to be expected in aggregates of non-cubic metals, a consideration which apparently is not properly appreciated.   For example, irregularity is to be expected in the thermal behaviour of such substances, because of the unequal thermal expansion in different directions; it may well be that something of this sort is the explanation of the polymorphic transition of Sb that has been claimed near $150°$, and the explanation of numerous other irregularities in the behaviour of such metals as Zn and Cd.

The resistance of single crystal Sb under pressure has many other abnormalities than that of sign.   At atmospheric pressure Sb is the only crystal yet measured in which the resistance across the planes of easiest cleavage is less than in the plane of cleavage.   The effect of pressure is to tend to make Sb like other metals, and in fact at 12,000 kg./cm.^{2} and $75°$ the resistance has become greater across the cleavage

plane than parallel to it, while at a temperature lower than 75° a somewhat higher pressure would be necessary to bring about this reversal. This reversal involves a great difference of pressure coefficient in different directions. Parallel to the basal plane the effect of pressure depends to a great extent on the temperature; at 30° resistance increases linearly with pressure by a comparatively small amount, while at 75° resistance at first increases slightly with the normal direction of curvature, but presently the direction of curvature reverses, and between 10,000 and 11,000 the resistance passes through a maximum, and then falls. This is the only known case of a maximum, and also nearly the only case of abnormal curvature. Perpendicular to the basal plane, on the other hand, resistance increases with pressure at both 30° and 75°, and with the normal direction of curvature over the entire pressure range. The mean coefficient at 30° for this orientation is $2 \cdot 80 \times 10^{-5}$.

Sb is also unusual in that there is considerable variation of temperature coefficient with direction, the average coefficient 0°–100° for flow across the basal plane being 0·0048, and for parallel flow 0·0057.

No measurements have yet been made on liquid Sb.

Arsenic, next in the chemical series above Bi and Sb, might be expected to show similar abnormalities. My measurements on As (B. unpublished) are entirely preliminary in character, but enough has been done to show that As is also abnormal. It is highly crystalline in properties, and therefore measurements on single crystals are demanded; I have not yet succeeded in getting single crystals of this of suitable dimensions. Arsenic is different from Bi and Sb in that there is a second modification stable at high pressures, the transition pressure at room temperature being in the neighbourhood of 5500 kg./cm.². This modification of As may be the analogue of the high-pressure modification of Bi, which has been suspected by analogy with water because of the abnormal expansion of Bi on freezing, but which has never been found. The results on As at pressure above the transition pressure were irregular. The probable interpretation of the results, for which the detailed paper must be

consulted, is that the pressure coefficient of resista   may
be positive for some directions in the single cry   and
negative for others, a possibility entirely consistent   the
behaviour of Sb at high pressures. The compres   ity
measurements on the low-pressure modification in   ed
that the compressibility varies greatly with direct   so
that variations in the pressure coefficient of resistan   are
not to be wondered at.   It was not possible to get sign   nt
measurements of the effect of pressure on the resistan   of
the high-pressure modification.   There is evidently room for
much further work with As.

Above As in the periodic table is vanadium; the pressure
effect on the resistance of this has not yet been measured.
Above V is phosphorus. The yellow and red modificatio s
of P are non-conductors of electricity, but the black moc
fication is a conductor of approximately the goodness
graphite. The effect of pressure on its resistance has bec
studied. Black P is anomalous in the enormous magnituc
of the effect of pressure; resistance is decreased by pressur
but the magnitude of the effect is quite without parall
in that at $0°$ the resistance at 12,000 kg./cm.^{2} is less tha
3 per cent. of its value at atmospheric pressure. There is r.
irregularity whatever in the behaviour of this substance, bu
resistance is a smooth single-valued function of both tem
perature and pressure. The direction of curvature of resis^t
ance against pressure is normal, as indeed it must be if the
curve is not to have a point of inflection. The specific
resistance of this substance at $0°$ at atmospheric pressure
is 1·00 ohm per cm. cube, and 3×10^{-2} at $0°$ at 12,000 kg./cm.^{2}.
The effect of pressure is thus to push the order of magnitude
of the resistance toward that of metals, but metallic resistance
is still far from reached, the resistance of metallic mercury,
for example, being in the neighbourhood of 10^{-4}. The
approach toward metallic resistance increases at a somewhat
accelerated pace as pressure increases; at $0°$ the first 6000
kg./cm.^{2} decreases the resistance to 20·4 per cent. of its initial
value, and the second 6000 produces a further decrease to
14·5 per cent. of the resistance at 6000. The non-metallic
character of the resistance under ordinary conditions is fur-

temperature coefficient, which is nega-
pressure tends to produce a displacement
viour of metals; at atmospheric pressure an
mperature from 0° to 100° reduces the resistance
its value at 0°, whereas at 12,000 kg./cm.² the
ding reduction is only 0·704.

hus see that the four members of the fifth column of
eriodic table, Bi, Sb, As, and P are highly anomalous
effect of pressure on resistance.   The alternate members
e column, Cb, Pr, and Ta do not seem to have any
al distinction.

the sixth column, perhaps Se should be called an
ormal element.   It has been measured by Montén [11] to
o kg./cm.², who found an enormous decrease of resistance
rom 0·005 to 0·02 of its initial value.   However, the pheno-
a in Se are known to be very much complicated by the
tence of two or more solid phases, which may be present
arying proportions.   Montén found a very large per-
ent decrease of resistance after his experiments, the final
nce varying from 0·5 to 0·2 of the initial resist-
o that it is obvious that he was not dealing with a
vell-defined substance, and the results are difficult
o in ret.

Th urth column in the periodic table contains several
abnorr substances.   C is difficult to get in a state of satisfactory urity, and my results can be regarded only as
prelimir ry.   In the amorphous condition, as arc carbon,
the resis nce was found to decrease under pressure.   Two
samples o raphite, on the other hand, one the Acheson
graphit mmerce, and the second a piece of specially
pure g the research laboratory of the Acheson
Co., b ncrease of resistance under pressure.
The tw ve different results, the first increasing
in resista it 4·5 per cent. for 12,000 kg./cm.², and
the second b 2 per cent.   Experiments should be made
on single crys f graphite.   Si, under C in the fourth
column, is appa tly normal, with a negative pressure
coefficient of −12 10⁻⁶.   However, the purity of the sample
of Si was probabl ot very high.   Ti, under Si, was available

only in a piece of inconvenient dimensio
inferior purity.   The effect of pressure on t
and probably positive, but there is no reaso
the effect of pressure on the pure metal may no
The only other element in this column which
measured having any claim to abnormality is Ce.
second modification at high pressures, the transition p
being about 7600 kg. at 30° and 9400 at 75°.   The res
of the low-pressure modification is abnormal in t
increases with pressure; the increase is linear wit
sure within the experimental error, with the coe
$+4\cdot42\times10^{-6}$ at 30° and $+2\cdot77\times10^{-6}$ at 75°.   The va
of the pressure coefficient with temperature is larger
usual.   The high-pressure modification is normal,
pressure coefficient of about -14×10^{-6}.   The abnorm
of the low-pressure modification are not confined
resistance, but the compressibility is also abnormal,
respect to its variation with pressure, becoming gre
higher pressures, and its temperature variation.

In the third column the only element measured tl
any claim to abnormality is Ga; here the abnormal
not in the behaviour of the resistance, but in the fact ti
is one of the few substances whose liquid is denser tha
solid, so that the melting temperature falls with
pressure.   In spite of this abnormality in the volume,
cating some unusual feature in the lattice structure of
solid, the resistance of both solid and liquid behaves regula
in that both the sign and the direction of curvature
normal.   The initial pressure coefficient of the solid at 0°
$-2\cdot5\times10^{-6}$, and that of the liquid at 30°
It is normal for the effect of pressure to be
liquid than on the solid phase, but so large
perhaps unusual.

In the second column there are several abnor elements.
Be, the first member of the column, was expected to be
abnormal because of its chemical proximity to other elements
which are abnormal, but this did not turn out to be the case.
Ca, the third element in the column, is abnormal in that the
resistance increases with pressure.   At 0° the average pressure

coefficient to 12,000 kg. is $+12\cdot9\times10^{-6}$. The curvature is also in the normal direction—that is, resistance increases at an accelerated pace at the higher pressures. The effect of temperature is about as usual; the mean temperature coefficient between $0°$ and $100°$ decreases from $0\cdot00333$ at atmospheric pressure to $0\cdot00311$ at 12,000, a decrease not large, but still perhaps somewhat larger than usual.

Sr, the next alkali earth below Ca, has a very large positive pressure coefficient of resistance, in fact the largest yet observed, the effect of 12,000 kg./cm.^{2} at $0°$ being to increase the resistance by $81\cdot6$ per cent. The direction of curvature is normal, the same as that of Ca. There is an unusually large change of pressure coefficient with temperature; the mean temperature coefficient $0°$–$100°$ is $0\cdot00383$ at atmospheric pressure and $0\cdot00275$ at 12,000 kg./cm.^{2}. The temperature variation takes the form of an unusually rapid drop of the initial pressure coefficient with rising temperature, so that it is not at all unlikely that at some temperature above $200°$ the initial pressure coefficient of resistance may be negative instead of positive; this would involve the phenomenon of a minimum of resistance with pressure above this temperature. The same sort of thing is shown by Ca, but to a very much less extent, so that the temperature at which the minimum might be expected is very much higher.

Ba is the next alkali earth below Sr. I expected a record-breaking positive coefficient, judging from the sequence Ca—Sr. Ba is not easy to obtain in a state of high purity as a massive metal, and it was only after several years of watchful waiting that I was fortunate enough to obtain a specimen of high purity from Dr. A. J. King of Syracuse University. My disgust can be imagined when my first measurements showed a decrease of resistance with pressure. I was then and there nearly cured of the temptation to attempt to predict the effect of pressure on the resistance of a metal from its position in the periodic table. However, Ba soon atoned for itself by showing the most interesting phenomenon of a minimum of resistance with pressure. At $0°$ this minimum occurs at 8100 kg./cm.^{2} and at $75°$ at 9600 kg./cm.^{2}. The scale of resistance changes is not great compared with either Ca or Sr; at 0° the resistance at the minimum is 0·9764. The effect of temperature is such that a temperature coefficient of 0·00584 at atmospheric pressure drops to 0·00527 at 12,000 kg./cm.².

In the first column of the periodic table is the group of five alkali metals; these have been long known to stand apart from the other elements by many properties, such as high atomic volume and high compressibility, the element Cs, for example, being considerably more compressible than water. It would be natural to expect unusual effects of pressure on resistance. The first measurements of the effect of pressure on resistance were made on Na and disclosed only an unusually large decrease of resistance, a fact in line with the high compressibility. Measurements on K next showed an even larger decrease, which was not surprising in view of the still larger compressibility of K. Then Li was tried, and it was found unexpectedly, in view of the negative effects already found, that its resistance increases under pressure. The direction of curvature was normal, and the pressure coefficient is nearly independent of temperature. The average coefficient to 12,000 is $+7·7 \times 10^{-6}$, and the initial coefficient is about 10 per cent. less than this. The coefficient of liquid Li was then investigated. For this a special technique was necessary, because of the great chemical activity of liquid Li. The method finally used was to enclose the liquid Li in a thin-walled capillary of a high-resistance Fe—Ni—Co alloy, and to measure the resistance of the capillary and Li together under pressure, and then correct for the effect of pressure on the resistance of the capillary, determined by separate experiment. The resistance of the liquid was found also to increase under pressure, the coefficient being $+9·27 \times 10^{-6}$, independent of pressure and temperature within the limits of error, which were less than 1 per cent. The effect on the liquid is thus greater than on the solid, which is in line with its greater compressibility. It is evident that the mechanism of the positive pressure coefficient must be different in Li from what it is in Bi, in that in the latter it is connected in some way with the lattice structure, and reverses sign when the solid melts, whereas in Li the coefficient is positive in both liquid and solid, and is larger in the liquid.

Cs was the next of the alkalies examined. There were at first difficulties of technique. In measuring the electrical properties of the alkali metals it has become the custom, because of their great chemical activity, to operate on them while enclosed in glass capillaries. When a solid metal is treated in this way there are apt to be serious internal strains developed because of unequal distortion of the metal and the glass. This turned out to be an important source of error in previous measurements of the temperature coefficient of resistance at atmospheric pressure of the alkali metals, but when the attempt was made to measure the pressure coefficient of resistance of these metals enclosed in glass capillaries the error proved to be so much greater as to be prohibitive. This made it necessary to work with bare wires of the alkali metals; these were formed by extrusion, the terminals were attached to them by threading fine wires of Ag through them, great care being necessary in the handling because of the mechanical softness and the chemical activity. The difficulties with Cs were greater than with the others, because it is by far the softest of the five metals, and it is also the most reactive chemically. There was further difficulty in that Cs is difficult to obtain in a state of sufficient purity. Cs, and to a less extent Rb, has the unusual property of dissolving its own oxide, so that unusual care is necessary in its preparation, demanding repeated distillation at the lowest possible temperature. The first attempts to measure the pressure coefficient of Cs disclosed an anomaly at high pressures which was interpreted as meaning a new modification. It was only after repeated attempts that clean-cut results could be obtained, and it could be established beyond doubt that the high pressure anomaly arose from a rather sharp minimum in the resistance at only 4000 kg./cm.². The initial rate of decrease of resistance with pressure is greater than for any other measured metal, being at the rate of about 22 per cent. for 1000 kg./cm.², a figure only slightly exceeded by black P. But the curvature is so great that at 4000 kg./cm.² a minimum is reached at a resistance 0·71 of the initial value, and from here it rises at a somewhat slower rate, recovering its initial resistance at about 12,000 kg./cm.². The pressure at which the minimum occurs is a function of the temperature, being 4000 kg./cm.² at 0° and 5600 kg./cm.² at 100°. Because of the location of the melting curve it was not possible to measure the resistance of the liquid metal at pressures high enough to reach the minimum, but simple extrapolation indicates that without much question the liquid will show the effect as well as the solid at temperatures above perhaps 140°, and there seems no reason to think that the mechanism responsible for the minimum has any essential connection with the lattice structure.

Rb was at first measured to pressures of about 12,000 kg./cm.². Its resistance decreases; its initial rate of decrease is greater than that of K, but the curvature is greater, so that the curves cross between 1000 and 2000 kg./cm.², and at 12,000 the resistance of Rb has dropped to 0·36 of its initial value, while that of K is 0·27. At this stage of the investigation enough data had been gathered so that a single generalisation began to stand out. With only one unimportant exception for a particular direction in an abnormal crystal, the curvature of resistance plotted against pressure was always convex toward the pressure axis, whether the resistance increases or decreases. This, together with the discovery of the minimum for Ba and Cs, naturally raised the question as to whether the resistance of all metals might not ultimately increase with pressure, if pressure could only be raised high enough. This meant that all those metals whose resistance initially decreases must pass through a minimum. The next most promising place to look for such a minimum was evidently Rb, because of its high compressibility and its close relationship to Cs. An examination of the experimental results to 12,000 kg./cm.² indicated that such a minimum was not an impossibility within realisable pressures. Accordingly, a special apparatus was constructed capable of reaching considerably higher pressures, and after several attempts a minimum was actually found for Rb at a pressure of 17,800 kg./cm.², where the resistance at 30° is 0·310 of its initial value. Re-examination of K up to 19,000 kg./cm.² did not give the minimum for it, but a short extrapolation, which can be made with considerable confidence by plotting first differences against pressure, indicates with almost certainty a minimum near 23,500 kg./cm.², where the resistance is probably 0·175 its initial value. A similar extrapolation of new measurements on Na to 19,000 kg./cm.² indicates with less certainty, but still with very high probability, a minimum near 28,000 kg./cm.², where the resistance is probably 0·442 of its initial value. The general conclusion is, therefore, that the ultimate behaviour of the five alkali metals at high pressures is an increase of resistance with pressure. Judging from the behaviour of Cs, the chances for another reversal at still higher pressures must be very small.

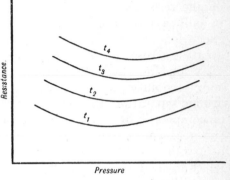

FIG. 71.—The relation between pressure, temperature, and resistance of all metals (elements) yet studied is consistent with a family of curves like this.

It is most tempting to extend this generalisation to all the other metals, and to expect that at sufficiently high pressures the resistance of all will increase with pressure. This expectation receives support from the behaviour of the three metals Ca, Sr, and Ba. The behaviour of these is consistent with the sort of connection between resistance, pressure, and temperature, indicated in fig. 71. The behaviour of different metals in the ordinary ranges of pressure and temperature may be thought of as given by small blocks cut out from the family of curves of the figure in different locations, depending on the metal.

It is interesting to speculate as to the pressures at which the minimum may be expected for other metals. A very rough answer is given by extrapolation with the second-degree formula which best represents the results within the experimental range. Of all the metals measured, indium indicates the lowest such pressure, which is about 40,000 kg./cm.², and Tl, Be, and Mg are next, with pressures in the neighbourhood of 50,000 kg./cm.². From the known departures of the measured results from a second-degree formula, it is fairly certain that the pressure of the minimum extrapolated from such a formula is too small, so that the chances of obtaining the minimum with any of the metals hitherto investigated are very small indeed.

Crystals. Characteristic crystal phenomena are to be expected only in the non-cubic metals, a cubic crystal being isotropic with respect to resistance. The effect of pressure on the resistance of non-cubic metals has been investigated in only a few cases: Zn, Cd, Sn, Sb, and Bi. With the exception of Sb, the resistance of these metals is always greatest across the plane of easiest cleavage—that is, in the direction in which the atoms are most separated. The compressibility is also greatest across planes of greatest atomic separation, this without exception. These differences of compressibility may be large, as with Zn, which is seven times as compressible perpendicular to the cleavage plane as parallel to it, or the difference of compressibility may be comparatively unimportant, as in Sn, where the excess perpendicular to the cleavage plane is only 12 per cent. One might expect differences of pressure coefficient of resistance in different directions similar to the differences of compressibility and specific resistance. These differences do, in fact, exist in the 'normal' metals Zn, Cd, and Sn. The pressure coefficient is negative in all directions and greatest numerically across the cleavage planes—that is, in the direction in which both compressibility and specific resistance are greatest. The pressure coefficient in different directions for these three metals seems more closely connected with compressibility than with specific resistance itself, although the variation of pressure coefficient with direction is not as great as the variation of compressibility. Thus in Zn and Cd, in which the specific resistance does not vary greatly with direction, but the linear compressibility varies by a factor of 7 or 6 respectively, the pressure coefficient of specific resistance varies from $-10\cdot87$ to $-6\cdot65$ and from

—13·1 to —8·7 respectively.   In Sn, on the other hand, in which the compressibility varies by only 12 per cent., although there is a variation of specific resistance of 50 per cent., the variation of pressure coefficient of resistance is only from —10·96 to —10·28.   In these three cases the effect of pressure is to tend to wipe out the distinction of different directions with respect to resistance.   In Bi the behaviour is the opposite.   Its compressibility perpendicular to the cleavage planes is 2·5 times as great as parallel to them, so that at high pressures the extra separation of the atoms perpendicular to the cleavage planes tends to disappear, but nevertheless the pressure coefficient, which is positive, is three times as great perpendicular to the cleavage planes as parallel, so that the excess of resistance in the direction of greatest atomic separation becomes accentuated at high pressures. In fact it has already been stated that the ratio increases from 1·275 at atmospheric pressure to 1·48 at 12,000.   In Sb the resistance is initially anomalously least in the direction of greatest atomic separation.   The linear compressibility is three times greater in this direction than at right angles. We have already seen that the effect of pressure is such as to make Sb behave like Bi at high pressures.

Liquid Metals.   Here again the effect of pressure has been measured in only a few cases, the difficulty being to combine the effect of high pressures and high temperatures.   In every case, except Li, the liquid decreases in resistance with increasing pressure.   There is in general a striking difference between the effect of pressure on the resistance of a liquid and a solid metal.   The positive pressure coefficient of liquid Li is about 33 per cent. greater than the positive pressure coefficient of the solid.   The negative coefficients of liquid Na and K are somewhat greater numerically than the corresponding coefficients of the solid, and for Rb and Cs the sequence is probably reversed, although exact statements are difficult here because of the uncertainty of converting measured into specific resistance.   In Ga the pressure coefficient of the solid is of the order of 2·5 times less than that of the liquid under comparable conditions, but in mercury the behaviour is reversed, and the pressure coefficient of the solid is numerically greater than that of the liquid. Where measurements could be made accurately enough to establish the direction of curvature, it was found that the pressure-resistance curve of liquids is convex toward the pressure axis like that of solids. There is no necessary connection between the curvature of the graphs of liquid and solid; thus the curvature of the graph for solid Ga is materially greater than that for the liquid, while the reverse is true for K. The temperature coefficient of resistance of liquid metals at atmospheric pressure is not invariably of the order of $\dfrac{1}{\tau}$, like that of solids, but in a number of cases is very much less. One would expect, therefore, greater variation than shown by solids in the pressure effect with temperature, or the temperature effect with pressure. The data on this point are very scanty, but it is known that the temperature coefficient of liquid Na decreases more with pressure than does that of the solid, whereas the temperature coefficient of liquid Li is approximately independent of pressure, and it is not improbable that the temperature coefficient of liquid Bi increases with rising pressure.

Summarising with regard to the pressure coefficients of liquid metals, it is to be said that although there does not seem to be any noteworthy tendency to a single type of behaviour, it is nevertheless remarkable that in many cases the resistance of the liquid responds more sluggishly to changes of pressure than does that of the solid, in that the coefficient of the liquid is either less than that of the solid, or else its pressure rate of change is less.

The way in which the relative resistance of solid and liquid changes along the melting curve is a matter of considerable interest. It will, in the first place, pay to recall the fact, already well known, that the direction in which the resistance changes when a metal melts is also the direction in which the volume changes. If the metal expands on melting, as is normal, the specific resistance increases on melting, and if the metal expands on freezing, the resistance of the liquid is less than that of the solid. This rule is true without exception. Ga and Bi are at present the only metals known in the second class; the data for antimony do not seem to be well established. I was able to add Li to the list of substances that obey this rule. This is of interest, because of the positive pressure coefficient of solid Li. With regard to the magnitude of the change of resistance on melting, there have been a number of theoretical proposals; the inaccuracy of the experimental results has allowed considerable latitude here. Thus theoretical considerations have been based on the assumption that the ratio of the resistance of the liquid to that of the solid is an integer. There is perhaps a tendency for the figures to cluster about the value 2, but it is now certain that within the limits of error the ratio is not integral.

The measurements under pressure bring out a fact that could not have been known before, namely, that the ratio of the resistance of solid to liquid is approximately a constant characteristic of the particular substance, which does not change greatly as pressure and temperature are changed along the melting curve. We now have the figures for the ratio of the resistance of liquid to solid for eight metals along the melting curve. For Li the accuracy was not great enough to permit more than the statement that the ratio does not change greatly in a pressure range of 8000 kg./cm.². For Na the ratio is 1·45 at atmospheric pressure, and has dropped to 1·36 at 12,000 kg./cm.²; for K the ratio of resistance of liquid to solid is 1·56 at 0 kg., and has dropped only to 1·55 at 9700 kg./cm.²; for Rb the ratio of resistances drops from 1·612 at 0 kg. to 1·571 at 3425 kg./cm.²; and for Cs the ratio of resistances rises from 1·660 at 0 kg. to 1·695 at 3790 kg./cm.². For mercury I determined the ratio of resistance of liquid to solid at the melting-point at 0° and 7640 kg./cm.² to be 3·345. I did not make measurements at any other temperature, but there are values by other observers. Onnes [14] finds 4·22, Cailletet and Bouty [15] 4·08, and Weber [16] gives 3·8 as the mean of six determinations, all for the ratio at the normal freezing-point at atmospheric pressure. The error is so large that it is not possible to say more than that the change in the ratio along the melting curve is not large, and is probably in the direction of a decrease with increasing pressure. The behaviour of the two abnormal metals Bi and

Ga is similar.   At 7000 kg./cm.^{2} I found the ratio of the resistance of liquid to solid Bi to be 0·45, and at atmospheric pressure Northrup and Sherwood [17] found 0·43.   The ratio is probably constant within the limits of error.   For Ga I found 0·58 for the ratio at 0 kg./cm.^{2}, and at 12,000 kg./cm.^{2} calculated the ratio to be 0·61.   This, again, is probably to be regarded as constant within the limits of error.   The results just given for Hg, Bi, and Ga are probably not as significant as for the alkali metals, because of the fact that they do not crystallise in the cubic system, so that very different results could be obtained if the melt crystallised as a single crystal with different orientations.

These results, which indicate a characteristic nearly constant increase of resistance for each metal when the regular arrangement of the lattice is replaced by the haphazard arrangement of the liquid, should be of some theoretical significance.

Resistance of Alloys under Pressure.   This subject has been barely touched in a systematic way.   Scattered through the literature there are miscellaneous observations of the effect of pressure on the resistance of various alloys, some of them made for ulterior purposes because of the utilisation of the alloy in some part of the pressure apparatus.   Thus there are early observations of Chwolson on brass, Lisell, Lafay, and Lussana on manganin, Lisell on Pt—Ir and German silver, and some observations of my own on two commercial Ni—Cr alloys, three commercial Ni—Cr—Fe alloys, and a commercial Cu—Mn—Al alloy.   Practically the only systematic work is that of Lisell on three Cu—Mn alloys up to 7·5 per cent. Mn; by Beckman on the Cd—Pb and Ag—Au series over the entire range of composition, and Au—Ni up to 14 per cent. Ni; by Ufford,[18] with my apparatus, on the complete series of Li—Sn, Ca—Pb, and Bi—Sn; by Oppenheimer (B. 65, p. 331), with my apparatus, on the Cu—Ni series up to 50 per cent. Ni; and by myself on the complete Fe—Co and Fe—Ni series.   The work done by Lisell and Beckman covered a pressure range of 2000 or 3000 kg./cm.^{2}, and was done at a single temperature; the work done with my apparatus covered the regular pressure range of 12,000 kg./cm.^{2}, and in most cases was done at two temperatures, giving some information about the temperature coefficients under pressure.

As might be expected, the results are very complicated, and few generalisations can be ventured with the small amount of material available. There does, however, seem to be one generalisation without present exception. The nature of this generalisation is suggested by the observation that positive pressure coefficients are much more common with the alloys than with pure metals. The early discovery of the positive coefficient of manganin is a case in point. Other examples are the entire central range of the Fe—Co series, and all the Cu—Mn series investigated by Lisell. The generalisation is that the initial effect of the addition of a foreign metal to any pure metal is to make the pressure coefficient greater algebraically. If the pressure coefficient of the pure metal is negative, as it usually is, this means that the coefficient of the impure metal is less numerically than that of the pure metal, a result consistent with all my experience. The result holds even when one of the pure metals has itself a positive coefficient. The alloys investigated by Ufford were chosen in the endeavour to obtain some light on the mechanism responsible for the positive pressure coefficient, and it will be noticed that in each pair in his three series one metal has a positive coefficient. Ufford found the rather surprising result that the coefficient of Bi becomes greater on adding small quantities of Sn, the coefficient of which by itself is negative. The range in which Bi shows this effect is exceedingly narrow, the addition of 0·25 per cent. of Sn being sufficient to depress the coefficient to less than the value for pure Bi. The same phenomena are also shown by Ca and Li, with the difference that the range of composition over which the alloy has a greater positive coefficient than that of the pure metal is very much wider, this range being up to 12 atomic per cent. Sn additional to Li, and up to 20 atomic per cent. Pb additional to Ca. This of itself rather tends to confirm the conclusion that we have already drawn that the mechanism of the positive coefficient in Ca and Li is different from that in Bi.

Going further in general description, one would expect specially simple behaviour in those alloys in which the two components are not miscible in each other, so that the alloy is merely a mechanical mixture of microscopic grains of fixed compositions. There are two well-marked examples of this; the entire Cd—Pb series of Beckman, and the Sn—Bi series of Ufford between 2 or 3 and 90 atomic per cent. of Sn. In each case the pressure coefficient of the intermediate members of the series is not far from a linear function in terms of the composition of the pressure coefficients of the end members of the series. If the components of the alloy form solid solutions, however, the coefficient is far from a linear function of composition, as other phenomena would suggest.

A further rough generalisation may be made, in that there is a crude parallelism between the pressure coefficient of resistance and the specific resistance; this is particularly striking for the Fe—Ni alloys. Beckman found that the pressure coefficient of the Au—Ni and Au—Ag alloys is a linear function of the electrical conductivity over considerable ranges of composition.

In one respect the behaviour of these alloys is quite understandable from the point of view of wave mechanics. Resistance, according to this picture, is a phenomenon of scattering of electron waves. In a perfectly regular lattice there is no scattering, and hence the resistance vanishes, but if irregularities are introduced by temperature agitation or by foreign atoms, then the scattering and hence the resistance increases. The same qualitative point of view was part of the theory in which I had been interested before the development of wave mechanics; here resistance was associated with a lack of perfect 'fit' between adjacent atoms. It is obvious that this picture, which associates resistance with irregularities in the structure, at once accounts for the fact that in general the resistance of an alloy is greater than that of either of its pure components. It goes further than this, and accounts for the algebraic increase of pressure coefficient when a foreign metal is added to a pure metal. This is a purely geometrical matter. At low pressures, a fair qualitative picture of various pressure phenomena is given by supposing that the atoms are relatively undeformable, but held apart in positions of equilibrium by interatomic forces, and that the first effect of an increase of pressure is to decrease the spaces between the atoms, so that they are pushed more or less into contact, with the result that at low pressures the deformability of the atoms plays a relatively less important part than at high pressures.   Now imagine a lattice of a pure metal in which the volume of the atoms is perhaps 75 per cent. of the total volume, and that of the free interatomic space 25 per cent. Replace some of these atoms by the atoms of another metal, with perhaps 10 per cent. greater volume.   Irregularities will be introduced into the structure by the foreign atoms, but this effect may be comparatively unimportant because of the large amount of free space available for mutual adjustment.   Now apply to the lattice an external pressure sufficient to squeeze out part of the free space.   The space available for smoothing out the irregularities becomes less, and the resistance rises. That is, on purely geometrical grounds, pressure tends to accentuate the relative irregularities arising from ill-matched atoms, and for this reason the pressure coefficient tends to become more positive.

It would appear, therefore, that those aspects of the problem of the positive pressure coefficient peculiar to the field of alloys can perhaps be understood.   It was a disappointment, however, that the measurements of Ufford on alloys of metals with positive coefficients did not throw any obvious light on the more fundamental question of why the scattering of electron waves in a pure metal may be increased by pressure.   The only suggested answer to this of which I know on the basis of the wave mechanics has been given by Houston,[19] who associates this effect with an abnormally small decrease of compressibility with rising pressure.   This suggestion failed to stand the test of experiment, however, in the case of invar, in which the change of compressibility with pressure was in the other direction from that indicated by theory.   However, this argument is obscured by the fact that invar is not a pure metal.

Each alloy considered alone shows behaviour with respect to pressure much like that of the pure metals.   The pressure coefficient is not much affected by temperature, and in general becomes less at the higher pressures.   There are, however, a comparatively large number of exceptions to the rule that the curve of resistance against pressure is convex toward the pressure axis.   It will be remembered that the only known exception to this rule among the pure elements was the low-pressure modification of Ce.   Among alloys, Ufford found that the curvature of the Bi—Sn alloys up to perhaps 6 per cent. Sn is abnormal, and I found the same phenomenon in the Fe—Ni series between perhaps 2 per cent. and 45 per cent. Ni.   This sort of thing is quite consistent with the picture presented above.   When pressure has pushed the atoms into complete contact, the previous accentuation of their lack of fit as they are brought closer together will disappear, so that we have here an exhaustion of an effect with high pressure, and so a law of diminishing returns, instead of the law of increasing returns, which appeared so surprising in connection with the pure metals.   Hence, depending on the relative magnitude of the atoms and the free spaces, we may have a law either of diminishing or of increasing returns.

One alloy investigated by Ufford merits special comment, the Bi—Sn alloy of 0·53 atomic per cent. Sn.   At 0° this alloy showed an initial increase of resistance, then a maximum, and finally decrease at 12,000 below the initial value.   The same sequence was found with decreasing pressure, but the exact path was not retraced, and the increase at the maximum was considerably greater than the increase with increasing pressure.   The maximum change of resistance was of the order of 2 per cent.   This sort of behaviour is not surprising in alloys in critical ranges of composition.   Under ordinary conditions Sn is soluble in Bi only up to a certain critical amount, and beyond this there is separation into two phases. The solubility limit is known to be a function of temperature, and it must also be a function of pressure.   It is obvious that an alloy which at low pressures is stable as a single phase may separate into two phases at higher pressures, and that because of the slowness of diffusion in the solid state, equilibrium will be attained only slowly, and measurements made under ordinary conditions will not be single valued.   From a detailed study of the pressure-resistance curves Ufford inferred that Sn becomes less soluble in Bi as pressure increases, but that Bi becomes more soluble in Sn.

Compounds. There is little to be said on this subject, which is practically untouched; there is an enormous field here, particularly with regard to intermetallic compounds. The only measurements of which I know on compounds with metallic conductivity are some unpublished ones that I have recently made on TiC and TiN. It has been shown by van Arkel that the specific resistance of these compounds is of the same order of magnitude as that of the metals, and, further-more, the temperature coefficient of resistance is positive. Up to 12,000 kg./cm.^{2} I found the resistance of both these compounds to decrease linearly with pressure, the coefficient of TiN being -9.35×10^{-7} at 30° and -9.97×10^{-7} at 75°, while that of TiC was -1.375×10^{-6} at 30° and -1.35×10^{-6} at 75°. These figures are of the same magnitude as for many metals, so that pressure does not bring out any peculiarities.

Three non-metallic compounds have been measured: AgS by Montén, and FeS and Fe_2O_3 by Beckman. These two investigators used the same apparatus, and their pressure range was about 3000 kg./cm.^{2}. Montén found for AgS a very large decrease of resistance, at 3000 kg./cm.^{2} and room temperature the resistance being 0.0765 of its value at atmo-spheric pressure, and at 0° and 3000 kg./cm.^{2} it was 0.0589 of its initial value. There was, of course, great curvature in the relation between pressure and resistance. Beckman found the resistance of FeS to decrease, but the decrease was comparatively small, and approximately linear with pressure, the mean value for the coefficient being -2.2×10^{-6}. Fe_2O_3 was investigated in directions perpendicular and parallel to the trigonal axis; in both directions the resist-ance decreases linearly with pressure by small and some-what different amounts, the coefficients being approximately -6.7×10^{-6} perpendicular, and -8.2×10^{-6} parallel.

[1] E. WARTMANN, *Phil. Mag.*, **17**, 441 (1859).
[2] O. CHWOLSON, *Bull. St. Pet. Acad.*, **27–28**, 187 (1881–1882).
[3] H. TOMLINSON, *Trans. Roy. Soc.*, **174**, 1–172 (1883).
[4] S. LUSSANA, *Nouv. Cim.*, **10**, 73 (1899) ; **5**, 305 (1903).

[5] E. LISELL, *Diss.*, Upsala (1902).

[6] R. LENZ, *Diss.*, Stuttgart (1882) ; *Beibl.*, **6**, 802 (1882).

[7] C. BARUS, *Amer. Jour. Sci.*, **140**, 219 (1890).

[8] A. DE FOREST PALMER, *ibid.*, **4**, 1 (1897) ; **6**, 451 (1898).

[9] W. E. WILLIAMS, *Phil. Mag.*, **13**, 635 (1907).

[10] A. LAFAY, *C.R.*, **149**, 566 (1909).

[11] F. MONTÉN, *Diss.*, Upsala (1909).

[12] B. BECKMAN, *Ups. Univ. Arsskrift* (1911), *Mat och Nat.*, **1**, 109.

[13] B. BECKMAN, *Ark. Mat. Ast. Fys.*, **7**, No. 42 (1912) ; *Proc. Amst.*, **15**, 947 (1913) (with K. ONNES) ; *Phys. ZS.*, **16**, 59 (1915) ; *Ann. Phys.*, **46**, 481 and 931 (1915) ; *Phys. ZS.*, **18**, 507 (1917).

[14] K. ONNES, *Kon. Akad. Wet. Proc.*, **4**, 113 (1911).

[15] CAILLETET et BOUTY, *C.R.*, **100**, 1188 (1885).

[16] WEBER, *Wied. Ann.*, **25**, 245 (1885) ; **36**, 587 (1888).

[17] E. F. NORTHRUP and R. G. SHERWOOD, *Jour. Frank. Inst.*, p. 477, (Oct. 1916).

[18] C. W. UFFORD, *Proc. Amer. Acad.*, **63**, 309 (1928).

[19] W. V. HOUSTON, *ZS. f. Phys.*, **48**, 449 (1928).

CHAPTER X

EFFECT OF PRESSURE ON THERMO-ELECTRIC
PROPERTIES

In 1891, Des Coudres [1] showed that the thermo-electric quality of mercury is affected by pressure. His apparatus was a simple arrangement of glass tubes, and pressures up to two atmospheres were obtained from the head of mercury in the tubes. He found that a thermo-couple consisting of one branch of mercury at a pressure of two atmospheres and the other at one atmosphere has an e.m.f. of $4·4 \times 10^{-10}$ volts per degree temperature difference between the junctions, driving the current from uncompressed to compressed mercury at the hot junction. In 1902 Agricola [2] extended the work of Des Coudres over the pressure range of 100 kg./cm.[2] and for the temperature range 0° to 100°. In this range the effect was linear with pressure and temperature, and practically the same coefficient was found as by Des Coudres. Agricola also investigated various dilute amalgams. Additions up to 2 per cent. of Sn, Cd, Sb, and Bi were found to make no appreciable difference in the behaviour, but similar additions of Zn, K, and Na increased the effect threefold. In 1908 Wagner [3] published investigations on fifteen metals and several alloys between 0° and 100° and up to 300 kg./cm.[2] pressure. The effects were linear in this range. For solids the magnitude of the effect is much smaller than for liquid mercury, and is roughly of the order of 5×10^{-12} volts per degree C. per kg./cm.[2]. Wagner drew various conclusions from his measurements as to the untenability of the classical electron theory of metals. Hörig [4] in 1909 investigated liquid mercury and the Na—K eutectic to 1400 kg./cm.[2] and 150°. He found the effect of pressure on both substances to be linear, and practically the same on both, $2·37$ and $2·32 \times 10^{-10}$ respectively. The figure found for Hg is thus somewhat larger than that found by Agricola or Des Coudres, but the difference is probably not significant in view of the magnitude of the experimental error. Siegel [5] in 1912 extended the temperature range to 400° in order to determine the effect on liquid Bi and Sn, but this extension of temperature range demanded the low-pressure range of only 50 kg./cm.². He was unable to establish any effect on liquid Bi and Sn, but did show that if it exists it is less than 2 per cent. of the effect on Hg. This is of the same order as the effect found by Wagner for solid metals, so that the presumption is that the pressure effect does not differ greatly on liquid and solid metals. Siegel was able to show that the effect on Hg somewhat more than doubles as temperature rises from 20° to 350°.

In 1918 I published (B. 29) an investigation of the effect of pressure on the thermo-electric quality of eighteen pure metals and two alloys for the temperature range 0° to 100° C. and the pressure range up to 12,000 kg./cm.². This extension of range proved to give essentially new information, because the effects for a number of metals are very far indeed from linear with pressure. The extension of range also made it possible, by allowing first and second differentiation with respect to temperature, to separate the total thermo-electric effect into the parts contributed by the Peltier heat and the Thomson heat, and so to find how each of these vary with pressure, which had not been possible with the results for a narrower range.

The method was a differential method much like that used by Wagner, as distinguished from that of Hörig, for example. Hörig's couple was composed of two different metals, one the metal to be investigated, and the other a standard metal, Pt for example, kept the same in all the experiments. The entire couple as exposed to hydrostatic pressure and the thermal e.m.f. for a given difference of temperature between the junctions measured with and without pressure. The difference, after a correction for the effect of pressure on Pt which had to be taken from Wagner's work, gave the effect of pressure on the unknown metal. In the method of Hörig, therefore, the effect sought is the difference between two comparatively large effects. The effect of pressure may be directly measured by making both arms of the couple of the same metal, and subjecting the one to pressure and leaving the other free.   If the two junctions between compressed and uncompressed metal are maintained at different temperatures, a thermal e.m.f. will be developed, just as if the couple were made in the ordinary way of two different metals, and, furthermore, the thermal e.m.f. of such a couple obviously gives directly the effect sought.

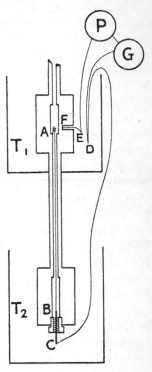

The experimental arrangements are indicated schematically in fig. 72. The two branches of the couple are the lengths of wire reaching from A to B and from C to D.   The length AB is stretched through the connecting pipe between two pressure cylinders, and is exposed to hydrostatic pressure over its entire length.   The cylinders at the two ends of AB are maintained at two different temperatures T_1 and T_2 by two thermostated baths.   The lower bath is kept at a constant temperature of $0°$ and the upper bath may vary from $20°$ to $100°$.   The other branch of the couple CD runs at atmospheric pressure between the same two temperature baths.   The terminals of AB and CD are connected in the upper bath with the measuring device, indicated by

FIG. 72.—Diagram of electrical circuits for determining effect of pressure on thermal e.m.f.

P and G in the figure to suggest a potentiometer and a galvanometer.   It is not necessary to describe here the details of the electrical measuring apparatus.   By taking all the precautions conventional in thermo-electric measurements, readings sensitive to 10^{-9} volts could be made, with an accuracy not greatly less than this.   The most essential precaution is that all parts of the circuit where there are dissimilar metals, as there must necessarily be in getting the leads into and out of the pressure cylinders, and all parts where there are stress gradients, must be as closely as possible at uniform temperature. This requirement was easily met in the lower ice-bath by rapid stirring; in the upper cylinder, which might rise to 100°, the requirement was more difficult, and the arrangement shown in fig. 72, in which the part of the couple reaching from F to E is of the same metal as AB, was adopted to minimise as much as possible the effect of slight fluctuations in the bath temperature, by surrounding the sensitive parts by heavy masses of metal.

The sign conventions are as follows: The thermal e.m.f. of the 'pressure couple,' that is a couple composed of compressed and uncompressed metal with junctions at different temperatures, is called positive if the ordinary positive current flows from uncompressed to compressed metal at the hot junction. The 'pressure' Peltier heat is called positive if heat is absorbed from the surroundings when positive current flows from uncompressed to compressed metal. The 'pressure' Thomson heat is positive if the amount of heat absorbed by positive current in flowing from cold to hot is greater in the compressed than in the uncompressed metal.

In the extended paper full details of the results are given ; for each metal a table is given of the thermal e.m.f. of the pressure couple for 10° temperature intervals and 2000 kg./cm.^{2} pressure intervals, and tables also of the pressure Peltier heat and the pressure Thomson heat for 20° temperature intervals and 2000 kg./cm.^{2} pressure intervals. The data of the tables are also plotted in diagrams. The thermal e.m.f. of the uncompressed metal at atmospheric pressure against lead is also given between 0° and 100°, and also the Peltier and Thomson heats against lead obtained by differentiation of the power series in temperature for the thermal e.m.f. The following metals were measured : Sn, Tl, Cd, Pb, Zn, Mg, Al, Ag, Au, Cu, Ni, Co, Fe, Pd, Pt, Mo, W, Bi, constantan, and manganin. A brief summary of the results is given here in Table XVI, abstracted from the summary in Smithsonian Tables, in which the thermal e.m.f. of 'pressure' couples of various metals is tabulated for wide intervals of pressure and temperature.

TABLE XVI

T.E.M.F. OF 'PRESSURE COUPLES' OF VARIOUS METALS

Signs positive, unless marked negative.

Metal	T.E.M.F., volts × 10⁹								
	2000 kg.		4000 kg.		8000 kg.		12,000 kg.		
	50°	100°	50°	100°	50°	100°	20°	50°	100°
Bismuth	53,000	85,000	110,000	185,000	255,000	425,000	185,000	452,000	710,000
Zinc	6,200	14,100	13,000	28,500	26,100	58,100	14,400	38,500	87,400
Thallium	4,930	10,870	9,380	20,290	17,170	37,630	8,780	23,750	52,460
Cadmium *	2,040	7,120	4,620	14,380	10,960	28,740	6,680	19,180	45,560
Constantan	2,850	5,950	5,800	11,810	11,530	23,790	6,750	17,200	35,470
Palladium	2,190	4,380	4,400	8,800	8,630	17,690	5,090	12,970	26,520
Platinum	1,810	3,600	3,600	7,310	7,370	14,350	3,880	11,030	21,570
Tungsten	1,190	2,530	2,360	4,990	4,690	10,120	2,700	7,050	15,140
Nickel	700	1,680	1,500	3,400	3,230	7,190	1,880	5,140	11,440
Silver	840	1,870	1,720	3,720	3,350	7,190	1,900	4,950	10,560
Iron	390	1,670	590	3,250	5,300	5,820	—990	220	7,680
Lead	460	1,050	920	2,120	1,860	4,210	880	281	6,330
Gold	456	1,052	905	2,051	1,791	3,974	990	2,627	5,760
Copper	292	584	580	1,216	1,124	2,420	596	1,616	3,546
Aluminium	— 70	101	— 91	294	32	929	— 68	312	1,962
Molybdenum	93	140	187	278	375	555	146	562	833
Tin	38	87	58	165	70	292	—182	10	390
Manganin	—123	—232	—242	—452	—489	—894	—308	—719	—1,314
Magnesium	— 84	—167	—181	—362	—395	—791	—259	—648	—1,296
Cobalt	—156	—348	—316	—692	—630	—1,360	—352	—937	—2,061

* Not comparable with ordinary Cd because of effect of pressure transitions. See p. 235.

Some of the substances show complicated effects. Two typical examples are reproduced. In fig. 73 is shown the thermal e.m.f. of a 'pressure' couple of Ag, which is perfectly regular, and in fig. 74 are the corresponding data for Fe, which are seen to be highly complicated.

It should be said that these pressure effects are apparently truly characteristic of the metal, and are not due to chance fluctuations in the quality or treatment of the metal, as other thermo-electric effects might lead one to fear. A proof of this is given by the fact that in those cases where the relation is linear, my coefficients agree suprisingly well with those of Wagner over his pressure range of 300 kg./cm.², and in cases like Fe, where there is great complication, the results can nevertheless be repeated rather well on replacing one sample of the metal by another, even when the different samples are known not to be of the same purity. This pressure effect is, therefore, not sensitive to impurity. On the other hand,

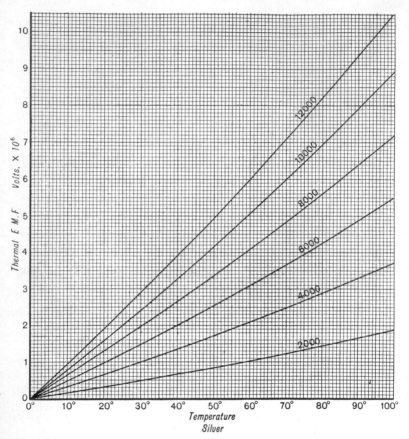

Fig. 73.—Thermal e.m.f. of a 'pressure couple' of silver as a function of pressure and temperature.

the effect of tension on thermo-electric quality is highly variable, and it is most difficult to get reproducible results and therefore to be sure that any such results are truly characteristic of the metal.

Only a rough summary can be attempted with results so complicated. Superposed on a bewildering mass of small scale complexity there are a few broad generalisations possible. The normal pressure effect turns out to be positive; that is, in a pressure couple the current normally flows from uncompressed metal to compressed at the hot junction. There are only three out of twenty cases, manganin, Mg, and

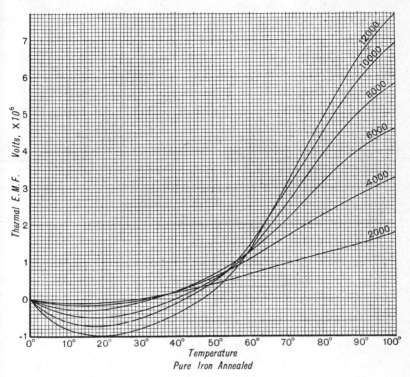

FIG. 74.—Thermal e.m.f. of a 'pressure couple' of iron as a function of pressure and temperature.

Co, in which the effect is negative over the entire range of pressure and temperature, and only three other metals, Fe, Al, and Sn, for which the effect is negative over any part of the range. The range of maximum values (that is, for 100° temperature difference and 12,000 kg. pressure difference) is from 710×10^{-6} volts for Bi to -20.6×10^{-6} for Co. The normal 'pressure' Peltier heat is also positive—that is, heat is absorbed by the current in flowing from uncompressed to compressed metal.   There are only four metals with negative pressure Peltier heat in part or the whole the range, namely, manganin, Sn, Mg, and Co.   The range of maximum values is from 2960×10^{-6} for Bi to $-90 \cdot 5 \times 10^{-6}$ for Co, measured in joules per coulomb.   Purely as a mnemonic device, it may be mentioned that the predominantly positive sign for the pressure Peltier heat is what would be suggested by the classical electron picture ;   the positive sign means that the electrons absorb heat in flowing from the compressed to uncompressed metal.   That is, as the electron gas expands in flowing from compressed to uncompressed metal it has to absorb heat to maintain itself isothermal, just like an ordinary gas.   Further implications of the classical electron picture, however, cannot be carried through in detail, as was brought out clearly in Wagner's discussion.

The 'pressure' Thomson heat is also usually positive, Co, Fe, and Bi being the only negative cases.   The range of values is from 220×10^{-8} for Zn to -280×10^{-8} for Bi, expressed in joules per coulomb per degree Centigrade.

When we try to correlate the thermo-electric behaviour under pressure with other properties of the metal we find complete lack of connection.   In Table XVII the metals are arranged in a number of columns, in each column the order being the order of relative magnitude of the property standing at the head of the column.   In the first place, the effect of pressure on thermal e.m.f. is not greatest for the most compressible metals, as we might expect, but the effect is apparently quite independent of the compressibility.   Compare, for example, the positions of Pb and Mg in the compressibility column with their positions in the pressure e.m.f. column. Furthermore, there is very little connection between the pressure effects on thermal e.m.f. and on resistance.   This can be seen on noting further that the effect of pressure on resistance runs approximately parallel to its effect on volume.   The conclusion suggests itself that there is no close connection between the mechanism which determines the thermo-electric behaviour of a metal and its electrical resistance.   This is further borne out by the lack of parallelism between the columns of pressure effect on resistance with those of pressure effect on Thomson and Peltier heat.

TABLE XVII

The Metals in Order of Decreasing Numerical Magnitudes of Various Properties

The horizontal bars in some columns show where the effect changes sign.

Compressibility	Specific resistance	Temp. coeff. of resis. 0°–100° at 0 kg.	Thomson heat at 0° and 0 kg.	Peltier heat against Pb at 0° and 0 kg.	Pressure effect on resistance	Pressure effect on E.M.F.	Pressure effect on Peltier heat	Pressure effect on Thomson heat
Mg	Bi	Fe	Cd	Fe	Pb	Bi	Bi	Zn
Bi	Cons	Tl	Mo	Cd	Tl	Zn	Zn	Pt
Tl	Man	Ni	W	Mo	Sn	Tl	Tl	Cd
Pb	Pb	Sn	Bi	Zn	Cd	Cd	Cd	Tl
Cd	Tl	Bi	Cu	Au	Mg	Cons	Cons	Sn
Sn	Fe	Al	Au	Cu	Zn	Pd	Pd	Pb
Zn	Ni	Mo	Ag	Ag	Al	Pt	Pt	Al
Al	Sn	Cu	Man	Tl	Ag	W	W	W
Ag	Pt	Cd	Al	W	Au	Ni	Ni	Cons
Man	Pd	Pb	Mg	Man	Fe	Ag	Ag	Ni
Cu	Co	Zn	Pb	Sn	Pd	Fe	Fe	Mo
Cons	Cd	Ag	—	Pb	Pt	Pb	Pb	Ag
Fe	Zn	Au	Sn	—	Cu	Au	Au	Pd
Au	W	Mg	Tl	Mg	Ni	Cu	Al	Au
Co	Mg	Pt	Zn	Al	Mo	Al	Cu	Cu
Ni	Mo	Co	Fe	Pt	W	Mo	Mo	Man
Pd	Al	W	Pt	Pd	Co	Sn	—	Mg
Mo	Au	Pd	Pd	Co	—	—	Man	—
Pt	Cu	Cons	Ni	Ni	Cons	Man	Sn	Co
W	Ag	Man	Co	Cons	Man	Mg	Mg	Fe
			Cons	Bi	Bi	Co	Co	Bi

There is almost exact parallelism between the columns of pressure effect on e.m.f. and on Peltier heat ; these two columns differ by only two single inversions, Al with Cu, and manganin with Sn. This means that nearly all the total e.m.f. in a pressure couple is provided by the difference of Peltier heat at the hot and cold junctions, and comparatively little by the effect of pressure on Thomson heat. Expressed otherwise, this means that the relation between e.m.f. and temperature at constant pressure is in most cases approximately linear.

The columns of pressure effect on Peltier and Thomson heats are not obviously related. Most striking is the transposition of Bi from the head of one column to the bottom of the other ; the pressure effect on Peltier heat of Bi is the maximum positive, and on the Thomson heat the maximum negative. One draws the conclusion that the Thomson heat mechanism and the Peltier heat mechanism are not closely related. The pressure effect on Thomson heat inclines to show parallelism to the pressure effect on resistance.

The thermo-electric behaviour at atmospheric pressure (Peltier heat against lead and Thomson heat) also shows little direct connection with the pressure effects. There is no obvious connection with other columns ; the thermo-electric mechanism seems to stand in a class by itself.

The magnitude of the pressure Thomson heat compared with the total Thomson heat under atmospheric conditions is of interest. The relative effect of pressure on the properties of most solids is small ; for example, the maximum effect on resistance was 14 per cent. for lead, and the effect on volume is only a few per cent. We would expect a similar state of affairs with the Thomson heat. This proves to be true, except for the anomalous metals. The maximum pressure Thomson heats of Al, Bi, Fe, Sn, and Zn are all of the order of magnitude of the whole Thomson heat under atmospheric conditions, but the other metals show the smallness of the effect to be expected. Lead needs special mention. It is usually assumed on the basis of Kelvin's work that the Thomson heat of lead is zero under normal conditions. However, the pressure Thomson heat for lead is of the same magnitude as that of other metals, being twice that of Ag, for example. This suggests that the Thomson heat mechanism of lead really has nothing unusual about it, but that the vanishing at atmospheric pressure is either a fortuitous result, or even that more accurate measurements may show that it is not exactly zero.

So far we have considered the effects only in broad outline, but when we come to consider the detailed variations with pressure and temperature we find great complexity. The normal behaviour of e.m.f. is a regular rise with pressure and temperature, the slope of the e.m.f. curves at constant temperature increasing with rising pressure. Fe, Al, and Sn are examples of complicated variations of e.m.f. with pressure and temperature, and there are also other examples where the slope at constant pressure may decrease with rising temperature, or the slope at constant temperature may increase with rising pressure. As for the detail of the variation of Peltier or Thomson heat with pressure and temperature, so many different types are presented that it is useless to try to enumerate them.

There is, however, one feature of perhaps sufficient significance to justify mention here, namely, that the effects with Zn and Cd were very irregular ; equilibrium after every change of pressure was set up only after a long time, and there were important hysteresis effects. None of the other metals showed these effects. Wagner also in his pressure range of 300 kg. found Zn and Cd unique in the length of time required to reach equilibrium after changes of pressure. I believe that an adequate explanation of these effects can be found in the very large differences of linear compressibility in different directions of single crystals of Zn and Cd, the compressibility along the axis being nearly six times as great as perpendicular. In a haphazard aggregate of small crystals the effect of hydrostatic pressure would therefore be to introduce large internal strains, which will give rise to time effects if they exceed the local elastic limit. This effect will not be shown at all by the cubic metals, which constituted the large majority of those investigated. Of the others, Mg, although hexagonal, has practically the same compressibility in all directions, and Sn also, which is tetragonal, has nearly the same compressibility in all directions. Thallium has not been investigated in the single crystal form. On the basis of this explanation the expectation is, therefore, that the effects will be by far the most irregular in Zn and Cd, which agrees with the facts.

The unexpected complications make results disappointingly meagre in their suggestions as to the nature of the thermoelectric mechanism. The conclusions are mostly negative in character ; the most unmistakable inferences may be drawn as to the untenability of the old gas-free electron theory of metals, but this is not surprising enough to be worth the trouble of extending the pressure range to 12,000 kg./cm.^{2} and, furthermore, the same inference was made years ago by

Wagner from his data up to 300 kg./cm.[2].   Further than this, the results suggest most strongly that the thermo-electric mechanism must be comparatively complicated, that it cannot be at all of the simplicity imagined in the free electron theory, and that most likely the effects which we measure are the resultant of different effects which sometimes, at least, work in different directions.   Houston (6) has recently given an expression for the thermal e.m.f. of a pressure couple on the basis of the wave mechanics, but when applied to Cu it gives a value ten times too small, and suggests none of the complications which we have found experimentally.

[1] T. Des Coudres, *Wied. Ann.*, **43**, 673 (1891).
[2] H. Agricola, *Diss. Erlangen* (1902).
[3] E. Wagner, *Ann. Phys.*, **27**, 955 (1908).
[4] H. Hörig, *ibid.*, **28**, 371 (1909).
[5] E. Siegel, *ibid.*, **38**, 588 (1912).
[6] W. V. Houston, *ZS. f. Phys.*, **38**, 449 (1928).

CHAPTER XI

THERMAL CONDUCTIVITY UNDER PRESSURE

CONSIDERABLE interest attaches to a determination of the effect of pressure on thermal conductivity because of the light which may be thrown on the mechanism of thermal conduction, which is not yet completely understood. The experiments may be conveniently classified according to whether the substance investigated is liquid or solid; as far as I know there are no measurements on gases in the pressure range of interest to us.

New methods had to be developed adaptable to use under pressure. The requirements are rather severe; the size of the apparatus is restricted by the necessity of enclosing it in the pressure cylinder, and there are further restrictions in that the entire apparatus must be immersed in the pressure transmitting liquid. This greatly increases the heat leak from various parts, so that many methods which might be applicable under atmospheric conditions cannot be used at all under pressure. It proved to be possible to meet these requirements much more successfully in the case of liquids (B. 46, 61) than of solids; in fact the apparatus finally adopted for liquids is capable of results in many cases more accurate than those previously found at atmospheric pressure, and the method should have a useful field of application under ordinary conditions. In the following, the results for liquids will first be discussed, and then those for solids, including metals.

The apparatus for liquids is shown in fig. 75. The liquid to be measured, shown shaded, is contained in the annular space between two copper cylinders, A and B. The outer cylinder, B, is placed inside the pressure cylinder, in which it is a fairly good fit. The pressure cylinder is filled with a liquid transmitting pressure equally to all parts of the apparatus. Thermal contact between B and the pressure cylinder is further secured by an arrangement of flat strips of spring copper (not shown in the diagram). The source of heat is at the axis H of the inner cylinder, where there is an insulated wire carrying a current of electricity. Heat flows radially from this wire through the cylinder A, the layer of liquid, the cylinder B to the pressure cylinder, and out through its steel walls to the temperature bath. The bath is rapidly stirred and maintained at a constant temperature by a regulator.

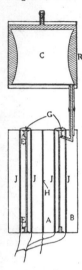

The thermal conductivity of the liquid may be calculated when the dimensions of the apparatus are known, the rate of heat input, and the difference of temperature between the outer and inner surfaces of the liquid. The formula for this is

$$k = \frac{Q}{2\pi(t_1 - t_2)} \log \frac{r_2}{r_1},$$

where k is the thermal conductivity of the liquid, Q the heat input per unit time per unit length along the axis, $t_1 - t_2$ the difference of temperature between inner and outer surfaces of the liquid, and r_1 and r_2 are the inner and outer radii of the layer of liquid.

FIG. 75. — Cross-section of apparatus for measuring the thermal conductivity of liquids under pressure.

Because the thermal conductivity of the copper is so much greater than that of the liquids (2500 times greater for the ordinary liquid), the temperature difference, $t_1 - t_2$, may be obtained with sufficient accuracy by measuring the temperature of the metal of the cylinders A and B at points near the liquid. Practically the entire drop of temperature which drives the heat flow takes place across the layer of liquid.

The heat input Q is measured in terms of the resistance of the wire per unit length and the current. The difference of temperature is determined from the e.m.f. of thermo-couples of copper-constantan, one of the junctions of which, J, is in A and the other in B. Three of these couples were used, situated at angular intervals of 120° and connected in parallel. In this way the mean of three temperature differences was obtained, thus correcting for any slight geometrical irregularities. For convenience in drawing, only two couples are shown in the figure.

The apparatus for measuring the heating current and the e.m.f. of the couples was a potentiometer arrangement, practically the same as that used for measuring thermal e.m.f. under pressure. The details need not be given; all measurements could be made to considerably better than 0·1 per cent.

Successful performance depends essentially on various details of mechanical construction. In the first place, and most important, in order to secure freedom from convection currents in the liquid, particularly since the cylinders are used in a vertical position, it is necessary that the annular space between the cylinders be small. This distance was only $\frac{1}{64}$ in. (0·040 cm.), the outer diameter of the inner cylinder being $\frac{3}{8}$ in. (0·95 cm.), and the inner diameter of the outer cylinder $\frac{13}{32}$ in. (1·03 cm.). In the early days of thermal conductivity measurements a number of radial flow methods between concentric cylinders were used, without, however, using an axial source of heat as above. These methods have been later more or less discredited because of errors due to convection. The layer of liquid used in this present apparatus is many times thinner than the thinnest used previously. Convection is further reduced by the unusually small temperature difference between the inner and outer cylinders, which was usually about 0·6° C. The heat conveyed convectively varies, other things being equal, as the square of the temperature difference, so that it is important to keep this difference down. That the performance of the apparatus in this regard was satisfactory may be seen by comparing the values of the absolute conductivity found with it with the best of previous measurements. A further advantage of the thin layer is that thermal equilibrium is very quickly attained with it; it was possible to make final readings ten minutes after every change of pressure. It is not unusual with apparatus built for use at atmospheric pressure for several hours to be necessary for the attainment of thermal equilibrium.

The inner cylinder was supported concentrically with the outer by rings of German silver G at top and bottom, 0·002 in. (0·005 cm.) thick. There is some heat leak from inner to outer cylinder at the two ends across the German silver rings, which, however, is small, because of the thinness of the German silver and its relatively poor thermal conductivity. The leak was further reduced by increasing the radial separation of the cylinders at the two ends, so that the conduction from one to the other through the liquid was decreased. By choosing the dimensions of the enlargement E correctly, it is possible to make the decreased conductivity through the liquid exactly compensate for the increased conductivity through the German silver. The dimensions were so chosen that this compensation should be exact for the average of the liquids used.

Because of the volume compression of the liquid as pressure is raised, it is necessary to provide a reservoir of the liquid outside the space between the concentric cylinders. This was accomplished as follows: the reservoir R was made of a thin collapsible tube of tin 0·75 in. (1·78 cm.) in diameter, soldered at top and bottom to end discs of brass. It was filled with liquid and connected with the annular space between the cylinders with a tube of German silver. In order to cut down the volume compression as much as possible, the interior of the collapsible tube was filled with a metal core C of the shape shown, of such dimensions that when the tube was entirely collapsed against the core there should be a loss of volume of perhaps twice the volume compression of the liquid between the cylinders under the maximum pressure. This scheme was entirely successful in separating the liquid to be measured from the surrounding liquid—petroleum ether, or kerosene—by which pressure was transmitted from one part of the apparatus to another, and also in transmitting the pressure freely to the liquid between the cylinders. The thin rings of German silver separating the inner from the outer cylinder never showed the slightest sign of deformation after a run, which, in view of their extreme thinness, is good evidence that the pressure in the liquid between the cylinders must have been very nearly the same as in the liquid outside.

There are a number of corrections, which are described in detail in the original paper. The only correction for the change of dimensions under pressure is for the change of length; there is a correction for the change of resistance of the heating unit with pressure and temperature; there are corrections for the changes of constants of the thermo-couples with pressure and temperature; there is a correction for heat leak through the transmitting liquid at the ends between inner and outer cylinder, and its variation with pressure. All these corrections are small, and together do not amount to more than 4 per cent. of the total effect. The smallness of the corrections constitutes a marked advantage of the method.

There is another sort of correction not peculiar to these measurements under pressure, but affecting the thermal conductivity measurements of transparent substances, which has received very little attention in previous measurements, namely, correction for radiation. Two sorts of radiation effect may be considered. The first is direct radiation through the liquid between the two walls of the containing vessel, which are maintained at different constant temperatures. Transfer of heat by radiation in this way is effective in all those cases where the radiation path through the liquid is not so long that it will be approximately all absorbed, and this is apparently the condition in all experiments hitherto made. An exact calculation of this effect is not possible without a determination of all the emissivity constants of the vessel. An upper limit can be found by assuming the walls of the vessel to act like a black body; with the pressure apparatus described above, this upper limit would be between 1 and 2 per cent. for most of the liquids used. The upper limit is doubtless much too high, and the correction was merely neglected in my work. A simple dimensional examination will show that the importance of the effect varies directly as the distance between walls. In the apparatus above, in which this was only 0·4 mm., the effect is therefore small, but in other forms of apparatus used at atmospheric pressure the distance frequently is 1 cm. or more, when it

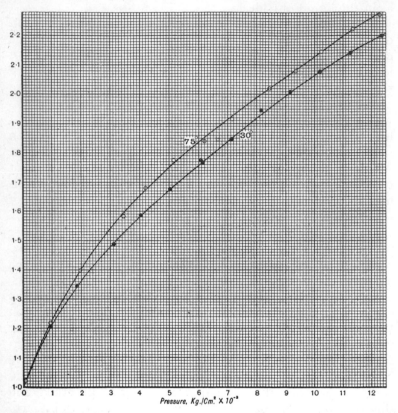

Fig. 76.—The effect of pressure on the relativity thermal conductivity of i-propyl alcohol at 30° and 75°. The points shown are observed points, with small corrections still to be applied.

may be important. Neglect of this effect may very easily result in false values for the temperature coefficient of conductivity, and this may be an explanation of the chaotic values for this scattered through the literature. A second way in which radiation may enter is by direct transfer from one part of the liquid to another. This is a matter which has been very little, if at all, discussed. I am not sure that it is possible in principle to separate a heat transfer by a mechanism of this kind from a transfer by the ordinary conduction mechanism. No correction was applied for anything of this sort in my experiments, but the possibility of such an effect should at least be mentioned.

The Results for Liquids. The liquids chosen for investigation were for the most part those whose compressibility I had already investigated, in order that it might be possible to correlate the effect of pressure on thermal conductivity with other pressure effects. Various details about the purity and handling of the different liquids may be found in the detailed paper. A set of measurements on a typical liquid of the set is shown in fig. 76, from which an idea may be formed of the regularity of the measurements. The average deviations of the single readings from smooth curves varied from 0·06 to 0·31 per cent., averaging about 0·17 per cent. The smooth curves drawn through the points should be considerably more accurate than this.

The results are summarised in Table XVIII. This contains the absolute conductivity at atmospheric pressure at 30° and 75° as determined by these measurements, and the relative conductivity at each of these temperatures as a function of pressure up to 12,000 kg./cm.². Both the temperature and the pressure effects shown in the table merit discussion.

Previous values of the temperature coefficient of thermal conductivity are quite contradictory, and only the most recent can be even considered. It will be noticed that at atmospheric pressure my temperature coefficients are negative for all liquids except water. The negative sign of the coefficient agrees with that found by Lees [1] and by Goldschmidt,[2] but my values are numerically considerably smaller than those of either. The temperature coefficient of the conductivity of water is exceptional in being positive; my value is an increase of 0·22 per cent. per degree rise of temperature. Practically the only previous value determined with proper care is that of Jakobs,[3] who finds for the coefficient 0·28 per cent. per degree. Considering the magnitude

TABLE XVIII

EFFECT OF PRESSURE ON THERMAL CONDUCTIVITY OF A
NUMBER OF LIQUIDS

Liquid	Temperature	Conductivity at 0 kg./cm.²	Relative conductivity as a function of pressure in kg./cm.²						
			0	1000	2000	4000	6000	9000	12,000
Methyl alcohol	30°	·000505	1·000	1·201	1·342	1·557	1·724	1·927	2·097
	75°	493*	1·000	1·212	1·365	1·601	1·785	2·007	2·191
Ethyl alcohol	30°	·000430	1·000	1·221	1·363	1·574	1·744	1·954	2·122
	75°	416*	1·000	1·233	1·400	1·650	1·845	2·083	2·278
Isopropyl alcohol	30°	·000367	1·000	1·205	1·352	1·570	1·743	1·963	2·150
	75°	363	1·000	1·230	1·399	1·638	1·812	2·030	2·211
Normal butyl alcohol	30°	·000400	1·000	1·181	1·307	1·495	1·648	1·842	2·008
	75°	391	1·000	1·218	1·358	1·559	1·720	1·923	2·099
Isoamyl alcohol	30°	·000354	1·000	1·184	1·320	1·524	1·686	1·893	2·069
	75°	348	1·000	1·207	1·348	1·557	1·724	1·934	2·126
Ether	30°	·000329	1·000	1·305	1·509	1·800	2·009	2·251	2·451
	75°	322*	1·000	1·313	1·518	1·814	2·043	2·316	2·537
Acetone	30°	·000429	1·000	1·184	1·315	1·511	1·659	1·864	Freezes
	75°	403*	1·000	1·181	1 325	1 554	1·738	1·960	2·137
Carbon bisulphide	30°	·000382	1·000	1·174	1·310	1·512	1·663	1·834	1·962
	75°	362*	1·000	1·208	1·366	1·607	1·789	1·998	2·154
Ethyl bromide	30°	·000286	1·000	1·193	1·327	1·517	1·657	1·815	1·928
	75°	273*	1·000	1·230	1·390	1·609	1·772	1·944	2·121
Ethyl iodide	30°	·000265	1·000	1·125	1·232	1·394	1·509	1·628	1·724
	75°	261	1·000	1·148	1·265	1·442	1·570	1·715	1·837
Water	30°	·00144	1·000	1·058	1·113	1·210	1·293	1·398	Freezes
	75°	154	1·000	1·065	1·123	1·225	1·308	1·412	1·506
Toluol	30°	·000364	1·000	1·159	1·286	1·470	1·604	1·768	(2·394†)
	75°	339	1·000	1·210	1·355	1·573	1·738	1·932	2·089
Normal pentane	30°	·000322	1·000	1·281	1·483	1·777	1·987	2·245	2·481
	75°	307*	1·000	1·319	1·534	1·855	2·122	2·440	2·740
Petroleum ether	30°	·000312	1·000	1·266	1·460	1·752	1·970	2·215	2·379
	75°	302*	1·000	1·268	1·466	1·780	2·026	2·324	2·561
Kerosene	30°	·000357							
	75°	333	1·000	1·185	1·314	1·502	1·654	1·839	2·054

* Extrapolated.
† Toluol freezes at 9900 kg./cm.² at 30°.   The figure is for the solid at 11,000.

of the corrections which have had to be applied in the previous methods, I believe that the apparatus developed here for use under pressure is capable not only of giving the pressure coefficient, but also of giving better values for the ordinary temperature coefficient than most previous methods.

In general characteristics the effect of pressure is the same on all liquids tried. The thermal conductivity increases under a pressure of 12,000 kg./cm.² by an amount which varies from 1·5-fold for water to 2·7-fold for normal pentane. In general, the effect is greater for those substances with the lower boiling or freezing temperatures (in general these are also the most compressible substances). The effect is not at all proportional to pressure, but at high pressures a given increment of pressure produces a much smaller effect, both absolutely and relatively, than at lower pressures. It may be seen from the results in the table that the second 6000 kg./cm.² produces an effect which, in general, is about one-half that of the first 6000 kg./cm.².

The relative effect of pressure is greater at 75° than at 30° by something of the order of 5 or 10 per cent. If one calculates from the relative values in the table the absolute conductivity as a function of pressure and temperature, it will be found that the change with temperature of the pressure effect is so large that at the upper end of the pressure range the temperature coefficient of conductivity is reversed, and all liquids are more conducting at the higher temperatures at the higher pressures. The pressure at which this reversal takes place varies somewhat with the liquid, but as a rough average it is at about 3000 kg./cm.². This reversal by pressure of a temperature derivative is not the first instance, but we have seen that at high pressures the thermal expansion of all these liquids is greater at low temperatures rather than at high temperatures, the latter being the normal behaviour at atmospheric pressure. It may be significant that the pressure of reversal of this effect also is roughly 3000 kg./cm.².

The conductivity of solid toluol at 11,000 kg./cm.² is about 28 per cent. greater than that for the liquid extrapolated to

the same pressure. We would in general expect the solid
to have the greater conductivity, because its specific volume
is less than that of the liquid, and the velocity of elastic
waves in it is greater. This relation of conductivity in liquid
and solid is also what has been found by Barus for thymol,
the only other organic substance that I know that has been
measured in both solid and liquid.

**Theoretical Considerations on the Thermal Conductivity of
Liquids**. The thermal conductivity of liquids is a subject
that has received very little theoretical attention. It has
been recognised that the mechanism of conduction in a
liquid must be different from that in a gas. In the gas there
is an intimate connection between thermal conductivity and
viscosity; each is connected with the length of free molecular flight, conductivity being concerned with the transfer
of energy and viscosity with the transfer of momentum.
The relation between these is exhibited by the elementary
formula

$$k = \eta C,$$

η being viscosity, and C specific heat. That there is no such
relation for liquids may be shown in the first place by substituting numerical values into this formula; the thermal
conductivity will be found to be of the order of ten times too
small. That there is no simple relation between thermal
conductivity and viscosity in liquids is also shown by the
known fact that the viscosity of solutions of varying strength
of gelatine in water varies by a factor of many fold, with only
slight changes in the thermal conductivity. The same is
suggested by the pressure experiments above; the change of
viscosity of all the liquids measured above, except water, is
many fold greater than the change in thermal conductivity.

The only relation that I know which has been suggested
connecting the thermal conductivity of liquids with their
other properties is an empirical relation of Weber,[4] first given
as $k/\rho c =$ const., and later modified to

$$\frac{k}{\rho c} \left(\frac{m}{\rho} \right)^{\frac{1}{3}} = \text{const.}$$

Here ρ is the density of the liquid, c its specific heat, and m its molecular weight. The factor $\left(\dfrac{m}{\rho}\right)^{\frac{1}{3}}$, by which the modified formula differs from the original one, is seen to be proportional to the mean distance of separation between the centres of the molecules. Weber's own data indicate a surprising constancy of the modified expression, but later discussion [5] has disclosed considerably greater variations than supposed by Weber. It seems probable that the relation discovered by Weber has no particular absolute significance. He gave no way of calculating the constant, which turns out to have the dimensions $M^{-\frac{1}{3}}L^3T^{-1}$. No simple relation between a constant of these dimensions and any of the fundamental constants suggests itself. It is to be expected *a priori* that it would not be difficult to discover relations like those of Weber, because the properties of ordinary organic liquids vary through only a comparatively small range.

In the course of speculations on the significance of the data above I have come across an expression for thermal conductivity which is different from that of Weber, in that it enables us to calculate the thermal conductivity completely, with no empirical constant unaccounted for. The relation is

$$k = 2av/\delta^2,$$

where a is the gas constant, v the velocity of sound in the liquid, and δ the mean distance of separation of the centres of the molecules, assuming a cubical arrangement on the average, and calculating δ by the formula $\delta = \left(\dfrac{m}{\rho}\right)^{\frac{1}{3}}$, where m is the absolute weight in grams of one molecule of the liquid.

In Table XIX are shown the computed and observed values of thermal conductivity of those substances for which the necessary data have been procured. The velocity of sound has been directly measured for only a few liquids. In order to treat all liquids consistently, I give in the table the velocity calculated from my values for the compressibility. The compressibility that enters the formula for the velocity of sound is the adiabatic compressibility. This is given in terms of the

TABLE XIX

COMPARISON OF COMPUTED AND OBSERVED THERMAL
CONDUCTIVITIES AT 30°

Liquid	Velocity of sound cm./sec.	δ^{-2} cm.$^{-2}$	Thermal conductivity Abs. C.G.S. units	
			Computed	Observed
Methyl alcohol .	$1 \cdot 13 \times 10^5$	$6 \cdot 00 \times 10^{14}$	$27 \cdot 4 \times 10^3$	$21 \cdot 1 \times 10^3$
Ethyl alcohol .	$1 \cdot 14$	$4 \cdot 74$	$21 \cdot 8$	$18 \cdot 0$
Propyl * alcohol	$1 \cdot 24$	$3 \cdot 94$	$19 \cdot 7$	$15 \cdot 4$
Butyl * alcohol .	$1 \cdot 05$	$3 \cdot 49$	$14 \cdot 9$	$16 \cdot 7$
Iso-amyl alcohol.	$1 \cdot 24$	$3 \cdot 13$	$15 \cdot 7$	$14 \cdot 8$
Ether . .	$0 \cdot 92$	$3 \cdot 19$	$11 \cdot 9$	$13 \cdot 7$
Acetone . .	$1 \cdot 14$	$4 \cdot 00$	$18 \cdot 5$	$17 \cdot 9$
Carbon bisulphide	$1 \cdot 18$	$4 \cdot 61$	$21 \cdot 9$	$15 \cdot 9$
Ethyl bromide .	$0 \cdot 90$	$3 \cdot 97$	$14 \cdot 5$	$12 \cdot 0$
Ethyl iodide .	$0 \cdot 78$	$3 \cdot 81$	$12 \cdot 1$	$11 \cdot 1$
Water . .	$1 \cdot 50$	$10 \cdot 4$	$63 \cdot 0$	$60 \cdot 1$

* Computed for normal propyl and iso-butyl alcohols, observed for iso-propyl and normal butyl.

isothermal compressibility, the thermal expansion, the specific heat, and the absolute temperature by the formula

$$\left(\frac{\partial v}{\partial p}\right)_s = \left(\frac{\partial v}{\partial p}\right)_\tau + \frac{\tau}{C_p}\left(\frac{\partial v}{\partial \tau}\right)_p^2.$$

Here the second term is a small correction; the values of thermal expansion and specific heat used in the computation of it were taken from tables of constants. The calculated velocity agrees with the directly determined velocity within a couple of per cent. where the comparison is possible.

In view of the simplicity of the relation, the agreement between observation and calculation shown in Table XIX is surprising. It is particularly surprising that the high conductivity of water is reproduced so closely. The high conductivity of water, therefore, does not appear to be due in any special way to the molecular peculiarities of water (two or more molecular species), but is directly referable to its low compressibility and the fact that the centres of the molecules are closer together than in the average liquid.

This expression for thermal conductivity may be obtained in terms of a very simple physical picture. Imagine the molecules in simple cubical array, the distance between molecular centres being δ. Let there be in the liquid a temperature gradient $d\tau/dx$. The energy of a molecule is $2a\tau$ (half-potential and half-kinetic), where a is the gas constant, $2\cdot02 \times 10^{-16}$, and τ is the absolute temperature. The difference of energy between adjacent molecules in the direction of the temperature gradient is $2a\delta\dfrac{d\tau}{dx}$. This energy difference is to be conceived as handed down a row of molecules with the velocity of sound. The total energy transferred across a fixed point of any row of molecules per unit time is the product of the energy difference and the number of such energy steps contained in a row v cm. long, or $\left(2a\delta\dfrac{d\tau}{dx}\right)\left(\dfrac{v}{\delta}\right)$. The total transfer across unit cross-section is the product of the transfer across a single row and the number of rows, or $2av\dfrac{d\tau}{dx}\delta^{-2}$. But by the definition of thermal conductivity this transfer is also $k\dfrac{d\tau}{dx}$. Whence, identifying coefficients, we have for the thermal conductivity

$$k = 2av\delta^{-2}.$$

The matter has been further discussed by Jeffreys,[6] who remarks with justice that this picture is over-simplified; evidently insufficient stress is put on the three dimensional aspect of the elastic waves which constitute the heat motion. He suggests an alternative account by supposing that in a liquid the irregularity of motion is so great that the microscopic elastic waves by which heat is transferred are completely scattered in the smallest distance possible—that is, the distance between molecular centres. He suggests that the formula may be applicable to glasses, and I have also found it roughly applicable to a solid like hard rubber; it would not be surprising if it were roughly applicable to all amorphous solids.

Not only does this very simple expression give approximately the absolute value of the thermal conductivity, but it also accounts for the sign, although not the numerical magnitude, of the temperature coefficient of conductivity. Thus it will be found that for the ordinary organic liquid both v and δ^{-2} decrease with rising temperature, giving a conductivity decreasing with rising temperature, as found experimentally. For water, on the other hand, v increases with rising temperature (both the isothermal compressibility and the thermal expansion of water changing abnormally with temperature) at a rate more than sufficient to compensate for the decrease of δ^{-2}, so that the net effect is to increase the conductivity, again agreeing with experiment.

In view of the success of this simple expression at atmospheric pressure, it was a surprise to find that it does not account at all well for the phenomena under pressure. It does give the correct sign for the pressure effect, for both v and δ^{-2} increase with pressure, so that an increase in conductivity is to be looked for. But the change given by the formula is greater than that actually found. The velocity of sound is the factor that changes most with pressure. We have already seen that the ratio of the compressibility at 12,000 kg. to that at atmospheric pressure for the normal organic liquid is of the order of 15 to 1. Combined with the change of density, this results in a change of the velocity of sound by a factor usually between 3 and 4. Combined with the factor δ^{-2} this demands a change of thermal conductivity at 12,000 of between three- and fourfold, whereas the experimental value is at most 2·7, and for the majority of liquids is in the neighbourhood of 2. The computed change of conductivity with pressure for water agrees no better with experiment than it does for other liquids.

If the formula becomes at high pressures a better approximation at high temperatures than at low temperatures, the reversal of sign of the temperature coefficient of conductivity at high pressures would be a consequence.

Thermal Conductivity of Solids. The problem of devising

a suitable method of measuring the thermal conductivity of solids under pressure proved to be much more difficult than for liquids, and several methods had to be used (B. 38) depending on the absolute magnitude of the conductivity of the solid. If the conductivity is comparatively low, a modification of the radial flow method used for liquids proved to be feasible. Wherever possible this method should be used, as it has the ideal advantage that the heat losses are small. By this method a number of minerals and similar substances were investigated (B. 48). There is evidently some geological interest in such measurements, because an analysis of the thermal condition of the earth demands a knowledge of the thermal conductivity of the minerals of the earth's crust at depths at which the pressure is high.

The minerals were worked either by machining, if they were soft enough, or by grinding if they were not, to the form of cylindrical tubes 1·27 cm. outside diameter, 1·02 cm. inside diameter, and 2·5 cm. long. Copper cylinders fitted inside and outside, provided with a heating wire along the axis, and thermal couples with junctions near the inner and outer surfaces of the cylinder of mineral. There is a complication in that the fit between copper and mineral cannot be made perfect, but there was of necessity a film of the pressure transmitting liquid. The dimensions of this film could be obtained by a suitable combination of measurements of dimensions and density of the other parts of the apparatus, and since the effect of pressure on the thermal conductivity of the transmitting liquid and on the dimensions of copper and minerals had been previously determined, the correction for the film of liquid could be calculated as a function of pressure. The correction for this film was of the order of 30 per cent. at atmospheric pressure; at high pressures it becomes less.

The following minerals were investigated: basalt, Solenhofen limestone, talc, pipestone (Catlinite), NaCl, and pyrex glass. Detailed analyses of these substances are given in the detailed paper. I am much indebted to Professor R. A. Daly of the Department of Geology of Harvard University for suggestions as to significant minerals.

The results are shown in Table XX.   In this table is given the absolute conductivity in grm. cal. per sec. per cm.² per degree C. per cm. as a function of pressure (measured in kg./cm.²) at 30° and 75° C., and also the percentage change

TABLE XX

SUMMARY OF THERMAL CONDUCTIVITY MEASUREMENTS
ON NON-METALLIC SOLIDS

Substance	Thermal conductivity as function of pressure	Percentage change of conductivity for 1000 kg./cm.²
Pyrex glass	$0.00261 + 0.000010 \left(\dfrac{p}{1000}\right)$, 30° and 75°	$+0.38$
Basalt	$0.00404 + 0.000019 \left(\dfrac{p}{1000}\right)$, 30°	$+0.47$
	$0.00414 + 0.0000089 \left(\dfrac{p}{1000}\right)$, 75°	$+0.22$
Solenhofen Limestone	$0.00523 + 0.000005 \left(\dfrac{p}{1000}\right)$, 30°	$+0.1$
	$0.00451 + 0.000030 \left(\dfrac{p}{1000}\right)$, 75°	$+0.67$
Talc	$0.00733 + 0.000115 \left(\dfrac{p}{1000}\right)$, 30°	$+1.57$
NaCl	$0.00880 + 0.000317 \left(\dfrac{p}{1000}\right)$, 30°	$+3.6$
	$0.00756 + 0.00027 \left(\dfrac{p}{1000}\right)$, 75°	$+3.6$
Pipestone	Pressure conductivity 30° 0 0.00438 2,000 0.00506 4,000 0.00544 6,000 0.00563 8,000 0.00574 10,000 0.00587 12,000 0.00596	Average $+3.0$

of conductivity at each of these temperatures for a pressure change of 1000 kg./cm.² (corresponding to a depth in the earth's crust of about 2·5 miles).   The results for pipestone, which do not vary linearly with pressure, are given as a function of pressure at 2000-kg. intervals.   In all cases the thermal conductivity is found to increase with increasing pressure, and for all of the materials except pipestone the increase with pressure is linear.

For the harder rocks, basalt and Solenhofen limestone, which presumably correspond more nearly to the average to be found in the crust, the change per 1000 kg./cm.[2] is 0·5 per cent. or less. It is greater for the softer substances, talc and pipestone, being 1·57 and 3·0 per cent. respectively, and is greatest for NaCl, for which it is 3·6 per cent. (It should be remarked with regard to NaCl that my value for the absolute conductivity is of the order of one-half that which Eucken has found. The absolute conductivity of NaCl is the highest of that of any of the solids measured by this method, and therefore the apparatus is least well adapted to measuring it.)

In general the change of thermal conductivity with pressure does not appear to be so large that it will introduce any important change into previous geological speculations which involve only the conductivity near the surface of the earth.

The effect of temperature on thermal conductivity, unlike that of pressure, does not always seem to be of the same sign. The conductivity of limestone and NaCl decreases with rising temperature by fairly large amounts, but that of basalt increases slightly. A positive temperature coefficient for basalt was also found by Poole,[2] as opposed to a numerically much larger negative coefficient for granite. The numerical value of Poole's coefficient agrees as nearly with mine as can be expected for material from different sources. For ordinary liquids we have seen that the thermal conductivity usually decreases with rising temperature, and this may perhaps be taken as the normal effect in rocks. But there is not any great certainty here, and the combined effect of pressure and temperature at a depth of several hundred miles, where the pressure is of the order of a hundred thousand atmospheres, may be such that we must be prepared for the possibility of the conductivity being several fold greater than the accepted value.

Pyrex glass does not depart from the average crystalline rock in the effect of pressure on conductivity, although the absolute conductivity is lower, as is usually the case for a glass.

Metals are the most interesting solids for which to determine the effect of pressure on thermal conductivity, because of the important part that thermal conductivity has played in the electron theories of metals and the connection with electrical conductivity which is formulated in the Wiedemann-Franz law.   Unfortunately, because of the high conductivity of metals, the pressure effect is very difficult to measure, and the results for them are very much less satisfactory than for liquids or for minerals.

Two methods were used.   The first was a simple modification of the radial flow method used for liquids.   If in fig. 75 the two copper cylinders with the intermediate annular ring of liquid are replaced by a single cylinder of metal, with heating element and thermo-couples unchanged, the essential features of the radial flow method for metals will be reproduced.   The mechanical difficulty is in getting into the cylinder of metal the long small holes that are required for the heating element and the thermo-couples.   In the case of the easily fused metals, these holes were cast into position, using cores of tungsten wire, but with the metals of high melting-point, such, for example, as iron, copper, and nickel, larger holes had to be drilled through the cylinder, which were then filled with small copper tubes, sweated into place. Measurements made with this method gave very good smooth curves;   the difficulty was that different specimens of the same metal did not give consistent results.   There proved to be important sources of error: difficulty of exactly controlling the temperature of the external surface, and in those cases in which copper tubes were necessary to carry the thermo-couples, difficulty of getting perfect thermal contact between the tubes and the body of the metal.   Both these difficulties increase very rapidly as the conductivity of the metal increases, and it proved impossible to obtain consistent results for the metals of high conductivity.   But for metals with low conductivity, such as lead and tin, consistent results were obtained, and I believe that there cannot be much question about the correctness of these.

For metals of high conductivity, an entirely different method had to be used.   This was a longitudinal flow method, in which heat flows from a source at one end of the rod, along the rod, to a sink at the other end.   The source was a diminutive heating coil let into a hole drilled in the end of the rod, and the sink was a massive block of copper, in good thermal contact with the walls of the pressure cylinder.   The temperature gradient in the rod is measured with a differential thermocouple attached at two points near the mid-length of the rod. The thermal conductivity can obviously be calculated in terms of dimensions, heat input, and temperature gradient. The chief complication with this method arises from the escape of heat from the rod laterally into the pressure-transmitting medium.   A simple dimensional examination will show that this effect can be minimised by making the dimensions of the specimen as small as possible; the rods were actually 3 mm. in diameter, and 1 cm. long.   Approximate calculations of the amount of heat escaping, both at atmospheric and at high pressure, can be made in terms of the dimensions and the known constants of the transmitting liquid.   The magnitude of this correction varies with the conductivity of the metal; for silver, the correction changes at the maximum pressure by an amount equivalent to a 1 per cent. change in the conductivity of the metal, for nickel the corresponding effect is 5 per cent., and in the extreme case of bismuth nearly 14 per cent.   These corrections are an important part of the total pressure effect, so that for this reason alone there is more uncertainty in these pressure results than in any other sort of pressure measurement.   But there was an additional difficulty in practice.   In order that the thermo-couple should accurately register the temperature of the metal, it was threaded through a fine hole which pierced the rod perpendicular to the axis.   This hole should be a close fit for the thermal-couple wire in order to secure good thermal contact and geometrical certainty in the location of the junction, but should be a loose fit in order to avoid strains in the thermo-couple wire arising from the viscosity of the transmitting medium.   The compromise that was adopted was to make the hole about 0·007 and the wire about 0·005 of an inch in diameter.   The resulting uncertainty of 0·002 in. in the location of the junction was about 5 per cent. of the total distance between junctions.   Under pressure the couple has the possibility of lateral play of this amount, with corresponding variation in the apparent thermal conductivity. An analysis of the irregularities made it possible to very much minimise the effect. Each couple apparently had one or two positions of stability in the holes; by careful manipulation long successions of readings could often be obtained with no change of position, but very often after a critical pressure had been reached the couple would flop to a new, more stable position, with the result that all the data showed a tendency to lie on three or four distinct straight lines. By taking many points, and by a careful study of the stability relations, I believe that I have in every case got essentially the correct result, but it must be admitted that an unsympathetic critic might find considerable ground for scepticism in the more unfavourable cases.

The results are shown in Table XXI, in which the percentage change of thermal conductivity produced by 12,000

TABLE XXI

EFFECT OF PRESSURE ON THERMAL CONDUCTIVITY
OF METALS AT 30°

Metal	Average pressure coefficient between 0 and 12,000 kg.	Pressure coefficient of Wiedemann-Franz ratio
Lead . . .	$+17 \cdot 3 \times 10^{-6}$	$+ 6 \times 10^{-6}$
Tin	$+12 \cdot 2$	$+ 3$
Cadmium . . .	$+ 7 \cdot 4$	$- 1 \cdot 7$
Zinc	$+ 2 \cdot 1$	$- 2 \cdot 5$
Iron	$- 0 \cdot 3$	$- 2 \cdot 6$
Copper . . .	$- 7 \cdot 5$	$- 9 \cdot 3$
Silver . . .	$- 3 \cdot 7$	$- 7 \cdot 0$
Nickel . . .	$-12 \cdot 0$	-13
Platinum . . .	$- 1 \cdot 6$	$- 3 \cdot 5$
Bismuth . . .	$-31 \cdot 0$	-10
Antimony . . .	$-21 \cdot 0$	-10

kg./cm.^{2} is listed. In only two cases was the accuracy sufficient to establish that the effect was not linear in the pressure; these two cases are lead and tin, in which the departure is in the normal direction, the effect becoming less at the higher pressures. However, the curvature in these two cases is not certain enough to tabulate. The experimental results for lead are shown in fig. 77, from which

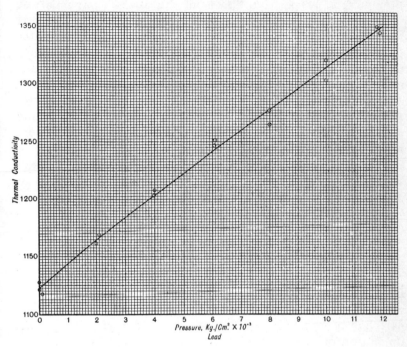

FIG. 77.—The thermal conductivity in relative units of lead as a function of pressure.

an idea may be obtained of degree of regularity possible in a favourable case.

It is to be noted that a number of the metals measured do not crystallise in the cubic system, so that the results for them do not have the unique significance of the results for cubic metals. But all the non-cubic metals were prepared in such a way as to be of fine-grained structure, and it is very probable that the results listed are the mean of the pressure effects for all the directions in the crystal. For example, four samples of antimony were measured; two were prepared by very rapid chilling from the melt, and two by extrusion. The changes found for 12,000 kg. were −23·9, −26·3, −24·8, and −23·9 per cent. respectively.

The most surprising feature of the results is that the effect is negative for about half the metals, contrasted with the universally positive effect in liquids and minerals. A negative effect is perhaps to be expected in bismuth and antimony, because the electrical conductivity also decreases under pressure, but the electrical conductivity of the other metals increases with pressure, and a corresponding increase in the thermal conductivity is to be expected in view of the close relation between the two conductivities expressed by the Wiedemann-Franz law. If the Wiedemann-Franz ratio were truly independent of the metal, then it must also be independent of pressure for any particular metal. The third column of the table gives the pressure coefficient of the Wiedemann-Franz ratio. It is not zero, and for nearly every metal except lead and tin is negative—that is, with these two exceptions, the increase of electrical conductivity with pressure is greater than the increase of thermal conductivity.

Theoretically a fairly satisfactory account has been given of the Wiedemann-Franz law, both on the basis of the classical electron theory and on the new wave mechanics. Both of these theories give only a partial account of the mechanism of thermal conduction, because only the electronic part has been considered, leaving out the part arising from the action of the neutral atoms. The contribution made by the atomic mechanism is almost certainly different in different metals. Furthermore, if the atomic mechanism of heat transfer is affected by pressure differently from the electronic mechanism, and there is no reason to think that it may not be, the Wiedemann-Franz ratio may be expected to vary with pressure. There is, however, no immediate correlation between the pressure coefficient of the Wiedemann-Franz ratio and its absolute value, so that it is obvious that the state of affairs is not entirely simple.

The effect of pressure on the thermal conductivity of six pure metals and four alloys had been previously measured by Lussana [8] up to 3000 kg./cm.². His method was that of

Despretz; this may be considered as a modification of the longitudinal flow method used above. Instead of allowing heat to flow between source and sink at two ends of a comparatively short rod, the rod is made infinitely long, and the heat input from a source at one end flows along the rod until it is all dissipated by lateral flow to the surroundings. The conductivity of the metal can be obtained in terms of the temperature readings at various parts of the length, provided the constants of lateral escape are known. It is evident that the thermal conductivity of the transmitting liquid plays a much larger part in this method than in mine, and that therefore the pressure effect on the conductivity of the transmitting liquid must be determined with proportionally greater accuracy. The effect of pressure on the thermal conductivity of the liquid is ten times greater than on the metal itself, yet Lussana determined it to only one significant figure. For these reasons it seems to me that much confidence cannot be placed in Lussana's results, which differ greatly from those given above. Lussana has, however, defended his measurements,[9] and the reader may form his own opinion, if necessary. In his paper of rebuttal I believe that he has misunderstood my method in several important particulars, in particular in the part which he supposes was played by heat leak along the thermo-couple wires.

[1] G. H. Lees, *Trans. Roy. Soc.*, **191**, 399 (1898).
[2] R. Goldschmidt, *Phys. ZS.*, **12**, 417 (1911).
[3] M. Jakob, *Ann. Phys.*, **63**, 537 (1920).
[4] H. F. Weber, *Wied. Ann.*, **10**, 101, 304, 472 (1880); *Berlin Akad. Sitzber.*, 809 (1885[2]).
[5] E. van Aubel, *ZS. Phys. Chem.*, **28**, 336 (1899).
[6] Harold Jeffreys, *Proc. Camb. Phil. Soc.*, **24**, 19 (1928).
[7] H. H. Poole, *Phil. Mag.*, **27**, 58 (1914).
[8] S. Lussana, *Nuov. Cim.*, **15**, 130 (1918).
[9] S. Lussana, *ibid.*, **25**, 115 (1923).

CHAPTER XII

VISCOSITY

THERE are comparatively few previous experiments on the effect of pressure on viscosity. Coulomb made an unsuccessful attempt by a swinging-disc method to detect a difference between the viscosity of water at atmospheric pressure and in vacuum. The first successful attempt was by Roentgen[1] in 1881 (published in 1884), who found a decrease of 1 per cent. or less in the viscosity of water at temperatures between 5° and 11° under a pressure of twenty atmospheres. Warburg and Sachs[2] found in 1884 at 20° and up to 150 kg./cm.² pressure a linear decrease of the viscosity of water by 1.7×10^{-4} per kg./cm.², and a much greater linear increase for ether, benzene, and carbon dioxide. Their method was a capillary flow method, in which a mercury column provided the driving pressure. R. Cohen[3] studied the effect of pressure up to 900 kg./cm.² between 0° and 25° on water, NaCl solutions of four different concentrations, and turpentine. The method was a modified form of the capillary flow method of Roentgen and of Warburg and Sachs. The viscosity of water was found to decrease with pressure, and by smaller amounts the higher the temperature. The experiments left uncertain whether there was any pressure effect at all at 40°. Furthermore, at the lower temperatures the effect of pressure became less at the higher pressures. The viscosity of concentrated NaCl and NH_4Cl solutions was found to increase with increasing pressure; as the solution becomes more dilute the abnormal behaviour of pure water is approached. Turpentine showed very large effects, an increase of 100 per cent. under 600 kg./cm.². Hauser[4] extended the temperature range of Cohen to 100°, although decreasing his pressure range to 400 kg./cm.², and found that above 32° pressure increases the viscosity of water like other liquids. Faust,[5] working in Tammann's laboratory, measured up to 3000 kg./cm.[2] at several temperatures the effect on ether, CS_2 and ethyl alcohol. Up to about 1500 kg./cm.[2] the viscosity increases linearly with pressure, but above this it increases more rapidly. E. Cohen and Bruins [6] measured the effect of a pressure of 1500 kg./cm.[2] on the viscosity of mercury at 20°. All of these experimenters used some form of a capillary flow method, with glass capillaries and mercury to provide the driving pressure, so arranged that only a single reading could be obtained with each set-up. Hyde [7] made measurements up to 1500 kg./cm.[2] on a number of lubricating oils, and found large increases of viscosity, rising to as much as fifteen-fold. The method was a very ingenious one, which permitted continuous readings; the oil could be made to run back and forth from one reservoir to another through a long steel capillary under a head of mercury, and the whole apparatus was mounted on knife edges so that the position of the liquid could be determined by a balancing operation. The viscosity of lubricating oils has been further measured, since the publication of my results to be described later, up to 4000 kg., and in the temperature range to 140° by Hersey and Shore,[8] by a method which also permits continuous readings. The work of Hersey was first done in the Jefferson Laboratory, and his pressure technique was in many respects the same as mine. The viscosimeter, which was adapted by Hersey to high-pressure measurements from a form originally due to Flowers, is a simple rolling-ball arrangement, in which a steel ball rolls from one end to another of an inclined smooth steel tube filled with the oil under pressure. The time of roll is determined electrically. By tipping the tube back and forth, the measurement may be repeated as often as desired, without opening the apparatus. Again large effects were found, running up to several hundred-fold. It was found that the logarithm of the viscosity was approximately linear in the pressure, as is also the case with the pure liquids to be described presently. These results on oils are evidently of much technical interest, but they are difficult to interpret because of the physical complexity of the materials, which are mixtures of many components, and they will not be discussed further here.

In addition to the quantitative measurements just discussed, a couple of phenomena may be mentioned which are intimately related to these, but which have been observed only qualitatively. Barus [9] observed a very great increase in the viscosity of marine glue under pressure; in particular the increase was so great that there was no appreciable increase in the velocity of extrusion of marine glue through a small crack on doubling the pressure. In my preliminary pressure work I had observed a similar enormous increase in the viscosity of ordinary paraffin with pressure. In one of the collapsing experiments on steel tubes, a hollow, heavy-walled steel tube, closed at both ends, was imbedded in a mass of paraffin, and pressure applied by a piston directly to the paraffin, with the idea that the paraffin would transmit the pressure approximately hydrostatically to the steel tube, which would be collapsed at some high pressure. The rigidity of the paraffin was increased by pressure so much, however, that no relative motion between steel and paraffin took place, but the steel tube was shortened lengthwise by an amount equal to the volume compression of the paraffin.

None of the methods previously used for the measurement of the effect of pressure on viscosity was adapted to the greater pressure range of this work, in general the apparatus being too bulky or too cumbersome in operation, requiring, for example, that the pressure chamber be reopened and the apparatus refilled for every new determination at a new pressure. The apparatus adopted for this work does not give the absolute viscosity, but does give the relative viscosity, and therefore the pressure coefficient of viscosity. Two different methods had to be used: one demanded that the liquid be a non-electrical conductor (B. 55), and was applied to the study of forty-three liquids, mostly organic, and the other was applied only to mercury (B. 61). The apparatus used with the non-conducting liquids is shown in fig. 78. The general idea of the method is very simple; in a steel cylinder of approximately 6 mm. internal diameter, filled with the liquid under investigation, there is a steel

cylindrical weight separated from the walls of the cylinder
by a narrow annular space. The time of vertical fall of the
weight from one end of the cylinder to the other is determined; the time is a measure of the viscosity.

The pressure-producing apparatus is so mounted and connected with the viscosity cylinder that it may be rotated

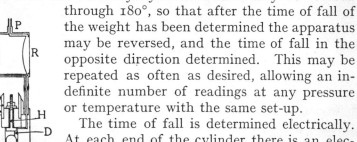

through 180°, so that after the time of fall of
the weight has been determined the apparatus
may be reversed, and the time of fall in the
opposite direction determined. This may be
repeated as often as desired, allowing an indefinite number of readings at any pressure
or temperature with the same set-up.

The time of fall is determined electrically.
At each end of the cylinder there is an electrically insulated terminal D against which
the weight rests at the end of its fall. In this
position electrical connection is made from
the walls of the cylinder through the weight
to the terminal, and the completion of this
connection is made to operate
a suitable timing device. The
weight, which is shown in
detail in fig. 79, was provided
with three small projecting
lugs at the top and bottom
ends, which act as guides to

Fig. 78.—Section
of the viscos-
ity apparatus
for insulating
liquids.

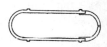

Fig. 79.—Detail of
the falling weight
of the apparatus
of fig. 78.

keep the weight concentric as it falls, and through which electrical contact is made with the cylinder. In order to secure sufficiently good contact, these lugs had to fit the cylinder within
better than 0·0025 cm. In order to keep the time of fall within
reasonable limits, different weights were used according to the
absolute viscosity of the liquid, the annular space between
weight and cylinder varying between 0·0125 cm. for water to
0·075 cm. for the more viscous liquids such as glycerine.
Further flexibility could be attained by changing the total
weight by placing within the cavity appropriate weights of
tungsten or gold.

The insulating plugs at the two ends of the cylinder are

shown in the figure; they must not only provide a terminal electrically insulated from the cylinder, but must prevent the liquid under investigation in the cylinder from mixing with the pump liquid in which the cylinder is immersed. Mechanical tightness was secured by making the central stem of the plug H of brass; this stem becomes relatively shorter and so tighter at high pressure, because of the differential compressibility between brass and steel. The insulating plug was soldered into the end of the cylinder; in this way the liquid under investigation came in contact only with metals or mica, and was kept pure. The entire apparatus was disassembled for each new filling with a fresh liquid; this proved to be necessary because of the extreme necessity for removing every particle of mechanical dirt of any kind. The soldering and unsoldering were facilitated by two German silver sleeves E and F permanently soldered to the cylinder and plug respectively. The soldered connection between cylinder and plug was made and broken at A.

The liquids under investigation were kept from contact with the surrounding liquid by which pressure is transmitted by means of a reservoir R of very thin pure tin, exactly as already described in connection with measurements of the effect of pressure on the thermal conductivity of liquids. The reservoir and cylinder were filled through the stem P, which was then sealed with a soldered plug not shown. After filling, the cylinder and reservoir were attached to an insulating plug, not shown, and the plug, with cylinder and reservoir as one self-contained unit, were screwed into the large pressure cylinder.

The large pressure cylinder was connected with a pipe to the pressure-generating apparatus and the pressure gauge, which were the same as that used in previous work, except that they were mounted horizontally, so that they could be rotated about the connecting pipe as an axis. During rotation the hydraulic press with which pressure was produced had to be disconnected from the hand-pumps, which were the source of pressure. This was made possible by valves mounted to rotate with the press, so that the pressure produced by the pump could be maintained after the transmitting pipe has been disconnected. The large pressure cylinder which contained the viscosity apparatus was T-shaped; there was a side connection at which entered the pipe connecting to the pressure-producing apparatus, which acted as the long arm of the T, which was horizontal and about which rotation took place. The main body of the cylinder acted as the cross-arm of the T, and by rotation was changed from one vertical position to the inverse. The falling weight within the viscosity apparatus fell along the axis of the cross-arm of the T. The viscosity cylinder was kept at constant temperature by the conventional stirred bath and regulators; the horizontal connecting pipe entered the bath through a simple stuffing-box.

The timing apparatus was a simple combination of a small synchronous 60-cycle A.C. motor with an ordinary clock movement, started and stopped with appropriate relays. The necessary error in the determination of a single time-interval was not over $\frac{1}{120}$ sec. Considerable difficulty was found in getting a proper source of current to operate the relay through the contact made by the falling weight. If the weight is to fall freely, there is considerable contact resistance between the lugs and the walls of the cylinder. Direct current is not suitable, for if the voltage is high enough to jump the gap, an arc follows, which may decompose the liquid or make the weight stick to the walls of the cylinder. After some trial, the alternating current delivered by a small bell-ringing magneto of the type used in insulation testing was found suitable.

Various possible sources of error in starting and stopping of the clock and the operation of the various relays made necessary the construction of a special calibrating timing device, which need not be described in detail. Error in timing was minimised by taking the average of 50 fall times, if the time of fall of the weight was less than 5 sec. Above 5 sec. fall time, the mean of 10 fall times was taken, until the time got to be 40 or 50 sec., when a smaller number of readings was taken, but never less than two even in the extreme case of one and a half hours. The error in the finally corrected fall time should not be greater than 0·002 or 0·003 sec. for the short times, and no more proportionally for the longer times.

In addition to the corrections of the timing device, there were a number of other corrections. There is a correction arising from the fact that when the apparatus is rotated for a new fall the weight begins to fall before the vertical position is completely reached. There is a correction due to the inertia effect at the beginning of fall in virtue of which a certain amount of time is needed to build up the final velocity. The largest correction is for the change of buoyancy of the liquid on the falling weight as the density of the liquid changes under pressure or with temperature. This correction is a percentage correction, and is naturally greatest at the highest pressures; it demands a knowledge of the compressibility of the liquid. The correction, which is complicated, need not be given in detail, but the most important term is $\frac{\rho_p - \rho_0}{\rho_0}$, where ρ_p is the density of the liquid at pressure p, and ρ_0 the density at atmospheric pressure. The compressibility of about a dozen of the liquids of this investigation had already been measured, so that for these liquids the pressure correction could be exactly computed. The compressibility of the other liquids has not been measured to high pressures, so that for them some sort of estimate had to be made of the correction. This estimate could be made with considerable confidence, because it is known that $\frac{\rho_p - \rho_0}{\rho_0}$ as a function of pressure does not vary greatly from liquid to liquid among such organic liquids as those used here. The procedure in calculating the correction for a liquid whose compressibility had not been measured was to substitute into the formula the value of ρ_0 of the liquid (obtained from the tables), and to use for $\frac{\rho_p - \rho_0}{\rho_0}$ the corresponding value for that one of the twelve liquids whose compressibility had been measured which was most nearly like the liquid in question. In selecting the most similar liquid, weight was given both to the chemical constitution and to the compressibility at low pressures, in those cases in which this had been determined by other observers.

It was not difficult to select a similar liquid in practically every case except that of glycerine. The low-pressure compressibility of glycerine is so much less than that of any other of the liquids that an estimate of the correction did not seem safe, and the compressibility was therefore specially determined.

The correction for change of buoyancy with pressure rises to as much as 5 per cent. in only a few extreme cases. The temperature correction for buoyancy in passing from 30° to 75° at atmospheric pressure was about 0·05 per cent. for nearly all the liquids, but for a substance with so extreme a density as ethylene dibromide rose to a maximum of 1·8 per cent. I believe that the uncertainty in the final result due to buoyancy is practically always less than 0·1 per cent. However, if there should be abnormal features in the compressibility, this limit may be exceeded. In the original paper enough data are recorded to permit allowance for this if future exact measurements of the compressibility of any liquid should disclose the necessity.

Finally, a correction has to be applied for the change of dimensions of the apparatus under pressure. The time of fall may be found by a dimensional argument to be proportional to the square of the linear dimensions. The correction for distortion, accordingly, turns out to be linear in the pressure, independent of temperature, and equal to 0·46 per cent. at 12,000 kg.

In addition to the corrections described above peculiar to these measurements of viscosity, there are other corrections common to all high-pressure measurements, which have been described in sufficient detail previously. All the corrections together do not amount to over 10 per cent. in the extreme case, and more often were of the order of 3 or 4 per cent.

The apparatus was subjected to a number of checks to assure its proper functioning. This included comparison of the times of fall at different pressures with varying weights and a comparison of the relative viscosities of different liquids and of their temperature coefficients with the values of other observers.   The performance was entirely satisfactory.

Water gave special difficulty; a great many experimental attempts were made, but an accuracy equal to that obtained with the other liquids was not achieved.   The difficulty was caused by the electrical conductivity of the water interfering with the operation of the timer through the electric contact arrangement.   It was not possible to use the cylinder of bessemer steel used with the other liquids, because after the water had stood in contact with the steel for a while, there was enough chemical action to produce a short-circuit in all positions of the weight.   Various special forms of apparatus were tried.   Finally, by using the proper grade of stainless steel for the cylinder, and pure nickel for the weight, results were obtained which without doubt are essentially correct, but which are still not as good as for the other liquids.

The original paper must be consulted for a detailed description of the liquids.   Many of these were of exceptional purity, namely, fourteen obtained from Professor J. Timmermans of the Bureau Belge d'Étalons Chimiques, and five from Professor F. Keyes of Massachusetts Institute of Technology.   Most of the others were from the Eastman Kodak Company, their purest grade.

The method so far described is, as mentioned, applicable only to insulators.   The measurements with this method on organic liquid were carried out first, and it appeared from that work that it would be of particular interest to extend the measurements to a substance with the simplest possible molecule, which naturally suggests mercury.   Unexpected difficulties were found in devising a method suitable for mercury, and many forms of apparatus were tried and discarded.   The greatest difficulty arises from irregular capillary action; it is practically impossible to obtain regular results with a method in which the position of the mercury meniscus in glass or steel or similar material must be determined, and no pressure apparatus should involve this as an essential feature if it is possible to avoid it.   Final success was attained only by making the apparatus of copper, with which mercury amalgamates, so that there is consequently no irregular capillary action. The use of copper involves possible error from the solution of copper in mercury, but Richards [10] has found that mercury dissolves less than 0·02 per cent. copper, so that any error from this effect must be very small.

The apparatus finally adopted for measuring the viscosity of mercury under pressure is shown schematically in fig. 80. It consists of two copper reservoirs connected by a fine copper capillary. The arrangement may be tipped back and forth through definite angles, and the time of flow of mercury from one reservoir to the other determined by electric contacts, through which a timing device is operated, as in the other method.

The corrections to be applied are much larger than in most of my pressure work, being equal in magnitude to the effect itself, and demand a rather elaborate mathematical discus-

Fig. 80.—Section of the flow part of the apparatus for determining the viscosity of mercury.

sion, for which reference must be made to the original paper. The principal complications arise from the fact that the head through which the mercury flows varies so much that it is not accurate enough to use a simple average, and also from the fact that the so-called 'kinetic energy' correction must be applied. The correction for the change of density of the transmitting liquid, petroleum ether, demanded a special determination of its compressibility.

A check on the performance of the apparatus was afforded by the fact that the temperature coefficient of viscosity at atmospheric pressure determined by it agrees very closely with that of other observers. Rough checks of the correctness of the pressure coefficient could also be obtained by comparison with the results given by some of the preliminary and less satisfactory forms of apparatus. The known difficulties in the performance of these preliminary forms of apparatus were of such a character that it was possible to say whether they would give too large or too small a value.

Thus one form of apparatus was known to give too large an effect because of capillary action, and this gave a 40 per cent. increase of viscosity under 12,000 kg. against 33·5 per cent. found with the final apparatus. Three other forms of preliminary apparatus, which were known to give too small values because of dirt sticking to the contacts, gave 24·5 per cent., 25 per cent., and 26 per cent. respectively against the final 33·5 per cent.

In the course of the measurements on organic liquids a number of new freezing observations were made of liquids not previously observed. Freezing is, of course, shown by the weight refusing to fall. There is no previous warning of the approach of freezing, but the viscosity curve of the liquid runs without change into the sub-cooled region. The method is not well adapted to give accurate values; in general, an observed melting-point gives the true equilibrium pressure more accurately than a freezing-point, because the solid may not be superheated while the liquid may be sub-cooled. The data for the effect of pressure on melting-points determined in this way have already been tabulated in the chapter on melting under pressure.

The numerical results for all the substances except water and mercury are now given in Table XXII. In the original paper there is a table giving the results in considerably greater detail. The table gives the logarithm of the relative viscosity as a function of pressure and temperature, the viscosity at 30° and atmospheric pressure being taken as unity. The logarithm of the viscosity, instead of the viscosity itself, is given because the variation with pressure of the viscosity is very rapid, and the curve of viscosity against pressure has rapidly varying curvature, whereas the curve of log (viscosity) against pressure approaches a straight line at high pressure, and is not too much curved at low pressures. The pressures tabulated in the table are 0, 1000, 4000, 8000, and 12,000 kg./cm.^{2}, the intervals being shorter at the lower end of the range because of the much more rapidly varying curvature. From the values of the logarithms the ratio of the viscosity at 30° to that at 75° at various pressures may be found and is tabulated. Also the absolute viscosities at

TABLE XXII

Substance		Pressure, kg./cm.^{2}					η_{30}
		1	1000	4000	8000	12,000	
Methyl alcohol	$\log\dfrac{\eta}{\eta_0}\begin{cases}30°\\75°\end{cases}$ η_{30}/η_{75}	0·000 9·769 1·702	0·167 9·933 1·714	0·471 0·208 1·832	0·750 0·448 2·004	0·998 0·655 2·203	0·00520
Ethyl alcohol	$\log\dfrac{\eta}{\eta_0}\begin{cases}30°\\75°\end{cases}$ η_{30}/η_{75}	0·000 9·657 2·203	0·200 9·873 2·123	0·617 0·289 2·128	1·023 0·634 2·449	1·390 0·919 2·958	0·01003
n-Propyl alcohol	$\log\dfrac{\eta}{\eta_0}\begin{cases}30°\\75°\end{cases}$ η_{30}/η_{75}	0·000 9·598 2·523	0·283 9·880 2·529	0·836 0·368 2·938	1·402 0·827 3·758	1·915 1·223 4·920	0·01779
n-Butyl alcohol	$\log\dfrac{\eta}{\eta_0}\begin{cases}30°\\75°\end{cases}$ η_{30}/η_{75}	0·000 9·548 2·845	0·321 9·867 2·858	0·934 0·312 3·343	1·609 0·941 4·679	2·208 1·396 6·518	0·02237
n-Amyl alcohol	$\log\dfrac{\eta}{\eta_0}\begin{cases}30°\\75°\end{cases}$ η_{30}/η_{75}	0·000 9·540 2·884	0·341 9·871 2·951	1·060 0·466 3·926	1·811 1·049 5·781	2·495 1·562 8·570	
i-Propyl alcohol	$\log\dfrac{\eta}{\eta_0}\begin{cases}30°\\75°\end{cases}$ η_{30}/η_{75}	0·000 9·505 3·141	0·343 9·851 3·120	0·982 0·425 3·624	1·640 0·957 4·844	2·311 1·424 7·748	0·01757
i-Butyl alcohol	$\log\dfrac{\eta}{\eta_0}\begin{cases}30°\\75°\end{cases}$ η_{30}/η_{75}	0·000 9·444 3·597	0·388 9·824 3·664	1·203 0·488 5·188	2·075 1·158 8·260	2·898 1·747 14·16	0·02864
i-Amyl alcohol	$\log\dfrac{\eta}{\eta_0}\begin{cases}30°\\75°\end{cases}$ η_{30}/η_{75}	0·000 9·424 3·805	0·386 9·787 4·012	1·185 0·492 4·970	2·069 1·168 8·042	2·952 1·780 15·76	
n-Pentane	$\log\dfrac{\eta}{\eta_0}\begin{cases}30°\\75°\end{cases}$ η_{30}/η_{75}	0·000 9·811 1·545	0·315 0·163 1·419	0·847 0·676 1·483	1·360 1·119 1·742	1·846 1·493 2·254	0·00220
n-Hexane	$\log\dfrac{\eta}{\eta_0}\begin{cases}30°\\75°\end{cases}$ η_{30}/η_{75}	0·000 9·803 1·574	0·332 0·171 1·449	0·914 0·701 1·633	1·514 1·198 2·070	1·646	0·00296
n-Octane	$\log\dfrac{\eta}{\eta_0}\begin{cases}30°\\75°\end{cases}$ η_{30}/η_{75}	0·000 9·810 1·549	0·327 0·153 1·493	1·088 0·763 2·113	1·363	··	0·00483

TABLE XXII—*continued*

Substance			Pressure, kg./cm.^{2}					η_{30}
			1	1000	4000	8000	12,000	
i-Pentane	$\log \dfrac{\eta}{\eta_0}$	$\begin{cases} 30° \\ 75° \end{cases}$	0·000	0·344	0·894	1·431	1·947	
			9·821	0·193	0·715	1·179	1·586	0·00198
		η_{30}/η_{75}	1·510	1·416	1·510	1·786	2·296	
i-Amyl decane	$\log \dfrac{\eta}{\eta_0}$	$\begin{cases} 30° \\ 75° \end{cases}$	0·000	0·435	1·354			
			9·772	0·178	0·925	1·727		
		η_{30}/η_{75}	1·690	1·807	2·685			
Ethylene di-bromide	$\log \dfrac{\eta}{\eta_0}$	$\begin{cases} 30° \\ 75° \end{cases}$	0·000					
			9·756	0·003	0·354 (3000)	..	..	0·01490
		η_{30}/η_{75}	1·754					
Ethyl chloride	$\log \dfrac{\eta}{\eta_0}$	$\begin{cases} 30° \\ 75° \end{cases}$	0·000	0·242	0·649	1·008	1·323	
			9·850	0·131	0·514	0·834	1·111	
		η_{30}/η_{75}	1·413	1·291	1·365	1·493	1·633	
Ethyl bromide	$\log \dfrac{\eta}{\eta_0}$	$\begin{cases} 30° \\ 75° \end{cases}$	0·000	0·222	0·631	1·043	1·400	
			9·806	0·072	0·472	0·816	1·123	0·00368
		η_{30}/η_{75}	1·567	1·413	1·442	1·687	1·892	
Ethyl iodide	$\log \dfrac{\eta}{\eta_0}$	$\begin{cases} 30° \\ 75° \end{cases}$	0·000	0·218	0 656	1·108	1·549	
			9·837	0·057	0·467	0·854	1·200	0·00540
		η_{30}/η_{75}	1·455	1·445	1·545	1·795	2·234	
Acetone	$\log \dfrac{\eta}{\eta_0}$	$\begin{cases} 30° \\ 75° \end{cases}$	0·000	0·226	0·605	0·987		
			9·895	0·113	0·445	0·762	1·031	0·00285
		η_{30}/η_{75}	1·274	1·297	1·445	1·679		
Glycerine	$\log \dfrac{\eta}{\eta_0}$	$\begin{cases} 30° \\ 75° \end{cases}$	0·000	0·260	0·936	1·741		
			8·810	9·023	9·529	0·094	0·628	3·8
		η_{30}/η_{75}	15·49	17·26	25·53	44·36		
Ethyl acetate	$\log \dfrac{\eta}{\eta_0}$	$\begin{cases} 30° \\ 75° \end{cases}$	0·000	0·258	0·818	1·393	1·974	
			9·836	0·081	0·517	0·992	1·416	0·0039
		η_{30}/η_{75}	1·459	1·503	2·000	2·518	3·614	
n-Butyl bromide	$\log \dfrac{\eta}{\eta_0}$	$\begin{cases} 30° \\ 75° \end{cases}$	0·000	0·269	0·816	1·408	2·018	
			9·832	0·090	0·564	1·040	1·484	0·00537
		η_{30}/η_{75}	1·472	1·510	1·786	2·333	3·420	
Cineole	$\log \dfrac{\eta}{\eta_0}$	$\begin{cases} 30° \\ 75° \end{cases}$	0·000					
			9·654	0·142				
		η_{30}/η_{75}	2·218					

TABLE XXII—continued

Substance		1	1000	4000	8000	12,000	η_{30}
		Pressure, kg./cm.²					
Oleic acid $\log \dfrac{\eta}{\eta_0}$	30°	0·000	0·616				
	75°	9·419	9·989	0·843			
	η_{30}/η_{75}	3·811	4·236				
CCl$_4$ $\log \dfrac{\eta}{\eta_0}$	30°	0·000	0·351				
	75°	9·760	0·100	0·542	··	··	0·00845
	η_{30}/η_{75}	1·738	1·782				
Chloroform $\log \dfrac{\eta}{\eta_0}$	30°	0·000	0·211	0·660			
	75°	9·858	0·094	0·480	0·914	··	0·00519
	η_{30}/η_{75}	1·387	1·309	1·514			
CS$_2$ $\log \dfrac{\eta}{\eta_0}$	30°	0·000	0·160	0·509	0·840	1·189	
	75°	9·875	0·051	0·372	0·671	0·946	0·00352
	η_{30}/η_{75}	1·334	1·285	1·371	1·476	1·750	
Ether $\log \dfrac{\eta}{\eta_0}$	30°	0·000	0·324	0·792	1·261	1·670	
	75°	9·878	0·149	0·601	0·986	1·311	0·00212
	η_{30}/η_{75}	1·324	1·496	1·552	1·884	2·286	
n-Amyl ether $\log \dfrac{\eta}{\eta_0}$	30°	0·000	0·401	1·230	2·091		
	75°	9·736	0·107	0·776	1·437	2·007	
	η_{30}/η_{75}	1·837	1·968	2·844	4·508		
Cyclohexane $\log \dfrac{\eta}{\eta_0}$	30°	0·000					
	75°	9·723	0·169	··	··	··	0·00828
	η_{30}/η_{75}	1·892					
Methyl cyclohexane $\log \dfrac{\eta}{\eta_0}$	30°	0·000	0·388	1·274	2·318		
	75°	9·747	0·154	0·900	1·756	2·582	0·00639
	η_{30}/η_{75}	1·791	1·714	2·366	3·648		
Benzene $\log \dfrac{\eta}{\eta_0}$	30°	0·000	0·347				
	75°	9·765	0·081	0·498 (3000)	··	··	0·00566
	η_{30}/η_{75}	1·718	1·845				
Chlorobenzene $\log \dfrac{\eta}{\eta_0}$	30°	0·000	0·253	0·867			
	75°	9·814	0·053	0·563	1·146	··	0·00711
	η_{30}/η_{75}	1·535	1·585	2·014			
Bromobenzene $\log \dfrac{\eta}{\eta_0}$	30°	0·000	0·262	0·897			
	75°	9·801	0·044	0·558	1·029 (7000)	··	0·00985
	η_{30}/η_{75}	1·581	1·652	2·183			

TABLE XXII—*continued*

Substance			Pressure, kg./cm.^{2}					η_{30}
			1	1000	4000	8000	12,000	
Aniline	$\log \dfrac{\eta}{\eta_0}$	$\begin{cases}30° \\ 75°\end{cases}$	0·000 9·551	0·376 9·847	0·560	··	··	0·0319
		η_{30}/η_{75}	2·812	3·381				
Diethyl-aniline	$\log \dfrac{\eta}{\eta_0}$	$\begin{cases}30° \\ \\ 75°\end{cases}$	0·000 9·690	0·394 9·984	1·070 (3000) 0·758	1·775		
		η_{30}/η_{75}	2·042	2·570				
Nitro-benzene	$\log \dfrac{\eta}{\eta_0}$	$\begin{cases}30° \\ 75°\end{cases}$	0·000 Decomposes	0·264				
		η_{30}/η_{75}						
Toluene	$\log \dfrac{\eta}{\eta_0}$	$\begin{cases}30° \\ 75°\end{cases}$	0·000 9·796	0·274 0·065	0·897 0·597	1·699 1·186	 1·832	0·00523
		η_{30}/η_{75}	1·600	1·618	1·995	3·258		
o-Xylene	$\log \dfrac{\eta}{\eta_0}$	$\begin{cases}30° \\ 75°\end{cases}$	0·000 9·767	0·311 0·057	0·689	··	··	0·00709
		η_{30}/η_{75}	1·710	1·795				
m-Xylene	$\log \dfrac{\eta}{\eta_0}$	$\begin{cases}30° \\ 75°\end{cases}$	0·000 9·799	0·290 0·079	0·967 0·637	1·333	··	0·00552
		η_{30}/η_{75}	1·589	1·626	2·138			
p-Xylene	$\log \dfrac{\eta}{\eta_0}$	$\begin{cases}30° \\ 75°\end{cases}$	0·000 9·797	 0·092	··	··	··	0·00568
		η_{30}/η_{75}	1·596					
p-Cymene	$\log \dfrac{\eta}{\eta_0}$	$\begin{cases}30° \\ 75°\end{cases}$	0·000 9·800	0·333 0·087	1·194 0·749	1·612		
		η_{30}/η_{75}	1·585	1·762	2·786			
Eugenol	$\log \dfrac{\eta}{\eta_0}$	$\begin{cases}30° \\ \\ 75°\end{cases}$	0·000 9·429	0·541 9·810	2·273 (3000) 0·805	2·343		
		η_{30}/η_{75}	3·724	5·383	29·38			
Petroleum Ether	$\log \dfrac{\eta}{\eta_0}$	$\begin{cases}30° \\ 80°\end{cases}$	0·00 ··	0·30 ··	0·93 0·56	1·59 1·06	2·18 1·49	
		η_{30}/η_{80}	··	··	2·34	3·39	4·90	
Kerosene	$\log \dfrac{\eta}{\eta_0}$	$\begin{cases}30° \\ 80°\end{cases}$	0·00 ··	0·46 ··	1·71 0·91	1·88	2·80	
		η_{30}/η_{80}	··	··	6·3			

atmospheric pressure at 30° are listed when these are known; many of these values were taken from the Smithsonian Tables, and others were given by Timmermans.

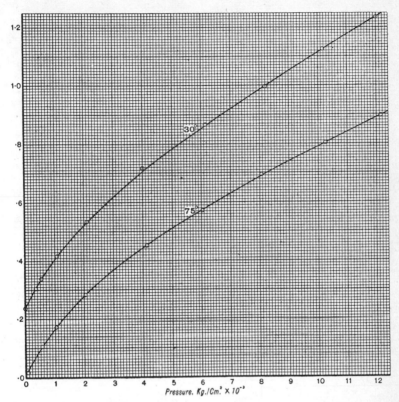

FIG. 81.—The common logarithm of the corrected time of fall of the weight plotted against pressure for methyl alcohol. The time of fall is proportional to viscosity.

A sample curve for methyl alcohol, giving the logarithm of the observed relative viscosity is shown in fig. 81.

Because of their comparatively small variation of viscosity with pressure, water and mercury are treated separately. In Table XXIII are given the relative viscosities of water (not log relative viscosity) as a function of pressure at 0°, 10·3°, 30°, and 75°, and the results are shown graphically in fig. 82.

TABLE XXIII

RELATIVE VISCOSITY OF WATER

Pressure, kg./cm.²	Relative viscosity			
	0°	10·3°	30°	75°
1	1·000	0·779	0·488	0·222
500	0·938	0·755	0·500	0·230
1,000	0·921	0·743	0·514	0·239
1,500	0·932	0·745	0·530	0·247
2,000	0·957	0·754	0·550	0·258
3,000	1·024	0·791	0·599	0·278
4,000	1·111	0·842	0·658	0·302
5,000	1·218	0·908	0·720	0·333
6,000	1·347	0·981	0·786	0·367
7,000	..	1·064	0·854	0·404
8,000	..	1·152	0·923	0·445
9,000	..	..	0·989	0·494
10,000	Freezes		1·058	..
11,000			1·126	..

The results for mercury are shown in Table XXIV and fig. 83, in which viscosity is plotted against pressure at two temperatures.

TABLE XXIV

VISCOSITY OF MERCURY PRESSURE UNDER

Pressure, kg./cm.²	Absolute viscosity	
	30°	75°
1	0·01516	0·01340
2,000	0·01585	0·01400
4,000	0·01663	0·01463
6,000	0·01742	0·01528
8,000	0·01822	0·01598
10,000	0·01912	0·01674
12,000	0·02007	0·01759

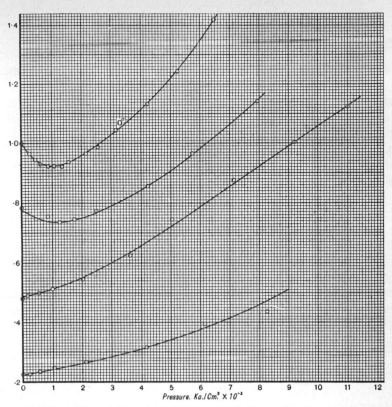

FIG. 82.—The relative viscosity of water at 0°, 10·3°, 30°, and 75° as a function of pressure.

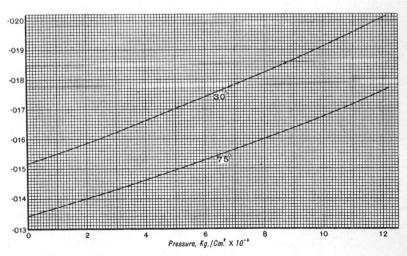

FIG. 83.—The absolute viscosity of mercury at 30° and 75° as a function of pressure.

Some comparison is possible between these results and those of other observers. For organic liquids there are practically only the results of Faust [5] to 3000 kg. on ethyl alcohol, ether, and CS_2. He finds for ethyl alcohol an increase of viscosity at 30° under 3000 kg. of 2·94-fold against my 2·31, for ether 3·96-fold against my 3·27, and for CS_2 3·43 against 2·03. The agreement certainly ought to be much closer. It is not evident from his paper whether all the corrections were applied, but it is hard to see how any corrections could be responsible for so large a difference. For mercury there are the results of Cohen and Bruins [6] to 1500 kg./cm.^{2}. Their result is equivalent to an increase of viscosity of 6·1 per cent. at 2000 kg./cm.^{2} at 30° against 5·1 per cent. found above. It is probable that some corrections should be applied to Cohen's results, which are discussed in my detailed paper, but when they are all applied there will probably be an outstanding difference of 11 per cent. of the effect—that is, an increase of 5·6 per cent. at 2000 kg. by Cohen against my 5·1 per cent.

In certain qualitative features, the behaviour of all the liquids investigated here, except water, is alike, although there are very large quantitative differences. The viscosity increases with pressure at a rapidly increasing rate, so that if viscosity is plotted against pressure, a curve of very rapid upward curvature is obtained. This is unusual; most pressure effects become relatively less at high pressure by a sort of law of diminishing returns. In fig. 84 is shown viscosity against pressure at 30° for CS_2 and ether, two substances with comparatively small pressure effect. It is seen that over the first two or three thousand kilograms the relation between pressure and viscosity is nearly linear, but above this the departure is extreme. If the logarithm of viscosity is plotted against pressure the curve obtained is in general concave toward the pressure axis. The curvature is much the greatest at low pressures; above 2000 or 3000 kg./cm.^{2} the curve approximates to a straight line in a little more than half the cases, while in the remaining cases it gently reverses curvature. This means that above 3000 viscosity either increases geometrically or else even more rapidly as pressure increases arithmetically. Among the substances investigated, eugenol and p-cymene have the most rapid increase.

The temperature coefficient of viscosity is shown by the rows in Table XXII giving η_{30}/η_{75}. Here again the effect is abnormal; most temperature effects become less at high

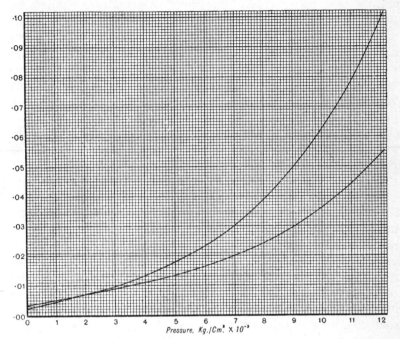

Pressure, Kg./Cm². x 10⁻³

FIG. 84.—The viscosity in abs. C.G.S. units at 30° of CS_2 and ether as a function of pressure. The curve for ether starts below that for CS_2, and rises above it beyond 2000 kg.

pressures, which is to be expected if the modification in the structure produced by temperature agitation becomes less under the greater constraints imposed by the high pressure. But here the relative change of viscosity with temperature becomes very markedly greater at high pressure, the ratio η_{30}/η_{75} changing under 12,000 kg./cm.² by a factor of as much as 4.

Apart from these qualitative resemblances, the most varied quantitative behaviour is shown by the various substances.

In fact, viscosity is a unique property in regard to the magnitude of the pressure effect and its variation from substance to substance. The compressibility at atmospheric pressure, for example, varies by a factor of not more than four- or five-fold for the substances investigated here, and under 12,000 kg./cm.^{2} the compressibility of any one substance diminishes by not over fifteen-fold. The thermal expansion changes by a factor of 2 or 3 under 12,000 for these liquids, and the specific heats and thermal conductivities do not vary more. Excepting water and mercury, the smallest effect of pressure on viscosity found above is that on methyl alcohol, which increases tenfold under 12,000, and the largest is by over 10^7 for eugenol (obtained by linear extrapolation, which gives too low a value).

In general, the largest pressure effects are for those substances with the most complicated molecules. This is very plainly shown by the series of the alcohols, or by the various compounds derived from benzene; the relative pressure effect is greater the more complicated the group substituted for hydrogen. There is also a very marked constitutive effect, the iso-compounds having a larger effect than the normal compounds, and a similar effect is seen in the three xylenes. A heavier atom substituted into a molecule produces in general a larger pressure effect, as is shown in the series ethyl chloride, bromide, iodide, or by chloro- and bromo-benzene. There appears, however, to be a tendency working in the other direction at low pressures. In the ethyl halogen series, the increase of viscosity produced by 500 kg./cm.^{2} is in the order Cl, Br, I, whereas the increase under 12,000 is in the order I, Br, Cl. It may well be that the effect at low pressure is due to the abnormally large compressibility of C_2H_5Cl, due to the neighbourhood of the critical point, at which the compressibility is infinite. Abnormal behaviour is also shown by methyl cyclohexane, the pressure effect being larger than for cyclohexane.

Water is quite different in character from the other liquids. Previous investigations have been made to 400 kg./cm.^{2} by Hauser.[4] He found that below 30° viscosity decreases with increasing pressure, and above 30° increases. At higher pressures we now find that at 0° and 10° there is a minimum viscosity, at a pressure roughly 1000 kg./cm.², the minimum being less pronounced at 10° than at 0°. At 30° and 75° there is a regular increase of viscosity with pressure over the entire range. (Not much weight must be placed on the precise numerical values given for 75°, there being much experimental uncertainty here because of electrical conductivity by the water.) It is natural to see in this abnormal behaviour of water an association effect; at low pressures and temperatures, water is strongly associated with large molecules and a large viscosity, but as pressure increases the association decreases, and the average size of the molecules decreases, giving a term in the viscosity which diminishes fast enough to more than compensate the normal increase of viscosity under pressure. At higher pressures the association effect is exhausted, and the behaviour becomes normal.

Except for absolute magnitude, the results for mercury are of the same general character as those for other liquids. Viscosity increases with increasing pressure at an accelerated rate, and the percentage increase at 75° is less than at 30°. It will be found on inspection of the diagram that the curve for 75° may be obtained by displacing bodily the 30°-curve along the pressure axis by about 5750 kg./cm.². This means that the fractional change of viscosity produced by a given increment of pressure is a function only of the viscosity, no matter whether a given value of the viscosity is found at a low pressure and low temperature or at a high temperature and a high pressure. The experimental data are not sufficient to allow a similar statement about temperature coefficients at equal viscosities. The result for the pressure coefficient was not found to hold in the case of other liquids, so that possibly in the case of mercury the result is only approximate, holding only over a comparatively narrow range.

On the theoretical side, I believe that these pressure results indicate very strongly that there is a feature in the mechanism of liquid viscosity not yet sufficiently considered. Comparatively little work has been done on the theory of the viscosity of liquids. It has, of course, been recognised that the mechanism is different in liquids and gases. A striking demonstration of this is to be found in the pressure effects, the reciprocal of viscosity and thermal conductivity of a gas change together with pressure, and are known to be simply related, whereas in a liquid, thermal conductivity increases with increasing pressure, but reciprocal viscosity decreases enormously. In most discussions of the viscosity of liquids, the volume has been looked on as the most important factor. Thus Philips [11] has a theory in which viscosity is a function of volume only, and Faust [5] thought that he could infer from his data that at high pressures viscosity in the limit becomes a linear function of volume only. As a matter of fact, his data for CS_2 and ether were favourable to this hypothesis, but ethyl alcohol was distinctly unfavourable.

These data show most definitely that viscosity cannot be a function of volume only. In figs. 85 and 86 are plotted the logarithm of relative viscosity at 30° and 75° as a function of volume for ether, CS_2, and i-amyl alcohol. If viscosity were a function of volume only, the curves for different temperatures would coincide. They do approximately coincide at the volumes corresponding to low pressures for ether and CS_2, but at smaller volumes (higher pressures) there is marked divergence. The discrepancy for ether at the upper end of the pressure range corresponds to a factor of 2 on the viscosity. The relation is not satisfied at any part of the range by i-amyl alcohol; even at the low pressure end of the range the viscosity at 30° is greater by a factor of 2·5 than at the same volume at 75°. The behaviour of mercury, which might be expected to be specially simple because it is monatomic, shows the same sort of departure from expectation. The volume of mercury at 75° and 2000 kg./cm.² is the same as at 30° and atmospheric pressure, but the viscosity at 75° and 2000 kg./cm.² is 0·0140 against 0·0152 at 30° at atmospheric pressure. The pressure coefficient of viscosity would have to be nearly three times as great as found above if viscosity were a function of volume only; such a change is entirely beyond the possibility of experimental error. In general, then, viscosity decreases when temperature and pressure are increased together in such a way as to keep volume constant.

The theory of Brillouin [12] does give a pure temperature effect at constant volume. The fundamental idea of this

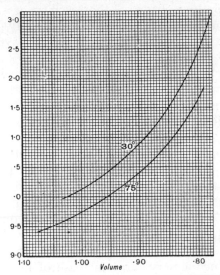

FIG. 85.—The common logarithm of relative viscosity at 30° and 75° of i-amyl alcohol as a function of volume.

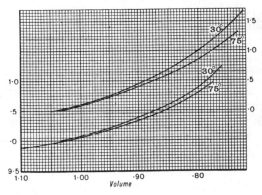

FIG. 86.—The common logarithm of relative viscosity at 30° and 75° of ether and CS_2 as a function of volume. The curves for ether are the upper curves with the scale of ordinates at the right.

theory is that in a liquid momentum is transferred by elastic waves in much the same way as the thermal energy is transferred by elastic waves in Debye's theory of heat conduction.

A simple expression is found by Brillouin for the temperature coefficient of viscosity at constant volume, which involves the thermal conductivity and the velocity of sound. Substitution of the numerical values, which may be obtained from the data of this chapter, shows, however, that the formula fails by a factor of nearly 5000.

It seems fairly evident, therefore, that there is some very important element in the situation not hitherto considered. This, I believe, to be an interlocking effect between the molecules, which prevents the free motion of one layer of molecules over another. Slipping of two interlocking molecules past each other can take place only when haphazard temperature agitation has so far separated them that the interlocking parts are free. According to such a picture, viscosity would be expected to decrease with rising temperature both at constant pressure and constant volume. When the volume is decreased at constant temperature by increasing pressure, a comparatively small decrease of total volume may evidently produce a very large increase of interlocking, and the effect would be expected to increase more and more rapidly as the pressure increases. Furthermore, it is evident that the magnitude of the effect may be very different for different substances.

Such an interlocking effect would be expected to be most important in the most complicated molecules, and strong evidence in favour of such a picture is the very marked tendency found experimentally for the pressure effect to increase as the molecule becomes more complicated, and conversely the comparatively very small effect in monatomic mercury. Evidently 'complication of the molecule' is a very hazy concept, and any numerical measure of it can be only very crude. It is evident, I think, that the molecule may be more complicated either because it contains more atoms or because the atoms themselves are more complicated. As a measure of the complication of the atom, I have taken the number of extranuclear electrons (the atomic number), and to measure the complication of the molecule I have multiplied the total number of extranuclear electrons in all the atoms which the molecule contains by the total number of atoms in the molecule. This evidently neglects many factors which we would like to include; for example, no distinction is made between an iso- and a normal alcohol, although our ordinary structural formulas would suggest that the molecule of an iso-alcohol is more likely to interlock with others of its kind than is a molecule of a normal alcohol.

If the complexity of the molecule, measured in this way, is plotted against the pressure effect on viscosity, a very definite correlation will be found. This plot will be found in the detailed paper (B. 55). Considerable experiment on my part failed to find any other properties with as definite a correlation with viscosity. Stronger evidence of the importance of molecular complexity may be found by confining oneself to a single related series of chemical compounds, and noticing the effect of making the molecule more complicated in the series. Striking examples of this are the increase of pressure effect with molecular weight in the series of alcohols, or of hydrocarbons, C_nH_{2n+2}, or the ethyl halogens.

If some such interlocking effect is an important part of the viscosity mechanism, this means that so far as viscosity phenomena are concerned, the molecule preserves its inviolability and continues to function as a unit when the volume is greatly decreased by high pressure. There are other phenomena, such as compressibility, in which the molecule seems to lose its significance at small volumes, and the atom becomes more significant. A reason for the difference may be seen in the different sorts of relative motion involved. If a molecule ceased to function as a whole during viscous shear it would be torn apart by the relative motion of the parts of the liquid, but the relative motion involved in a hydrostatic compression is not such as to destroy the molecule, even under comparatively large changes of volume. At present it would probably be prohibitively complicated to attempt to put into mathematical form an exact analysis of any such interlocking effect.

After the galley proofs of this book had been read I was most kindly informed by Professor Andrade that he has had considerable success in extending the formula which he had previously deduced for the variation of viscosity with temperature [13] to include the variation with pressure also. The extended formula is :

$$\eta v^{\frac{1}{3}} = A e^{\left(p + \frac{r}{v^2}\right)\frac{s}{T}},$$

where v is the specific volume as given by experiment, T the absolute temperature, and A, r, and s are empirical constants. A and rs are fixed by the variation of viscosity with temperature at atmospheric pressure, so that there is effectively only a single constant available to reproduce the variation of viscosity with pressure.   This can be done with much success over the entire pressure range and at both temperatures, 30° and 75°, for such liquids as C_2H_5Br, C_2H_5I, and CS_2. Whether the formula would be as successful for the liquids which have a much larger pressure coefficient, such as eugenol, for which the relation between pressure and volume was not available, was not evident at the time of writing.   The fundamental point of Andrade's theory, namely, that viscosity involves a temporary freezing together of the molecules of the liquid into larger aggregates more nearly crystalline in character, is somewhat like the interlocking effect suggested above.   The theory will probably be published in detail before the appearance of this book.

[1] W. C. ROENTGEN, *Wied. Ann.*, **22**, 518 (1884).
[2] WARBURG und SACHS, *ibid.*, **22**, 518 (1884).
[3] R. COHEN, *ibid.*, **45**, 666 (1892).
[4] L. HAUSER, *Ann. Phys.*, **5**, 597 (1901).
[5] O. FAUST, *ZS. Phys. Chem.*, **86**, 479 (1914).
[6] E. COHEN and H. R. BRUINS, *ibid.*, **114**, 441 (1925).
[7] J. H. HYDE, *Proc. Roy. Soc.*, **97**, 240 (1920).
[8] M. D. HERSEY and H. SHORE, Paper presented at the December Meeting of American Society of Mechanical Engineers (1927).
[9] C. BARUS, *Proc. Amer. Acad.*, **27**, 13 (1891–92).
[10] T. W. RICHARDS, *Carnegie Inst. Wash.*, Pub. No. 118 (1909).
[11] H. B. PHILLIPS, *Proc. Nat. Acad. Sci.*, **7**, 172 (1921).
[12] L. BRILLOUIN, *Jour. Phys. et Rad.*, **3**, 326, 362 (1922).
[13] E. N. DA C. ANDRADE, *Nat.*, 1st March and 12th April 1930.

CHAPTER XIII

MISCELLANEOUS EFFECTS OF PRESSURE

THIS chapter of miscellanies will be largely devoted to work of other observers on topics not already discussed, in which the pressure range has been low, seldom as much as 3000 kg./cm.². The chapter will also include several investigations done with my high-pressure apparatus and more or less under my supervision, namely, on dielectric constant, magnetic permeability, optical absorption, and irreversible chemical reactions, mostly polymerisations. The pressure range of these investigations is 12,000 kg./cm.². Finally, an account will be given of my own work on the transformation of yellow to black phosphorus, and on the effect of pressure on rigidity, for which there has been no place in the earlier chapters.

The Effect of Pressure on Solubility. As already stated in the chapter on melting, this effect may properly be regarded as a special topic in the behaviour of binary mixtures under pressure, but since no systematic examination has been made of the general field, and the technique of these experiments is special to itself, they may conveniently be described in this chapter of miscellanies.

The effect of pressure on solubility satisfies an equation which is a simple extension of Clapeyron's equation for one-component systems:

$$\left(\frac{\partial L}{\partial p}\right)_\tau \Big/ \left(\frac{\partial L}{\partial \tau}\right)_p = -\tau \frac{\Delta V}{H},$$

where $\left(\dfrac{\partial L}{\partial p}\right)_\tau$ is the pressure coefficient of solubility, $\left(\dfrac{\partial L}{\partial \tau}\right)_p$ the ordinary temperature coefficient of solubility, ΔV the total change of volume when 1 molecule of the solute goes into solution under equilibrium conditions (that is, into an

357

infinite amount of the already saturated solution), H the heat of solution of 1 molecule under the same conditions, and τ the absolute temperature. In this equation everything may be regarded as known at atmospheric pressure except $\left(\dfrac{\partial L}{\partial p}\right)_\tau$ and the equation therefore affords a means of calculating the pressure coefficient of solubility. It would appear that there would be no particular point in multiplying experimental verifications of the formula at pressures so low that $\left(\dfrac{\partial L}{\partial \tau}\right)_p$ and the other factors are approximately the same as at atmospheric pressure. As a matter of experiment, however, the equation has proved unexpectedly difficult of verification. The matter is discussed at considerable length in the book of Cohen and Schut, where the early experiments are described, which were, however, so inaccurate that verification of the formula failed by large amounts; these early experiments will not be described here. By far the most careful work in this field has been done by Cohen and his students, and the experimental verification may now be regarded as satisfactory. The pressure range is that usual in the work of Cohen, about 1500 kg./cm.²; this range was sufficient to show departures of L from linearity with pressure. No attempt was made, however, to find the variation of the other factors in the thermodynamic equation, so that the experimental verification of the formula can be regarded as made only at atmospheric pressure. There would seem to be no particular point in trying merely for a verification of the formula at higher pressures.

One of the chief technical difficulties is in ensuring that the solution shall be actually saturated under pressure, and in properly measuring the concentration when known to be saturated. Stirring of the solution under pressure may be accomplished either with some sort of electromagnetic device operating within the pressure chamber, or by arranging the apparatus so that it can be tipped back and forth, with mercury in the bottom of the pressure chamber, the solution being swashed about by the motion of the mercury. The concentration of the solution at saturation under pressure may be determined by arranging so that a valve may be opened, giving a small sample, the analysis being made rapidly so that the salt has no time to deposit from the supersaturated condition. Or the concentration may be determined electrically, by measuring the resistance of the saturated solution under pressure, and then from measurements of the resistance under pressure of unsaturated solutions of various known concentrations extrapolating to the concentration of the unknown saturated solution.

Cohen was able to reproduce many of his results with a second degree expression in pressure. Some of the results follow ; p is in atmospheres, and C is the number of grams of salt in 100 grm. of solution.

Thallo-sulphate at 30°:

$$C = 5.831 + 3.295 \times 10^{-3}p - 1.09 \times 10^{-7}p^2.$$

Naphthalene in tetrachlorethane at 30°: $C = 35.07$, $p = 0$; $C = 30.26$, $p = 250$; $C = 26.40$, $p = 500$; $C = 23.33$, $p = 750$; $C = 20.89$, $p = 1000$.

m-dinitrobenzene in ethyl acetate at 30°, p to 500 atmospheres:

$$C = 52.54 - 2.674 \times 10^{-2}p + 9.825 \times 10^{-6}p^2.$$

NaCl in water at 24.05°, to 1500 kg.:

$$L = 35.898 + 1.647 \times 10^{-3}p - 3.268 \times 10^{-7}p^2. \begin{cases} (L = \text{grm. of salt/} \\ 100 \text{grm. of water}). \end{cases}$$

Mannite in water at 24.05°:

$$L = 20.65 + 9.31 \times 10^{-4}p - 1.806 \times 10^{-7}p^2.$$

The solubility may thus either increase or decrease with increasing pressure, and by largely varying amounts. In the case of naphthalene in tetrachlorethane, various empirical formulæ which were fitted to the experimental results indicated a minimum of solubility at pressures varying from 2200 to 3500 kg./cm.². Such a minimum has not been found experimentally, however, and Cohen is apparently doubtful as to whether such a minimum is actually to be expected.

In addition to these experiments on the solubility of solids in water, measurements have been made by Timmermans [1] on the critical temperature of complete mixing of nearly seventy pairs of organic liquids. The pressure range reaches an extreme of 1000 kg./cm.^{2}, but is usually of the order of a few hundred kg. The possibilities here are very complicated, as indicated by the theory of van der Waals. In particular, the domain of stable existence of the two components as a single homogeneous liquid phase may either increase or decrease with rising pressure. The results of Timmermans may be briefly summarised in the statement that the theory of van der Waals gives an adequate anticipation of the experimental possibilities.

Electrical Conductivity of Solutions. The measurements in this field cover an extensive range of materials, but the range of pressure is comparatively low, the highest pressure reached being 3700 kg./cm.^{2} in the investigations of Tammann and some of his collaborators. 1000 kg./cm.^{2} was the range of the work of Lussana, and the range of most of the other work was 500 or 1000 kg./cm.^{2}. The subject is expanded to the extent of 66 pages in Cohen's book, where most of the numerical data are reproduced. Practically no work has been done in this field since the publication of Cohen's book, so that a brief discussion here will suffice.

The first successful measurements were those of Fink [2] in 1885, who found a decrease of the resistance of solutions of HCl, $ZnSO_4$, and NaCl up to 500 kg./cm.^{2}. The effect of pressure decreases with increasing temperature and increasing concentration. Barus,[3] in 1890, showed from measurements to 150 kg./cm.^{2} that if pressure is raised simultaneously with temperature so as to keep volume constant, the resistance of $ZnSO_4$ solutions decreases with rising temperature, and he showed that the same is true for liquid mercury. The important later investigations in this field are by Roentgen,[4] Fanjung,[5] Lussana,[6] Stern,[7] Piesch,[8] Tammann,[9] and collaborators, particularly Körber [10] for aqueous solutions and Schmidt [11] for non-aqueous solutions.

In general, the effect of pressure on resistance is very complicated, as might be expected from the number of factors involved. At ordinary temperatures and in dilute solutions of strong electrolytes the effect of pressure is to decrease the resistance at a rate becoming less at higher pressures—that is, if relative resistance is plotted as ordinate against pressure as abscissa, the curve is concave upwards. As the concentration increases, the pressure effect becomes less numerically and the concavity increases, until a minimum appears; at still higher concentrations the position of the minimum moves toward lower pressures while the whole curve rises, until at high enough concentrations the effect of pressure is to increase the resistance from the beginning. Fig. 87, taken from Körber, shows a typical set of curves for NaCl. The details depend very much on the specific electrolyte; thus for HCl the curve for a solution of concentration $13 \times$normal still has a very pronounced minimum, although the minimum has entirely disappeared at a concentration of $5 \times$normal for NaCl.

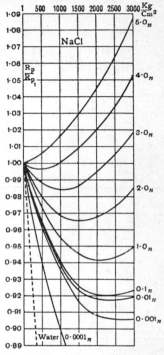

FIG. 87.—Electrical resistance of solutions of sodium chloride of various strengths as a function of the pressure. The resistance at atmospheric pressure is taken as unity for each strength.

The limiting pressure effect at very low concentrations in aqueous solutions of all electrolytes is due to the effect of pressure on the conductivity of the water itself; this cannot be determined with any certainty, because of the large effect of minute impurities. This feature of the situation was first pointed out by Tammann.

For weak electrolytes, that is electrolytes only slightly dissociated in ordinary concentrations, the initial pressure effect rapidly increases with increase of concentration until at something less than 0·001 normal it is four times as great as the initial effect at concentrations much lower, but still high enough so that the effect on the water itself is negligible;

from here on it decreases with increasing concentration, qualitatively like solutions of strongly dissociated electrolytes. For solutions of medium dissociation the effect is between these two, the initial pressure effect passing through a pronounced maximum at concentrations of the order of magnitude of normal.

In analysing these results there are at least four effects to be distinguished: (1) the pure volume effect, arising from the compression of the solution, thereby bringing more ions into unit volume; (2) the effect of pressure on the mobility of the ions, which one would expect as a first approximation to run parallel with the effect of pressure on the viscosity of the solution; (3) the effect of pressure on the degree of dissociation; and (4) the effect of pressure on the dissociation, and so the conductivity of water itself. The fourth effect may be disregarded for all except the most highly dilute solutions. If the solution is a moderately dilute one of a strongly dissociated electrolyte, the third effect may be disregarded, since the dissociation may be regarded as complete at all pressures, and the two first effects are left. The first effect may be calculated from the known compressibility, leaving only the second. This calculation has been made by Körber, and not with the expected results. The initial effect of pressure on mobility turns out to be negative, and there is a reversal to positive at higher pressures, as would be expected, since the initial effect of pressure is to decrease the viscosity of water, and then at higher pressures to increase it. But here the resemblance ceases; the pressure of reversal is in most cases much higher than the pressure of reversal of the viscosity of pure water. For example, LiCl and HCl show no reversal at all up to the maximum pressure of the experiment, 3000 kg./cm.². The pressure effect on mobility varies greatly in magnitude with the nature of the dissolved electrolyte. I do not see how these effects can be explained without supposing some specific effects on the ions themselves, such as a distortion under pressure, so that the net effect on mobility is a combination of the effect of the changing viscosity of water and that arising from the change of effective figure of the ions. This change of effective figure may involve the attraction of the ions for the surrounding molecules of water, and will involve the dielectric constant of water and the effect of pressure on it.   In one respect the pressure coefficient of mobility acts as would be expected, namely, the effect on a specific ion is independent of the other ions with which it is associated.   This is shown by the fact that the difference of the pressure effect on the mobilities of KCl and NaCl and KI and NaI are the same, within experimental error.   This is also consistent with the suggestion that there must be a deformation effect on the ions themselves.

If the salt is only feebly dissociated, the pressure effect contains, in addition to the pure volume and mobility effects, an effect due to the change of dissociation with pressure. This may be calculated in some cases from data on other salts in which the same radicals occur in other strongly dissociating combinations, so that only the first and second effects occur.   The relative change of dissociation for 500 kg. is a function of concentration, and for a typical case discussed by Tammann this may be as much as 1 per cent. at very small concentrations, but rise to as much as 10 per cent. for concentrations from 1/1000 to ten times normal.

The effect of a combination of pressure and temperature on the resistance of electrolytes has been investigated by Lussana and Körber.   If the resistance isotherms are plotted against pressure at constant concentration a family of curves will be obtained very much like fig. 85, which shows the curves of resistance against pressure for different concentrations at constant temperature, the progression in the family, from lower to higher temperatures being qualitatively the same as in the other family from lower to higher concentrations.   It appears, however, that the highest temperature yet reached, nearly 100°, is not high enough to make the pressure coefficient of resistance positive at all pressures, although this is possible in the other family of curves by going to concentrated enough solutions.

Non-aqueous solutions have been investigated only by Schmidt, who measured with Tammann's apparatus the effect of pressures up to 3000 kg./cm.² on the conductivity of several strongly dissociated electrolytes and one feebly dissociated electrolyte when dissolved in methyl, ethyl, iso-amyl alcohol, glycerine, and a number of other organic solvents. At low concentrations of the strongly dissociated salts the effects are as might be expected. The viscosity of these organic solvents increases with rising pressure, so that for this reason the resistance would be expected to increase. The compressibility effect works in the other direction to decrease the resistance, but numerically the relative change of volume under pressure is much less than the relative change of viscosity, so that the net effect would be expected to be an increase of resistance. This turns out to be the case; for this class of substance the plot of resistance against pressure looks very much like the curves at high concentrations for aqueous solutions. The upward curvature is what might be expected in view of the accelerated effect of pressure on viscosity at high pressures.

In non-aqueous solutions of weakly associated electrolytes, on the other hand, the effect of pressure in increasing the dissociation and so decreasing the resistance may exceed the effect on viscosity in increasing the resistance, with the net result that the resistance of these substances decreases with rising pressure. The direction of curvature is the same— that is, concave upward for weakly as well as strongly dissociated electrolytes; in some cases the curvature is sufficient to result in a minimum resistance at pressure below 3000 kg./cm.^{2}.

The effect of temperature at constant concentration on dilute solutions of strong electrolytes is opposite in non-aqueous and aqueous solutions. In non-aqueous solvents the effect of increasing temperature is to diminish the pressure effect on viscosity, whereas in water the pressure effect assumes larger positive values as temperature increases, the net effect being that the order of isotherms in the two families representing the pressure effect at different temperatures is the reverse for aqueous and non-aqueous solutions. There are greater irregularities in non-aqueous than in aqueous solutions, for the reason that even the strongly dissociated electrolytes in non-aqueous solutions are not usually more than from 0·5 to 0·8 dissociated, so that in no case is the effect of pressure on dissociation entirely to be neglected.

In comparing the pressure effects in different non-aqueous solutions of a single strongly dissociated electrolyte, Schmidt discovered the empirical rule that the logarithm of resistance is very nearly a linear function of pressure. Schmidt had at that time no measurements of the effect of pressure on the viscosity of his pure solvents, but his empirical rule is obviously closely connected with the result of my viscosity measurements, namely, that to a first approximation log viscosity is linear in pressure. Schmidt measured the effect in ten solvents; of these, five have been measured by me at high pressures. The order of the effect found by Schmidt on pressure coefficient of resistance is the same as the order of the pressure effect on viscosity for the three solvents with the smallest pressure coefficient of resistance, methyl and ethyl alcohols and acetone, but the parallelism is by no means close, and in fact, while the pressure effect on viscosity at low pressure is the same on acetone, nitrobenzene, and glycerine, the effect on resistance is more than twice as great for glycerine as for acetone, with nitrobenzene intermediate.

The Effect of Pressure on Dielectric Constant. Before any experiments were made on this subject there was the theoretical expression of Clausius-Mosotti, $\dfrac{\epsilon-1}{\epsilon+2} \cdot \dfrac{1}{d}=$const., to indicate what might be expected. The first measurements were by Roentgen [12] in 1894, who showed that the dielectric constant of water and alcohol is changed by less than 1 per cent. by 500 kg./cm.², although the formula would demand very much larger changes. Ratz,[13] in 1895, was able to make quantitative measurements to 250 kg./cm.² on water and a number of organic liquids, and found the dielectric constant to be increased by pressure, but by an amount very much smaller than is demanded by the Clausius-Mosotti expression, the discrepancy sometimes being as much as tenfold. Ortvay,[14] in 1911, measured a number of organic liquids to 500 kg./cm.² and found ϵ to increase with pressure as a quadratic function of pressure ($\epsilon=\epsilon_0(1+\alpha p+\beta p^2)$), α being

positive and β negative. The Clausius-Mosotti expression
$\frac{\epsilon-1}{\epsilon+2} \cdot \frac{1}{d}$ was found to decrease under 500 kg./cm.[2] by 1·3 per
cent. in the case of ether, and 1·17 per cent. in the case of
benzene.   The investigation of the effect of pressure on gases
was taken up at about this time in several papers by Occhialini [15] and Occhialini and Bodareu,[16] who established that up
to 350 kg./cm.[2] $\frac{\epsilon-1}{\epsilon+2} \cdot \frac{1}{d}$ is constant within experimental error
for H_2, N_2, O_2, and CO_2, but that $\frac{\epsilon-1}{d}$, another expression of
considerable historical importance, is not constant.   In 1920
Falckenberg [17] used a high-frequency method, which enabled
him to work with liquids of higher conductivity than had
been possible by previous methods, and measured up to 200
kg./cm.[2] the dielectric constant of water, ethyl alcohol,
methyl alcohol, and acetone.   It was found that $\frac{\epsilon-1}{\epsilon+2} \cdot \frac{1}{d}$ is
distinctly not constant, but that $\frac{\epsilon-1}{d}$ is more nearly so.
This is surprising in view of the results of Occhialini on gases.
In 1923 Waibel [18] used a resonance method of high precision
with high frequencies to measure the effect up to 130 kg./cm.[2]
on CS_2, hexane, C_6H_6, and air.   His results for the latter
checked those of Occhialini, and for the others the compressibility is not known accurately enough to permit a check of
the Clausius-Mosotti relation.   In 1925 Charlotte Franck [19]
measured benzene, hexane, pentane, and CCl_4 to 800 kg./cm.[2].
ϵ was found to increase with pressure, with the curvature
found by other observers.   Only for benzene are the compressibility data known accurately enough to permit a check
of the Clausius-Mosotti relation; here the 'constant' of this
relation decreases from 0·338 at atmospheric pressure to
0·330 at 700 kg./cm.[2].   Also in 1925 Grenacher [20] used a
high-frequency method of great precision on a number of
organic liquids, not all of them good insulators, but his
pressure range was only 60 kg./cm.[2].   The method was
delicate enough to detect departures from linearity in this

range. The Clausius-Mosotti 'constant' decreases for toluol by an amount estimated to be beyond experimental error; the other liquids could not be tested because of insufficient knowledge of compressibility.

By far the most extensive published work in this field is by Kyropoulos [21] in 1926, who measured with Tammann's pressure apparatus the effect up to 3000 kg./cm.[2] on ethyl alcohol, methyl alcohol, ethyl ether, CS_2, $CHCl_3$, petroleum ether, C_6H_6, CCl_4, acetone, water, and pyridine. When ϵ is plotted against pressure, a curve with the same curvature as that found by other observers is obtained. The Clausius-Mosotti expression applies very approximately to ether and CS_2, but fails for water, acetone, and the two alcohols. The other liquids could not be checked because of ignorance of the compressibility. $\dfrac{\epsilon - 1}{d}$ is approximately constant for CS_2, but is distinctly not constant for the others. The liquids for which the Clausius-Mosotti relation does not hold are known to be associated; Kyropoulos sees a possible explanation of the failure of the formula in the formation of complex molecules with smaller polarity. If $\dfrac{\epsilon - 1}{\epsilon + 2} \cdot \dfrac{1}{d}$ is plotted against pressure, a curve will be found of such a shape that a horizontal asymptote is not impossible at high pressure, where the effect of association may be supposed to be completed.

In addition to these published investigations, there are experiments as yet unpublished made by Z. T. Chang with my high-pressure apparatus. The results are preliminary in character; the materials examined were toluene, CS_2, n-pentane, n-hexane, ethyl ether, and i-amyl alcohol, and measurements were made at 30° and 75° up to 12,000 kg./cm.[2].

The change of $\dfrac{\epsilon - 1}{\epsilon + 2} \cdot \dfrac{1}{d}$, $\dfrac{\epsilon - 1}{d}$, and $\dfrac{\sqrt{\epsilon} - 1}{d}$ produced by pressure was calculated. The first of these expressions was by far the most nearly constant; it always decreased with increasing pressure and by about 3 per cent. for toluene and CS_2, but by 18·7 per cent. for i-amyl alcohol.

This investigation is at present being continued by another of my students, Mr. W. E. Danforth, jun.

Pressure on Index of Refraction.   This was a favourite subject for investigation fairly early, because the sensitiveness of interference methods was so great that comparatively large effects could be obtained with small pressures, and therefore no special demands were made on the pressure technique.   The earliest experiments are probably those of Jamin,[22] who in 1857 and 1858 measured with his refractometer the effect of pressures up to 2 atmospheres on the index of refraction of water, and found that within his limits of error $\dfrac{n^2-1}{d}$ was constant.   Mascart,[23] in 1874, also observed the effect on water up to a few atmospheres in a Jamin interferometer.   His results do not agree with those of Jamin and are intrinsically improbable, because he found the effect of pressure to increase at higher pressures.   Mascart incidentally made an interesting observation of the thermal effect of compression by counting the fringes passing after a sudden release of pressure.   In the same year Mascart obtained results on nine gases up to 8 atmospheres, and believed that he had established the law $\dfrac{n^2-1}{d}$ =const. within his limits of error, which were, however, rather large.   Quincke,[24] in 1883, obtained results on a number of liquids to a few atmospheres, which were later questioned by Roentgen, and which probably are not very reliable.   Between 1883 and 1887 Chappuis and Rivière [25] published several determinations of the effect of pressure up to about 20 kg. on the refraction of air, CO_2, and cyanogen.   Both the expressions $\dfrac{n-1}{d}$ =const. and $\dfrac{n^2-1}{n^2+2} \cdot \dfrac{1}{d}$ =const. reproduce within experimental error the results for these three gases.   In 1888–91 Zehnder and Roentgen [26, 27] published the most extensive work yet done on liquids.   Their method offered no novel features, and the pressure range was only a few atmospheres.   Measurements were made on water, CS_2, C_6H_6, ethyl alcohol, methyl alcohol, n-propyl alcohol, i-propyl alcohol, n-butyl alcohol, and amyl alcohol. In order to compute the 'refraction constant,' special measurements were made of the compressibility of a number of these liquids. It was found that neither of the expressions $\frac{n-1}{d}$ or $\frac{n^2-1}{n^2+2} \cdot \frac{1}{d}$ is constant; the variations of both are about the same on the average, and the directions of the discrepancies are opposite.

Since Roentgen and Zehnder, practically all the measurements in this field have been restricted to gases. Carnazzi,[28] in 1897, claimed to have found consistent departures from Gladstone and Dale's law (which states the constancy of $\frac{n-1}{d}$) in air, H_2, and CO_2 up to 5 atmospheres. In 1902 Gale [29] could find no variation in $\frac{n-1}{d}$ for air up to 19 atmospheres. Kaiser,[30] in 1904, investigated air, CO_2, SO_2, and H_2 at pressures below 1 atmosphere, and found the change of index to be linear in pressure—that is, $\frac{\partial n}{\partial p} = A(1 + ap)$. The constant a was not the same as that in the equation which expressed the dependence of density on pressure, so that neither Gladstone and Dale's nor the Lorentz-Lorenz expression holds. In 1904 Magri [31] measured the index of air to 176 atmospheres. n increases more rapidly than indicated by the formula $\frac{n-1}{d} = $const., and above 30 atmospheres $\frac{n^2-1}{n^2+2} \cdot \frac{1}{d} = $const. reproduces the results, whereas below 30 atmospheres the experimental accuracy is too low to answer the question as to which expression is better. Between 1911 and 1913 Siertsema and de Haas [32] published several papers on the refraction and dispersion of air, H_2, and CO_2 up to 100 kg./cm.². They found that the Lorentz-Lorenz expression, $\frac{n^2-1}{n^2+2} = \sum \frac{Ne_1^2}{3m_1(\nu_1^2 - \nu^2)}$, reproduces the results over the range of the visible spectrum and over their pressure range. Finally, in 1917, Posejpal [33] published results on air up to

20 atmospheres. He finds that if his results are expressed in the form $n-1 = \kappa p(1+\beta p)$, β must be a function of pressure, varying nearly inversely as the pressure up to 4 atmospheres, and from here on decreasing more slowly. Neither $\dfrac{n-1}{d}$ nor $\dfrac{n^2-1}{n^2+2} \cdot \dfrac{1}{d}$ is constant in his range, but both increase up to 20 atmospheres by about 0·25 per cent. In the range common to his work and that of Kaiser there are large discrepancies.

Evidently there is a fruitful field here for future investigation. It would be most interesting to compare the pressure effect on index with that on dielectric constant to really high pressures. The question of method is a difficult one. The simple interference methods which are adequate at low pressures can no longer be applied, because of the very great distortion of the glass windows.

The Effect of Pressure on the Rotation of the Plane of Polarisation by Solutions. This effect has apparently been attempted only once, by Siertsema,[34] who measured the effect of pressures up to 100 kg./cm.² on the rotation of sugar solutions of various strengths. The method was to almost entirely compensate for the rotation of the solution by a quartz plate, made very slightly wedge-shape. If the combination of quartz plate and solution is placed between crossed nicols, the quartz plate will be crossed by a black band, the location of which depends on the total rotation of the solution; if the rotation of the solution changes, the dark band will be displaced. In order to eliminate the effect of the very large double refraction in the windows produced by pressure, it was necessary to mount both the quartz plate and the nicols in the pressure chamber between the windows, where they were exposed only to hydrostatic pressure. The shifted black band was brought back to zero by rotating the nicols through a known angle; the amount of rotation measured the effect of pressure on the net rotation of quartz plate and solution together. The amount of rotation required was so small that it could be produced merely by twisting one end of the tube with respect to the other. The rotation produced by pressure in the quartz was not directly determined, but was calculated from measurements of Voigt of the compressibility, a somewhat questionable procedure. In any event this effect was small compared with the effect on the solution. The final result was that the rotation of sodium light is increased by pressure by a fractional amount independent of concentration over the range investigated (up to something of the order of 30 grm. of sugar per 100 grm. of water), the amount being about 0·26 per cent. per 100 kg./cm.².

Siertsema discussed his results from the point of view of Tammann's theory of solutions, namely, that the effect of an increase of external pressure is equivalent to the effect of adding solute and so increasing the internal pressure, and found agreement in some cases, but in other cases considerable discrepancies.

The Effect of Pressure on the Break - down Strength of Dielectrics. Considerable work has been done on this subject from a purely commercial point of view. The general result of increasing the pressure on either a liquid or a gaseous dielectric is to increase the voltage necessary for producing a spark between terminals a fixed distance apart. It will be sufficient to mention only one investigation, that of F. Koch,[35] in which the various questions are examined from a wider point of view than necessary for the purely engineering applications in mind. His measurements were made up to a maximum of 100 atmospheres. He was able to observe no effect on solids up to 50 kg./cm.², the maximum pressure to which they were examined. For liquids he found in the first 10 kg./cm.² a large increase of dielectric strength, which was nearly linear with pressure; above 10 kg./cm.² the increase was slower, and for many liquids an asymptotic value was apparently reached at a pressure of 70 or 80 kg./cm.². In liquids completely free from water the maximum effect was an increase of dielectric strength by about fivefold. There is a time effect involved, as shown by the fact that the break-down strength was greater for alternating than for direct current. Koch is of the opinion that the maximum strength is set by the detachment of electrons from the metal electrodes by the high field. The asymptotic value of the field is of the order of 900 KV/cm.; it is to be noticed that this is materially lower than the field necessary to produce cold electron emission in vacuum tubes.

The Effect of Pressure on Electro-motive Force. It was known comparatively early that the e.m.f. of a cell depends on the state of stress of the metals of the electrodes. Apparently Bichat and Blondlot,[36] in 1883, were the first to find a positive effect of a hydrostatic pressure acting uniformly on all parts of the cell. They found that the e.m.f. of the combination (Cu—$CuSO_4$ solution—Pt) was changed by about 0·001 volt by a pressure of 100 kg./cm.[2], and that of the cell (Pt—$AgNO_3$ solution—Ag) by about one-half as much. Gilbaut,[37] in 1891, found that a pressure of 500 kg. alters by appreciable amounts the e.m.f. of several types of cell, of which the Daniel cell (Cu—$CuSO_4$ solution—$ZnSO_4$ solution—Zn) was the most important. In 1901 Rolla Ramsay [38] applied pressure up to 300 kg./cm.[2] to the Clark and the Weston cells, and measured the change of e.m.f.

There is a thermodynamic relation connecting the effect of pressure on the e.m.f. of the cell with the change of volume when one coulomb of electricity passes through it. This is:

$$\left(\frac{\partial E}{\partial p}\right)_{\tau.e} = -\left(\frac{\partial v}{\partial e}\right)_{\tau.p}.$$

Here E is the e.m.f. of the cell, p the external pressure, t the absolute temperature, v the total volume of the cell, and e denotes the quantity of electricity which has flowed through the cell, all measured in consistent units. The subscript e in the derivative $\left(\frac{\partial E}{\partial p}\right)_{\tau.p}$ means that no current flows, or the e.m.f. is to be taken on open circuit. The derivative $\left(\frac{\partial v}{\partial e}\right)_{\tau.p}$ means the change of volume of the cell after unit quantity of electricity has flowed through it, pressure and temperature remaining constant.

Both Gilbaut and Ramsay discussed their results from the point of view of this formula, but without much success, as shown by a detailed critique by Cohen. The requirements in the physical system to which the formula may be applied are rather exacting; the cell must be reversible, it must be perfectly reproducible, and the chemical nature of the reactions taking place must be perfectly understood. Cohen and his collaborators have made an elaborate study of the effect of pressure up to 1500 kg. on the e.m.f. of a number of cells carefully selected to satisfy the above requirements, and their work in this field must be recognised as by far the most important that we have. In Cohen's book on Piezo-Chemistry will be found detailed results for eighteen different cells of various types. The volume changes are also determined in order to permit a comparison with the thermodynamic formula. The pressure coefficient, $\dfrac{\partial E}{\partial p}$ (pressure expressed in atmospheres) varied from $+8 \times 10^{-6}$ for the cell (Cd amalgam —$CdSO_4$ solution—$HgSO_4$—Hg) to $-2 \cdot 5 \times 10^{-6}$ for the cell (Zn amalgam—ZnCl solution—AgCl—Ag). $\dfrac{\partial E}{\partial p}$ varies somewhat irregularly with pressure, in general decreasing numerically at higher pressure. The agreement between the observed value and the value computed from the volume change may be characterised as fair, the discrepancies sometimes rising to as much as 10 per cent. Cohen, in his Baker lectures at Cornell, says that the agreement leaves much to be desired, and that new investigations should be made in which the compressibility of the material of the cells is taken into account.

The Effect of Pressure on Magnetic Permeability. In 1883 Tomlinson [39] made an unsuccessful attempt to detect an effect of hydrostatic pressure on magnetisation, and was able to show that if the effect exists it is of a smaller order of magnitude than the effect of a one-sided tension or compression. The existence of the effect was first established by Nagaoka and Honda [40] in 1898, who were able to measure the effect produced by 225 kg./cm.[2] on the magnetisation of Fe and Ni. They found the magnetisation of Fe to be decreased by pressure, while that of Ni was increased. The volume changes accompanying magnetisation were also measured and compared with the magneto-striction theory of Kirchhoff.   Theory and experiment failed to check, there being sometimes discrepancies even of sign.

In 1905 Miss Frisbie,[41] at the University of Chicago, measured the effect of pressure to 1000 kg./cm.² on the permeability of annealed and unannealed iron.   The effect of pressure on annealed Fe is to increase the permeability at all fields up to the maximum of 917 Gauss, but the effect on unannealed Fe was to decrease the permeability up to 5 Gauss, where the effect reversed, the permeability increasing above this value of the field.   Except for the anomalous effects on unannealed iron at low fields, the sign of the effect agrees with the demands of the thermodynamic theory of Kirchhoff.

In 1925 C. S. Yeh [42] published the results of measurements on Fe, Co, Ni, and two grades of steel up to 12,000 kg./cm.² made with my pressure apparatus.   Yeh's method was a ballistic method on toroidal specimens.   In order to attain regular results it was found that the specimens must be very carefully demagnetised before each reading; imperfect demagnetisation was doubtless responsible for the anomalous results of both Nagaoka and Honda and Miss Frisbie.   Over the entire pressure range up to 12,000 kg./cm.² Yeh found the percentage change of magnetisation at constant H to be linear with pressure for Fe, steel, and Co, but in the case of Ni the effect increases at a slightly accelerated pace at the higher pressures.   The nature of the effect depends very markedly on the field.   Yeh's maximum was about 100 Gauss.   For pure Fe magnetisation decreases under pressure. The change $\Delta B/B_0$ has a sharp minimum at -5.5 per cent. per 1000 kg. at $H = 1.2$, and at higher pressures appears to approach the value 0 asymptotically.   The effect on Ni is of the opposite sign; its magnetisation is increased by pressure, and the effect has a sharp maximum at about 4.5 per cent. per 1000 kg./cm.² at $H = 1.3$.   The effects are qualitatively the same for Fe with carbon as for pure Fe, but the minimum becomes somewhat less sharp and moves toward higher fields with increasing carbon content.   The effects for Co are more complicated.   At fields under 30 Gauss $\Delta B/B_0$ per 1000 kg./cm.² is negative, nearly independent of the field, and equal approximately to -0.2 per cent.   In the neighbourhood of 30 Gauss $\Delta B/B_0$ rises rapidly, and increases from 0·1 per cent. at 35 Gauss to 0·4 per cent. at 80 Gauss.

The retentivity of Fe was found to decrease under pressure.

Yeh also made a theoretical analysis of the situation, and showed that the volume change accompanying magnetisation may be analysed into two parts; one is prominent at low fields and involves the pressure effect just discussed. The second involves the elastic constants and the total H, and continues to increase in magnitude after saturation has been reached. These two effects may act in opposite directions. The experimental results agree qualitatively with this analysis; the volume changes were not measured, so that no quantitative comparison was possible.

Since Yeh, Mr. R. L. Steinberger has made similar measurements with similar apparatus, on the series of Fe—Ni alloys. The results, which are complicated, are as yet unpublished.

The Effect of Pressure on Reaction Velocity and Related Effects.
The first successful attempt to detect an effect of pressure on a reaction velocity was by Roentgen,[43] who, in 1892, discovered that a pressure of 500 kg./cm.^{2} decreases the velocity of the inversion of cane sugar by dilute HCl in aqueous solution. He explained this effect by a decrease of dissociation of HCl produced by pressure, an assumption contrary to the demands of thermodynamics that dissociation should be increased because of the volume relations. Rothmund,[44] in 1896, verified the results of Roentgen, and found quantitatively a decrease of inversion velocity of about 1 per cent. per 100 kg./cm.^{2}. On the other hand, Rothmund found that the velocity of saponification of methyl and ethyl acetate by dilute HCl is increased by 500 kg./cm.^{2} by about 20 per cent. The effect was apparently proportional to pressure over this range. Stern,[45] also in 1896, studied the effect of a pressure of 500 kg. on the velocity of inversion of sugar solutions of a variety of strengths by a number of kinds of acid. The inversion velocity is decreased by pressure when the acid is HCl, H_2SO_4, or oxalic acid, and the effect is smaller the smaller the concentration of the acid. On the other hand, the velocity is increased by pressure with phosphoric and acetic acid, and the increase is greater the

smaller the concentration of the acid. The pressure effect produced by HCl is nearly independent of the strength of the sugar solution, while in phosphoric acid the pressure effect increases with decreasing sugar concentration. The pressure effect is nearly independent of temperature between 15° and 25° C., although temperature itself has a very large effect on the velocity of inversion.

The most extensive work in this field has been done by Cohen and his collaborators, de Boer, Valeton, Kaiser, and Moesveld, over their regular pressure range of 1500 kg./cm.². Reference may be made to Cohen and Schut, *Piezochemie*, and to Cohen's Cornell lectures for some of the details of this work. There is a recent paper by Moesveld and de Meester [46] dealing with some of the puzzling theoretical questions, in which a compact summary will be found of all the experimental work at Utrecht. Twenty-eight different reactions have been studied; besides the inversion of sugar solution and the reaction of bromic and hydrobromic acid, all the other reactions have been saponifications of one sort or another in liquid solutions of various compositions. In most cases the effect of pressure is to increase the reaction velocity; the maximum effect is an increase of 4·5-fold produced by 1500 kg. on the saponification of bornylacetate in a 19 per cent. solution of propyl alcohol. In those cases where the effect is not proportional to the pressure, it was found that the effect becomes accelerated at high pressures, an effect the reverse of usual.

The great complication of the effect has been emphasised by Cohen, who feels that the results compel the conclusion that pressure exerts a specific effect on velocity. No ordinary considerations of the reaction velocity as affected by the number of collisions, which in turn may be affected by volume changes or changes of viscosity or association, are competent to explain the variety of the effects. The paper of Moesveld and de Meester is the beginning of an attempt to explain the specific effects. It is recognised that the deformation of the molecules is one of the elements which affect their ability to enter into a reaction, and this deformation may be a complicated function of the pressure and the medium in which the molecule is suspended. The whole question of activation of the molecules is also concerned. In view of the difficulty which has been experienced until quite recently in dealing with such a comparatively simple situation as monomolecular reactions in gases, it is evident that we may have to wait some time for a satisfactory untangling of the relations in this very much more complicated situation.

Closely connected with this question of reaction velocity are some measurements carried out by Professor J. B. Conant [47] with my high-pressure apparatus. I had already found early in my high-pressure work (B. 18) that egg-white and the proteids of meat may be coagulated by the action of pressures of the order of 6000 kg. at room temperature or lower. The nature of these processes is not at all understood, but the results at any rate suggest that pressure may produce irreversible reactions in organic compounds in a number of other cases. Professor Conant has begun a systematic attack on the whole problem, and already has a number of results. One of the most interesting cases is that of isoprene, which is polymerised by the action of pressure to a transparent solid that is essentially rubber. The amount of polymerisation increases very rapidly with increasing pressure. Thus, in one series of experiments the application of 6000 kg./cm.^{2} for 48 hours produced only 10 per cent. polymerisation, whereas polymerisation was practically complete after 12,000 kg. had been applied for 50 hours. The effect of pressure on this reaction has been studied up to 20,000 kg./cm.^{2}, where 70 per cent. polymerisation was observed in one experiment after 3 hours. On the other hand, at pressures of the order of a few hundred kilograms, there is no perceptible effect after the lapse of weeks. Dimethyl butadiene polymerises much like isoprene. Styrene and indene showed similar effects. Isobutyraldehyde and n-butyraldehyde were converted to a soft waxy solid of unknown nature by the action of 12,000 kg./cm.^{2} for 40 hours. This reaction was not permanent, but the solid slowly reverted to the original liquid on standing. This is probably a case of displaced internal equilibrium under the action of pressure. The effect of pressure on carboxyhemoglobin was studied in particular detail; this is converted by the action of pressure into a form insoluble in water, but soluble in dilute alkalies. Within experimental error the reaction runs as a first order reaction, the velocity of the reaction being a strong function of pressure; the rate of reaction also depends on the acidity of the solution.

The explanation of these effects is still obscure. They are not intimately connected with the volume effect. Thus the polymerisation of isoprene at 20,000 kg./cm.^{2} takes place with practically no volume change, although at atmospheric pressure the volume of the rubber is much less than that of the isoprene. Several liquids were tried which are known to have isomeric or polymeric forms of much smaller volume, but in no case was the form of smaller volume produced by pressure. The most striking of these is fumaric acid, which is an isomer of maleic acid, with 2·8 per cent. smaller volume.

There are other related effects of pressure on certain biological phenomena. It has been known for some time that pressures of the order of several thousand kilograms will kill micro-organisms. Experiments made at the West Virginia Experiment Station [48] showed this for a number of organisms, and the effect has recently been confirmed by Stuart Ballantine in unpublished experiments made with my apparatus. This effect has not been fully analysed, but is doubtless connected with the coagulation of the proteids of the organisms. Different organisms differ greatly in their resistance to the action of pressure.

Another biological effect was studied by P. A. Davies [49] at my suggestion. It is well known to seedsmen that certain seeds are very slow to germinate. White clover is a good example; it is not uncommon for only 20 per cent. of the seeds to germinate the first year after sowing, and for germination to continue at a roughly uniform rate for five years. This difficulty of germination probably entirely resides in the impermeable outer covering of the seed. Methods have been devised for destroying the impermeability of the coating by blowing the seeds against coarse sand-paper, or by the action of strong sulphuric acid. But the injury to the seeds produced by these methods is often so great that there is no resulting advantage. Davies found that very large increases in the germination of white clover (*Melilotus alba*) and alfalfa (*Medicago sativa*), rising in one case to as much as 90 per cent. germination for the treated seeds, against 25 per cent. germination for the untreated seeds, could be obtained by exposing the seeds to hydrostatic pressure in water. Two pressures were investigated, 500 and 2000 kg./cm.^{2}. The best results were obtained with 2000 applied for something of the order of 10 minutes. Pressures much higher than 2000 kill the seeds. The effect is apparently permanent, in that the seeds may be dried after exposure to pressure, and germination tests made up to ten months after treatment show the superiority of the treated seeds. The precise nature of the effects was not discovered. Microscopic examination disclosed no cracks in the external covering of the seed, which was unaltered in appearance. After the termination of these experiments, I found that De Vries had obtained an increased germination in the seeds of the evening primrose by the application of the few atmospheres' pressure that could be obtained from the city water-mains. His explanation of the effect was that in the seed-covering there are very fine capillary passages filled with air, which by capillary action resist the entrance of water. The air may be removed from these channels by the application of pressure, making the interior accessible to water. This can hardly be the explanation of the effect studied by Davies, for if this were the explanation, the optimum pressure should not be as high as 2000 kg./cm.^{2}. The experiments of Conant suggest the possibility that there may be a chemical reaction involved.

Optical Absorption under Pressure. The effect of pressure on the optical absorption of various materials has been measured by Miss Frances G. Wick,[50] up to about 3500 kg./cm.^{2} in my laboratory. The technique for mounting the glass windows is described in that paper; since it has now been improved it is not necessary to describe it in detail. It would probably have been possible to exceed 3500 kg./cm.^{2}, but as it was there were breakages of the windows, which would doubtless have been more frequent if the pressure range had been extended. If the apparatus is to be kept simple, it is not possible to measure the effect of pressure on absolute absorption, because of the unknown effect of pressure on the absorption of the windows and the transmitting liquid. To eliminate these would have demanded the use of different pieces of apparatus of different dimensions. Miss Wick confined herself to the easier observation of such effects on absorption bands as displacement or broadening or narrowing or intensification. The observations were mostly qualitative in character. None of the effects found were striking, but demanded rather careful examination of the photographic plates to establish them with certainty. The most marked effects were found in aqueous solutions of the salts of neodymium. In general, the effect of increasing pressure is similar to that of decreasing temperature or decreasing concentration, in that many of the absorption bands become narrower and unsymmetrically sharper. Not all bands are affected; the same ones are affected by pressure, temperature, and concentration. In one respect pressure has an effect not shown by temperature or concentration, in that some of the bands are shifted in position; this shift is usually toward the red, but one band, that at 4272 Å. units, is shifted toward the violet. The shifts produced by 3500 kg. are of the order of 10 Å. or less. To give an idea of the magnitudes involved, the description of the effect on the fine blue band at 4272 Å. may be quoted. "The total width of the band in a solution 0·287 normal at ordinary pressure is about 15 Å. The effect of a pressure of 1500 kg. is to narrow the red side about 8 Å., to shift the violet edge about 2 units towards the longer wave-lengths, and to reduce the width of the whole band to about 5 units." As a rough qualitative indication, the effect of increasing pressure to 3500 kg. was much the same as that of lowering temperature to −60° C. Other aqueous solutions of the praseodymium salts, erbium nitrate, the uranyl salts, and cobalt chloride showed smaller effects, in some cases so small that the existence of an effect could not be regarded as definitely established. The pressure effect was also measured on a number of glasses of different colours of the Corning Glass Company. Several of these, including one coloured with didymium salts, were not perceptibly changed by 1400 kg./cm.^{2}. Two orange glasses, G34 and G36, were made distinctly more yellowish by the application of 3700 kg./cm.^{2}, the transmission band in the red being shifted toward the violet. A red glass, G20, showed a similar displacement and also a very noticeable increase in the total amount of light transmitted. The effect of pressure on these three varieties of glass is similar to that of lowering temperature. The transmission band in the far red of synthetic ruby was somewhat sharpened and shifted toward the violet by 1600 kg., an effect similar to that of lowering temperature.

Recently Collins [51] has studied in my laboratory the effect of pressure on certain absorption bands in water, methyl and amyl alcohols, and toluene. The pressure range was 5000 kg./cm.^{2} for the first three liquids and 8000 for toluene. A considerable effect was expected, because these bands are molecular bands, and theory indicates that the effect of pressure on strongly polar liquids like these should be to very materially increase the polymerisation—that is, to change the molecular structure. However, no change could be found either in the spectral position or the intensity; the matter obviously needs further theoretical study.

The Effect of Pressure on Surface Tension. There appears to have been only a single attempt to measure the effect of pressure on surface tension, by Lynde [52] in 1906. The method was to measure the difference of height of the surface of separation of two liquids in the two arms of a U-tube of different diameters. The lower part of the U-tube contained one liquid, often mercury, and the upper part and all the rest of the pressure apparatus the other liquid. The level of the mercury in the two branches was observed through windows, which limited the range to 400 kg. The surface tension between the following pairs of liquids was measured; water-mercury, mercury-ether, water-ether, chloroform-water, and CS$_2$-water. It is difficult to know how much significance to attach to the results, which were quite irregular, varying sometimes by two- or threefold for different measurements of the same pair. The relative displacement of the surfaces under the highest pressure was 1 mm. at the maximum, and usually about one-third of this. However, it was possible

to establish with fair certainty that the effect was independent of the absolute size of the capillary, so that there is
this reason to believe that the effects measured were genuine.
The following are the percentage changes produced by 1000
lb./in.² in the relative surface tensions of pairs of liquids in
the order above: $-.15$ per cent., $-.25$ per cent., -5.2 per
cent., -0.15 per cent., $-.59$ per cent. Probably no significance can be attached to the last three, because of the
phenomena of mutual solubility, in virtue of which the
composition of the liquids separated by the surface of discontinuity varies with the pressure. With respect to the two
measurements, water *versus* mercury and ether *versus* mercury,
it is to be noted that 3 per cent. HNO_3 had to be added to
both water and ether in order to make the motions of the
capillary regular.

The subject is one which should be cultivated further,
because of the theoretical significance of the pressure coefficient of surface tension. It may be shown by a thermo-
dynamic argument that $\left(\dfrac{\partial \sigma}{\partial p}\right)_{A.\tau} = \left(\dfrac{\partial v}{\partial A}\right)_{p.\tau}$, where σ is the
surface tension and A is the area. That is, the pressure
coefficient of surface tension is a measure of the difference
of effective volume of a molecule in the interior and at the
surface of a liquid. In many liquids the surface molecules
are known to assume definite orientations with respect to
the surface, so that there is here a possible method of getting
information about the shapes of the molecules. If the molecules have the shape of long chains, and if these chains at
the surface stand parallel to each other, then one would expect
$\dfrac{\partial v}{\partial A}$ to be negative and hence $\dfrac{\partial \sigma}{\partial p}$ to be negative. Since the
effect is probably small in mercury, the measurements of
Lynde would suggest that in water and ether the effect is
of the other sign, and the molecules of these substances
demand more space when carried from the interior to the
surface.

The Effect of Pressure on Radioactive Disintegration. It is
stated in many books that the rate of radioactive disintegration cannot be appreciably affected by any known

physical agency, and in particular neither by high temperatures nor by high pressures. There are two experiments on the effect of pressure, both published in 1907, on the same page of *Nature*. The first was by A. Schuster,[53] who subjected a radioactive salt to a pressure of 2000 kg./cm.^{2} for five days, and at the end of the time could not observe a change of as much as $\frac{1}{3}$ per cent. in the amount of disintegration. The second was by Eve and Adams,[54] who enclosed a radioactive salt in a capsule of lead, which was subjected to a pressure of about 22,000 kg./cm.^{2}. No difference of as much as 1 per cent. could be detected in the amount of γ radiation emitted while exposed to pressure and while free.

Black Phosphorus. The transitions produced by pressure discussed in Chapter VIII were all reversible; in addition to reversible transitions, irreversible transitions may be produced by pressure. The number of these which have been studied up to the present is comparatively small; included here are the effects now being investigated by Professor Conant already described, and also an irreversible transition by which ordinary white or yellow phosphorus is transformed to a black modification, not known before (B. 16, 22). Up to the present, this black modification has been produced only under rather sharply limited conditions, by the simultaneous action of pressure between 12,000 and 13,000 kg./cm.^{2} and a temperature of 200° C. The material thus produced is much like graphite in appearance; its density is 2·67 against 1·9 for yellow phosphorus, it is a conductor of electricity, the effect of pressure on its conductivity having already been described, and it is indefinitely stable under atmospheric conditions. If left exposed to the air it very slowly oxidises, and becomes covered with a wet film, probably phosphoric acid, but if sealed into glass to prevent oxidation, it is apparently indefinitely stable, specimens having been kept for fifteen years with no apparent change. If heated, it vaporises, and the vapour condenses to ordinary red or yellow. I have not succeeded in condensing black phosphorus out of the vapour phase. The vapour pressure of the black is less than that of red or yellow; if it is heated in a closed container to about 650° C. under its own vapour pressure, it completely changes to ordinary red or violet, there being a point analogous to a triple point. Professor A. Smits [55] has investigated at some length this behaviour in the vapour phase, and has come to no definite conclusion, but believes the black form to be a system in incomplete internal equilibrium, and unstable with respect to the red. It does, however, appear to have a definite crystal form, and is probably hexagonal, as shown by the investigation of Linck and Jong.[56]

The most interesting features of the behaviour of black phosphorus are connected with its manner of formation.   It can be produced only from the solid phase; in preparing it, pressure must first be increased at room temperature beyond the melting curve before temperature is raised.   If temperature is first raised and then pressure applied to the liquid, only ordinary red phosphorus will be produced.   Black phosphorus is not produced immediately on carrying yellow phosphorus into the region beyond 12,000 kg./cm.^{2} and 200°, but a preliminary process of preparation is necessary before the transition runs.   This process of preparation takes about one half-hour for completion; it is accompanied by a comparatively small volume change, which proceeds at a gradually accelerating rate until suddenly the whole structure becomes unstable and the yellow phosphorus collapses into black with explosive rapidity.   No inoculation of the yellow phosphorus has the slightest efficacy in hastening the formation of the black, or in shortening the preliminary process of preparation. The whole affair is quite mysterious, and should be studied for its own sake, something that I have not yet had a chance to do.   There are two obvious things to do at once in such an investigation; first, to find whether the pressure necessary for formation does not become less at higher temperatures, and second, to try to stop the process of formation some time before the final collapse to black phosphorus, in order to find the nature of the intermediate preparatory form.

This black phosphorus is evidently partly metallic in its properties; one is reminded of the unstable non-metallic forms of the elements below phosphorus in the periodic table —arsenic, and antimony.

The Effect of Pressure on Rigidity. It would evidently be of great significance to determine the effect of pressure on all the elastic constants of single crystals, but this is a matter of great experimental difficulty, and at present no general method appears feasible. One such effect has already been measured, namely, the effect of pressure on linear compressibility. In addition to this, I have been able to measure the effect of pressure only on the rigidity or shearing modulus of such isotropic substances as may be produced in the form of helical springs (B. 68, 71).

For such substances, two methods are available, an absolute and a relative method. The absolute method has been applied only to steel. A helical spring of steel is suspended in a vertical position in the pressure chamber, and stretched to perhaps twice its original length with a weight. Attached to the weight is a high-resistance manganin wire sliding over a contact, by means of which the position of the weight, and so the extension of the spring, may be obtained by a potentiometer measurement of resistance, exactly as in the methods for determining the compressibility of solids or liquids. When pressure is applied, the stiffness of the spring changes in part because of the effect of pressure on rigidity, and the extension therefore changes and is measured. Obviously various corrections must be applied—for the change of dimensions, and for the change of buoyancy of the liquid, for example, but these can all be determined with the requisite precision.

Once the absolute effect has been determined on steel, the relative effect on other metals may be determined by stretching against each other two springs, one of steel, and the other of the other metal. If the relative stiffness of steel and the other metal changes in virtue of a change of hydrostatic pressure, there will be a shift in the position of the coupling point of the two springs, which can be determined electrically by a sliding contact device like that used in the absolute measurement of the steel.

In terms of the effect on steel, measurements were made on

a number of other metals, and also on several varieties of glass. The results are collected in Table XXV. The experimental accuracy was not great enough to justify giving a departure of the effect from linearity, although in several

TABLE XXV

EFFECT OF PRESSURE ON RIGIDITY AND INCOMPRESSIBILITY

Substance	Percentage change of rigidity under 10,000 kg./cm.²	Percentage change of incompressibility under 10,000 kg./cm.²
Steel	+2·2	+ 7·2
Tungsten	−0·3	+ 8·8
Tantalum	+0·3 to +1·5	+ 1·1
Molybdenum	+0·15	+ 6·9
Zirconium	−0·17	+13·6
Platinum	+2·4	+10·0
Thorium	+5·7	+28·4
Palladium	+1·1	+ 8·1
Nickel	+1·8	+ 7·9
Glass A	−0·62	+16·0
,, B	−8·45	− 4·0
,, C	−2·15	+17·8
,, D	−8·02	− 3·1
,, E	−8·80	− 1·8
,, F	−3·86	+ 5·2

cases it was evident that the effect does depart from linearity in the direction to be expected—that is, a smaller proportional effect at the higher pressures. The table shows that in general the rigidity of the metals increases under pressure, as probably one would expect. Furthermore, the effect on rigidity is in general less than on the compressibility. This, again, is what might be expected, and is doubtless connected with the very rapid increase in the force of repulsion between atoms when they are brought closer together than their normal distance of separation. The forces resisting volume compression are for the most part contributed by the pairs of atoms in closest contact. The repulsion between these increases rapidly when the distance between them is made less, so that the compressibility decreases by a comparatively large amount as the volume decreases under increased pressure. The forces resisting shear, on the other hand, contain a larger contribution from the more distant atoms, the component of the force effective in resisting shear contributed by the atoms directly in contact being of the second order. But the forces between distant pairs of atoms do not increase with decreasing distance as rapidly as those between adjacent pairs, and hence the effect of decrease of volume (or of increase of pressure) must be less on rigidity than on compressibility.

The rigidity of all the varieties of glass, on the other hand, decreases with increasing pressure. There is a very close connection between the decrease of rigidity in glass and the pressure coefficient of compressibility, the decrease being greatest for those glasses in which the abnormal increase of compressibility with pressure is greatest. Consultation of the original paper, in which the compositions of the glasses are given, will show that there is a further parallelism between the increase of compressibility with pressure and the SiO_2 content of glass.

Since an isotropic substance has only two independent elastic constants, it is possible to calculate from the pressure coefficients of compressibility and rigidity the pressure coefficient of any other elastic constant, such as Young's modulus or Poisson's ratio. It will be found for the metals that Young's modulus increases under pressure by an amount intermediate between the increase of rigidity and incompressibility. For the glasses, Young's modulus increases under pressure by something of the order of 2 per cent. for 10,000 kg./cm.^{2} in the case of those glasses which are most nearly normal with respect to the pressure coefficient of compressibility, but may decrease by 7 or 8 per cent. for the more abnormal glasses. Poisson's ratio increases with pressure for all varieties of glass, the increase for 10,000 kg./cm.^{2} varying from 1·5 to 18·1 per cent. By a combination of these data it may be found that the velocity of a wave of shear for all varieties of glass decreases under pressure by amounts running up to 5 per cent. for 10,000 kg./cm.^{2}. This would appear to be a matter of some geological interest.

1 J. TIMMERMANS, *Jour. Chim. Phys.*, **20**, 491 (1922).

2 J. FINK, *Wied. Ann.*, **26**, 481 (1885).

3 C. BARUS, *Amer. Jour. Sci.*, **140**, 219 (1890).

4 W. C. ROENTGEN, *Gött. Nachr.*, p. 505 (1893).

5 I. FANJUNG, *ZS. Phys. Chem.*, **14**, 673 (1894).

6 S. LUSSANA, *Nuov. Cim.*, **2**, 263 (1895) ; **5**, 1 (1897) ; **18**, 170 (1909); *ZS. Phys. Chem.*, **76**, 420 (1911) ; **79**, 677 (1912).

7 STERN, quoted in COHEN and SCHUT, without reference.

8 B. PIESCH, *Sitzber. Wien Akad.*, **103**, 784 (1894).

9 G. TAMMANN, *ZS. Phys. Chem.*, **17**, 725 (1895) ; *Wied. Ann.*, **69**, 767 (1899) ; *ZS. f. Elek. Chem.*, **16**, 592 (1910).
A. BOGOJAWLENSKY und G. TAMMANN, *ZS. Phys. Chem.*, **27**, 457 (1898).

10 F. KÖRBER, *ZS. Phys. Chem.*, **67**, 212 (1909) ; **77**, 420 (1911) ; **80**, 478 (1912).

11 E. W. SCHMIDT, *ibid.*, **75**, 305 (1910).

12 W. C. ROENTGEN, *Wied. Ann.*, **52**, 593 (1894).

13 FL. RATZ, *ZS. Phys. Chem.*, **19**, 94 (1895).

14 R. ORTVAY, *Ann. Phys.*, **36**, 1 (1911).

15 A. OCCHIALINI, *Nuov. Cim.*, **10**, 217 (1905) ; **7**, 108 (1914) ; *Atti Accad. Lin.*, **22**, 482 (1913).

16 A. OCCHIALINI et E. BODAREU, *Ann. Phys.*, **42**, 67 (1913).
A. BODAREU, *Atti Accad. Lin.*, **22**, 480 (1913).

17 G. FALCKENBERG, *Ann. Phys.*, **61**, 145 (1920).

18 F. WAIBEL, *ibid.*, **72**, 161 (1923).

19 CHARLOTTE FRANCK, *ibid.*, **77**, 159 (1925).

20 M. GRENACHER, *ibid.*, **77**, 138 (1925).

21 S. KYROPOULOS, *ZS. Phys. Chem.*, **40**, 507 (1926).

22 J. JAMIN, *C.R.*, **45**, 892 (1857) ; *Ann. chim. phys.*, **52**, 163 (1858).

23 M. E. MASCART, *C.R.*, **78**, 617, 801 (1874) ; *Pogg. Ann.*, **153**, 154 (1874).

24 G. QUINCKE, *Wied. Ann.*, **19**, 401 (1883) ; **44**, 774 (1891).

25 J. CHAPPUIS et CH. RIVIÈRE, *C.R.*, **96**, 699 (1883) ; **102**, 1461 (1886) ; **103**, 37 (1886) ; **104**, 1433 (1887) ; *Ann. chim. phys.*, **14**, 5 (1888).

26 L. ZEHNDER, *Wied. Ann.*, **34**, 91 (1888).

27 W. C. ROENTGEN und L. ZEHNDER, *ibid.*, **44**, 24 (1891).

28 P. CARNAZZI, *Nuov. Cim.*, **6**, 385 (1897).

29 H. G. GALE, *Phys. Rev.*, **14**, 1 (1902).

30 W. KAISER, *Wied. Ann.*, **13**, 210 (1904).

31 L. MAGRI, *Atti Accad. Lin.*, **13**, 473 (1904) ; *Phys. ZS.*, **6**, 629 (1905).

32 L. H. SIERTSEMA and M. DE HAAS, *Phys. ZS.*, **14**, 568, 574 (1913).
L. H. SIERTSEMA, *Proc. Amst.*, **15**, 925 (1913).

33 V. POSEJPAL, *Ann. Phys.*, **53**, 629 (1917).

34 L. H. SIERTSEMA, *Proc. Amst. Acad.*, **5**, 305 (1896–97) ; *Arch. Neerl.*, **3**, 79 (1900).

35 F. KOCH, *Elektrot. ZS.*, **36**, 85, 99 (1915).

[36] E. BICHAT et R. BLONDLOT, *Ann. Chim. Phys.*, **2**, 503 (1883).

[37] H. GILBAULT, *Lum. Elec.*, **42**, 7, 63, 175, 220 (1891); *C.R.*, **113**, 465 (1891); *Electrician*, **27**, 711 (1891).

[38] ROLLA R. RAMSAY, *Phys. Rev.*, **16**, 105 (1903).

[39] H. TOMLINSON, *Trans. Roy. Soc.*, **174**, 1 (1883).

[40] H. NAGAOKA and K. HONDA, *Phil. Mag.*, **46**, 261 (1898).

[41] F. C. FRISBIE, *Phys. Rev.*, **18**, 432 (1905).

[42] CHI-SUN YEH, *Proc. Amer. Acad.*, **60**, 503 (1925).

[43] W. C. ROENTGEN, *Wied. Ann.*, **45**, 98 (1892).

[44] V. ROTHMUND, *Öfver. af. Kong. Vet. Akad. Förhand. Stock.*, **53**, 25 (1896).

[45] O. STERN, *Wied. Ann.*, **59**, 652 (1896).

[46] A. L. T. MOESVELD and W. A. T. DE MEESTER, *ZS. Phys. Chem.*, **138**, 169 (1928).

[47] P. W. BRIDGMAN and J. B. CONANT, *Proc. Nat. Acad. Sci.*, **15**, 680 (1929).

J. B. CONANT and C. O. TONGBERG, *Jour. Amer. Chem. Soc.*, **52**, 1659 (1930).

[48] B. H. HITE, N. J. GIDDINGS, and CHAS. E. WEAKLEY, Jr., West Virginia University Agricultural Experiment Station, *Bull.*, p. 146 (1914).

[49] P. A. DAVIES, *Amer. Jour. of Bot.*, **15**, 149 (1928).

[50] FRANCES G. WICK, *Proc. Amer. Acad.*, **58**, 555 (1923).

[51] J. R. COLLINS, *Phys. Rev.*, **36**, 305 (1930).

[52] C. J. LYNDE, *ibid.*, **22**, 181 (1906).

[53] A. SCHUSTER, *Nat.*, **76**, 269 (1907).

[54] A. S. EVE and F. D. ADAMS, *ibid.*, **76**, 269 (1907).

[55] A. SMITS, G. MEYER, and R. PH. BECK, *Proc. Amst. Acad.*, **18**, 992 (1915).

[56] G. LINCK and H. JONG, *ZS. Anorg. Chem.*, **147**, 288 (1925).

APPENDIX

LIST OF PAPERS BY P. W. BRIDGMAN DEALING
WITH HIGH-PRESSURE EFFECTS

**In the body of the text these are referred to by their number
prefixed by B**

1. The Measurement of High Hydrostatic Pressure : I, " A Simple
   Primary Gauge," *Proc. Amer. Acad.*, **44**, 201–217 (1909) ; II, " A
   Secondary Mercury Resistance Gauge," *ibid.*, **44**, 221–251 (1909).
2. " An Experimental Determination of Certain Compressibilities,"
   ibid., **44**, 255–279 (1909).
3. " The Action of Mercury on Steel at High Pressures," *ibid.*, **46**,
   325–341 (1911).
4. " The Measurement of Hydrostatic Pressures up to 20,000 Kilograms
   per Square Centimeter," *ibid.*, **47**, 321–343 (1912).
5. " Mercury, Liquid and Solid, under Pressure," *ibid.*, **47**, 347–438
   (1912).
6. " Water in the Liquid and Five Solid Forms, under Pressure,"
   ibid., **47**, 441–558 (1912).
7. " The Collapse of Thick Cylinders under High Hydrostatic Pressure," *Phys. Rev.*, **34**, 1–24 (1912).
8. " Breaking Tests under Hydrostatic Pressure and Conditions of
   Rupture," *Phil. Mag.*, July 1912, pp. 63–80.
9. " Thermodynamic Properties of Liquid Water to 80° and 12,000
   Kilograms," *Proc. Amer. Acad.*, **48**, 309–362 (1912).
10. " Verhalten des Wassers als Flüssigkeit und in fünf festen Formen
    unter Druck," *ZS. Anorg. Chem.*, **77**, 377–455 (1912).
11. " Thermodynamic Properties of Twelve Liquids between 20° and
    80° and up to 12,000 Kilograms per Square Centimetre," *Proc.
    Amer. Acad.*, **49**, 1–114 (1913).
12. " The Technique of High Pressure Experimenting," *ibid.*, **49**, 627–
    643 (1914).
13. " Über Tammann's vier neue Eisarten," *ZS. Phys. Chem.*, **89**, 513–
    524 (1914).
14. " Change of Phase under Pressure "—I, *Phys. Rev.*, **3**, 126–141, 153–
    203 (1914).
15. " High Pressures and Five Kinds of Ice," *Jour. Frank. Inst.*, March
    1914, pp. 315–332.
16. " Two New Modifications of Phosphorus," *Jour. Amer. Chem. Soc.*,
    36, 1344–1363 (1914).

17. " Nochmals die Frage des unbeständigen Eises," *ZS. Phys. Chem.*, **89**, 252–253 (1914).
18. " The Coagulation of Albumen by Pressure," *Jour. Biol. Chem.*, **19**, 511–512 (1914).
19. " Change of Phase under Pressure "—II, *Phys. Rev.*, **6**, 1–33, 94–112 (1915).
20. " Polymorphic Transformations of Solids under Pressure," *Proc. Amer. Acad.*, **51**, 55–124 (1915).
21. " The Effect of Pressure on Polymorphic Transitions of Solids," *Proc. Nat. Acad. Sci.*, **1**, 513–516 (1915).
22. " Further Note on Black Phosphorus," *Jour. Amer. Chem. Soc.*, **38**, 609–612 (1916).
23. " Polymorphic Changes under Pressure of the Univalent Nitrates," *Proc. Amer. Acad.*, **51**, 581–625 (1916).
24. " The Velocity of Polymorphic Changes between Solids," *ibid.*, **52**, 57–88 (1916).
25. " Polymorphism at High Pressures," *ibid.*, **52**, 91–187 (1916).
26. " The Electrical Resistance of Metals under Pressure," *ibid.*, **53**, 573–646 (1917).
27. " The Resistance of Metals under Pressure," *Proc. Nat. Acad. Sci.*, **3**, 10–12 (1917).
28. " Theoretical Considerations on the Nature of Metallic Resistance, with Especial Regard to the Pressure Effects," *Phys. Rev.*, **9**, 269–289 (1917).
29. " Thermo-Electromotive Force, Peltier Heat, and Thomson Heat under Pressure," *Proc. Amer. Acad.*, **53**, 269–386 (1918).
30. " The Failure of Cavities in Crystals and Rocks under Pressure," *Amer. Jour. Sci.*, **45**, 243–268 (1918).
31. " Stress : Strain Relations in Crystalline Cylinders," *ibid.*, **45**, 269–280 (1918).
32. " A Comparison of certain Electrical Properties of Ordinary and Uranium Lead," *Proc. Nat. Acad. Sci.*, **5**, 351–353 (1919).
33. " An Experiment in One-piece Gun Construction," *Mining and Metallurgy*, Feb. 1920, pp. 1–16.
34. " Further Measurements of the Effect of Pressure on Resistance," *Proc. Nat. Acad. Sci.*, **6**, 505–508 (1920).
35. " Electrical Resistance under Pressure, including certain Liquid Metals," *Proc. Amer. Acad.*, **56**, 61–154 (1921).
36. " The Electrical Resistance of Metals," *Phys. Rev.*, **17**, 161–194 (1921).
37. " The Electron Theory of Metals in the Light of New Experimental Data," *ibid.*, **19**, 114–134 (1922).
38. " The Effect of Pressure on the Thermal Conductivity of Metals," *Proc. Amer. Acad.*, **57**, 77–127 (1922).
39. " The Compressibility of Metals at High Pressures," *Proc. Nat. Acad. Sci.*, **8**, 361–365 (1922).
40. " The Effect of Pressure on the Electrical Resistance of Cobalt,

Aluminum, Nickel, Uranium, and Cæsium," *Proc. Amer. Acad.*, **58**, 151–161 (1923).

41. "The Compressibility of Thirty Metals as a Function of Pressure and Temperature," *ibid.*, **58**, 166–242 (1923).

42. "The Compressibility of Hydrogen to High Pressures," Schreinemaker's Feestbundel, *Rec. Trav. Chim.*, July 1923, pp. 1–4.

43. "The Thermal Conductivity of Liquids," *Proc. Nat. Acad. Sci.*, **9**, 341–345 (1923).

44. "The Volume Changes of Five Gases under High Pressures," *ibid.*, **9**, 370–372 (1923).

45. "The Compressibility and Pressure Coefficient of Resistance of Rhodium and Iridium," *Proc. Amer. Acad.*, **59**, 109–115 (1923).

46. "The Thermal Conductivity of Liquids under Pressure," *ibid.*, **59**, 141–169 (1923).

47. "The Compressibility of Five Gases to High Pressures," *ibid.*, **59**, 173–211 (1923).

48. "The Thermal Conductivity and Compressibility of Several Rocks under High Pressure," *Amer. Jour. Sci.*, **7**, 81–102 (1924).

49. "Some Properties of Single Metal Crystals," *Proc. Nat. Acad. Sci.*, **10**, 411–415 (1924).

50. "Properties of Matter under High Pressure," *Mech. Eng.*, March 1925, pp. 161–169.

51. "Certain Aspects of High Pressure Research," *Jour. Frank. Inst.*, August 1925, pp. 147–160.

52. "The Compressibility of Several Artificial and Natural Glasses," *Amer. Jour. Sci.*, **10**, 359–367 (1925).

53. "Certain Physical Properties of Single Crystals of Tungsten, Antimony, Bismuth, Tellurium, Cadmium, Zinc, and Tin," *Proc. Amer. Acad.*, **60**, 305–383 (1925).

54. "Various Physical Properties of Rubidium and Cæsium, and the Electrical Resistance of Potassium under Pressure," *ibid.*, **60**, 385–421 (1925).

55. "The Viscosity of Liquids under Pressure," *Proc. Nat. Acad. Sci.*, **11**, 603–606 (1925).

56. "Linear Compressibility of Fourteen Natural Crystals," *Amer. Jour. Sci.*, **10**, 483–498 (1925).

57. "The Five Alkali Metals under High Pressure," *Phys. Rev.*, **27**, 68–86 (1926).

58. "The Effect of Pressure on the Viscosity of Forty-three Pure Liquids," *Proc. Amer. Acad.*, **61**, 57–99 (1926).

59. "The Breakdown of Atoms at High Pressures," *Phys. Rev.*, **29**, 188–191 (1927).

60. "Some Mechanical Properties of Matter under High Pressure," *Proc. Second International Congress for Applied Mechanics, Zürich* (1926).

61. "The Viscosity of Mercury under Pressure," *Proc. Amer. Acad.*, **62**, 187–206 (1927).

62. " The Compressibility and Pressure Coefficient of Resistance of Ten Elements," *ibid.*, **62**, 207–226 (1927).
63. " The Linear Compressibility of Thirteen Natural Crystals," *Amer. Jour. Sci.*, **15**, 287–296 (1928).
64. " The Pressure Transitions of the Rubidium Halides," *ZS. Krist.*, **67**, 363–376 (1928).
65. " The Effect of Pressure on the Resistance of Three Series of Alloys," *Proc. Amer. Acad.*, **63**, 329–345 (1928).
66. " The Compressibility and Pressure Coefficient of Resistance of Zirconium and Hafnium," *ibid.*, **63**, 347–350 (1928).
67. " Thermo-Electric Phenomena and Electrical Resistance in Single Metal Crystals," *ibid.*, **63**, 351–399 (1929).
68. " The Effect of Pressure on the Rigidity of Steel and Several Varieties of Glass," *ibid.*, **63**, 401–420 (1929).
69. " Thermische Zustandsgrössen bei Hohen Drucken," *Handbuch der Experimental Physik*, vol. viii, part 2, pp. 245–400, Leipzig (1929).
70. " General Survey of the Effects of Pressure on the Properties of Matter," *Proc. Phys. Soc. Lond.*, **41**, 341–360 (1929).
71. " The Effect of Pressure on the Rigidity of Several Metals," *Proc. Amer. Acad.*, **64**, 39–49 (1929).
72. " The Compressibility and Pressure Coefficient of Resistance of Several Elements and Single Crystals," *ibid.*, **64**, 51–73 (1929).
73. " The Minimum of Resistance at High Pressure," *ibid.*, **64**, 75–90 (1930).

INDEX

The names in parenthesis indicate joint authors

394